SOIL-WATER INTERACTIONS

BOOKS IN SOILS, PLANTS, AND THE ENVIRONMENT

Soil Biochemistry, Volume 1, edited by A. D. McLaren and G. H. Peterson
Soil Biochemistry, Volume 2, edited by A. D. McLaren and J. Skujiņš
Soil Biochemistry, Volume 3, edited by E. A. Paul and A. D. McLaren
Soil Biochemistry, Volume 4, edited by E. A. Paul and A. D. McLaren
Soil Biochemistry, Volume 5, edited by E. A. Paul and J. N. Ladd
Soil Biochemistry, Volume 6, edited by Jean-Marc Bollag and G. Stotzky
Soil Biochemistry, Volume 7, edited by G. Stotzky and Jean-Marc Bollag
Soil Biochemistry, Volume 8, edited by Jean-Marc Bollag and G. Stotzky

Organic Chemicals in the Soil Environment, Volumes 1 and 2, edited by C. A. I. Goring and J. W. Hamaker
Humic Substances in the Environment, M. Schnitzer and S. U. Khan
Microbial Life in the Soil: An Introduction, T. Hattori
Principles of Soil Chemistry, Kim H. Tan
Soil Analysis: Instrumental Techniques and Related Procedures, edited by Keith A. Smith
Soil Reclamation Processes: Microbiological Analyses and Applications, edited by Robert L. Tate III and Donald A. Klein
Symbiotic Nitrogen Fixation Technology, edited by Gerald H. Elkan
Soil-Water Interactions: Mechanisms and Applications, edited by Shingo Iwata, Toshio Tabuchi, and Benno P. Warkentin
Soil Analysis: Modern Instrumental Techniques, Second Edition, edited by Keith A. Smith
Soil Analysis: Physical Methods, edited by Keith A. Smith and Chris E. Mullins
Growth and Mineral Nutrition of Field Crops, N. K. Fageria, V. C. Baligar, and Charles Allan Jones
Semiarid Lands and Deserts: Soil Resource and Reclamation, edited by J. Skujiņš
Plant Roots: The Hidden Half, edited by Yoav Waisel, Amram Eshel, and Uzi Kafkafi
Plant Biochemical Regulators, edited by Harold W. Gausman
Maximizing Crop Yields, N. K. Fageria
Transgenic Plants: Fundamentals and Applications, edited by Andrew Hiatt
Soil Microbial Ecology: Applications in Agricultural and Environmental Management, edited by F. Blaine Metting, Jr.
Principles of Soil Chemistry: Second Edition, Kim H. Tan

Water Flow in Soils, edited by Tsuyoshi Miyazaki
Handbook of Plant and Crop Stress, edited by Mohammad Pessarakli
Genetic Improvement of Field Crops, edited by Gustavo A. Slafer
Agricultural Field Experiments: Design and Analysis, Roger G. Petersen
Environmental Soil Science, Kim H. Tan
Mechanisms of Plant Growth and Improved Productivity: Modern Approaches, edited by Amarjit S. Basra
Selenium in the Environment, edited by W. T. Frankenberger, Jr., and Sally Benson
Plant–Environment Interactions, edited by Robert E. Wilkinson
Handbook of Plant and Crop Physiology, edited by Mohammad Pessarakli
Handbook of Phytoalexin Metabolism and Action, edited by M. Daniel and R. P. Purkayastha
Soil–Water Interactions: Mechanisms and Applications, Second Edition, Revised and Expanded, Shingo Iwata, Toshio Tabuchi, and Benno P. Warkentin
Stored-Grain Ecosystems, edited by Digvir S. Jayas, Noel D. G. White, and William E. Muir
Agrochemicals from Natural Products, edited by C. R. A. Godfrey

Additional Volumes in Preparation

Seed Development and Germination, edited by Jaime Kigel and Gad Galili

Nitrogen Fertilization in the Environment, edited by Peter Edward Bacon

Phytohormones in Soils: Microbial Production and Function, W. T. Frankenberger, Jr., and Muhammad Arshad

SOIL-WATER INTERACTIONS

Mechanisms and Applications

Second Edition,
Revised and Expanded

Shingo Iwata
Ibaraki University
Ibaraki, Japan

Toshio Tabuchi
Tokyo University
Tokyo, Japan

Benno P. Warkentin
Oregon State University
Corvallis, Oregon

Marcel Dekker, Inc. New York • Basel • Hong Kong

Library of Congress Cataloging-in-Publication Data

Iwata, Shingo
Soil-water interactions: mechanisms and applications / by Shingo Iwata. Toshio Tabuchi, Benno P. Warkentin. — 2nd ed. rev. and expanded.
p. cm. — (Books in soils, plants, and the environment)
Includes bibliographical references and index.
ISBN 0-8247-9293-9
1. Soil moisture. 2. Soil physics. 3. Soils—Japan.
I. Tabuchi, Toshio. II. Warkentin, Benno P. III. Title. IV. Series
S594.I86, 1994
631.4'32—dc20 94-22883
CIP

The publisher offers discounts on this book when ordered in bulk quantities. For more information, write to Special Sales/Professional Marketing at the address below.

This book is printed on acid-free paper.

Marcel Dekker, Inc.
270 Madison Avenue, New York, New York 10016

Current printing (last digit):
10 9 8 7 6 5 4 3 2 1

PRINTED IN THE UNITED STATES OF AMERICA

Preface to Second Edition

Soil has three unique physical characteristics. The first and most conspicuous is that soil is a porous medium, and consequently possesses a filtering function. Soil consists of particles with various shapes and sizes—clay, silt, sand, and larger gravel and stones. Different mechanical composition and different arrangement of the same particles result in different pore structures, and these pore structures determine the water retentivity and permeability of soils. Moreover, natural soils have individual soil profiles, including macropores such as cracks and pores formed by soil fauna and plant roots, each of which has a complicated effect on water movement through soil. Second, soil has a very large specific surface area, which leads to a large interaction between ions and water molecules with the soil particles. Accordingly, soil possesses the ability to adsorb electrically neutral solutes. Both the first and second characteristics produce a high water retentivity in soil. Third, soil contains a high concentration of electrical charges due to clay and organic matter, a characteristic that allows soil to retain a high concentration of ions. This characteristic produces a repulsive force, which plays an important role in the prevention of an irreversible combination between particles. Swelling is also induced by the existence of this force.

These characteristics present particular problems that are different from those in other scientific fields. The first problem is related to physical heterogeneity, such as nonuniform distribution of ions in the soil solution. The second is concerned with complicated physical and physicochemical phenomena such as frost heaving and soil swelling. The third is connected with peculiarities of water movement such as infiltration into soil with macropores, unsaturated flow in soil under a near-saturated condition, and unsaturated flow at low water content.

This book provides a stimulating discussion of these problems, based on new viewpoints developed in the very useful interdisciplinary study of the interactions between soil and water. It will be of interest to the many scientists dealing with water in the vadose zone—soil scientists, agricultural, civil and environmental engineers, geologists, geochemists, hydrologists, agronomists, and pollution control scientists. Few books discuss advanced topics in soil water from different viewpoints. This book will serve that need for scientists from many fields who find they need an understanding of the interactions of soil and water. The modeling of these interactions, for example in descriptions of contaminant transport in soils, requires an understanding of the mechanisms that are discussed in this book. The interaction of soil particles with water is involved in most of the fundamental processes in soils. It is relevant not only to soil science, but also to the many technologies that deal with clays.

Each topic in this book is developed from fundamental concepts. No prior knowledge of soil science is assumed, although the reader will need some background in mathematics, physics, chemistry and thermodynamics. The authors begin at the atomic level and proceed to the field scale. For example, water adsorption on clay surfaces is developed from consideration of the nature of forces holding water at the surface, and then applied to phenomena such as water flux during frost heaving. The discussions of mechanisms of salt sieving or of clay swelling begin with the nature and characteristics of charges on clay particles, then are carried to the field scale in the management of water.

Concepts and measurements on interaction of water with soil particles date from about 50 years ago, when modern descriptions of soil water potential and physicochemical interaction of charged surfaces were developed. During subsequent years our understanding of the mechanisms of interaction has improved and applications have been made to various problems. Applications have been made in different scientific disciplines. The authors evaluate this work and then draw attention to those areas where our understanding is very incomplete, for example, flow of water and solutes through clay layers.

Some special features add to the value of this book. There is a thorough discussion of physical and chemical properties of allophane. The electric potential distribution around allophane particles is used to explain characteristics such as the development of a highly stable soil structure or fast flocculation. The differences between variable charge and constant charge mineral surfaces are discussed. There is a discussion of both capillary and surface forces in the description of water retention and water flow in soils and clays. The discussion of water fluxes shows the need to consider different physical processes for flow at different water saturation levels in the soil. The water regimes of paddy fields are discussed—a technology not familiar to most people working with soil water. Much of the material is based in part on the authors' own research.

Another distinct strength of this book is that it draws on the Japanese and Russian literature as well as English language research papers. Too many books written in English ignore the findings and ideas published in other languages. There is an extensive bibliography on many of the topics.

This revised edition, six years after the appearance of the first edition, gave us the opportunity to treat some topics now being developed in the soil physics literature, and to delete some topics available in other textbooks. About a third of the book has been rewritten, and about a quarter of the figures are new. Some of the main revisions are:

1. The importance of the physical meaning of the concept of "representative elementary volume" (REV) has been developed recently. A soil property can be treated as a continuum only if the REV is much smaller than the volume under consideration. Discussions based on this concept are included in new sections on "Heterogeneity of Hydraulic Conductivity in Soil" (6.2) and "Solute Movement in Field Soils" (6.3), and in a revised section 3.14 on "Other Forces of Interaction Between Charged Surfaces."
2. The existence in nature of heterogeneities in physical quantities is a barrier blocking development of research studies related to natural phenomena. While the best way to deal physically with heterogeneities has not yet been found, some considerations on this topic are included in sections 6.2 and 6.3.
3. The application to soil science of theoretical developments in a fundamental science has long been a way of understanding processes in soils. A new section 6.4 on "Application of Fractal Theory to Solute and Water Flows in Field Soil" is included.
4. Reviews have been made in research areas in which considerable understanding has accumulated in the past five years. Thus, new sections 2.5 on "Physicochemical Properties of Adsorbed Water" and 5.6 on "Recent Developments in Fingering Research" are added.
5. Material available in other textbooks that has been deleted in this second edition includes discussions on "Heat of Immersion," "Application of Darcy's Law to Unsaturated Flow," "Dependence of Unsaturated Soil Water Transport on Temperature," and "Application of Thermodynamics of Irreversible Processes."
6. Updated materials are added in many other sections, e.g. topics on electrical potentials at clay surfaces; effect of clay dispersion and swelling on permeability to water; state of the soil solution; forces between clay particles; solute and water flows in soil; and field water regimes.

We hope this revised edition will continue to serve a need in the literature of soil physics.

The authors acknowledge the important assistance of Pam Wegner, who prepared the manuscript for publication.

Shingo Iwata
Toshio Tabuchi
Benno P. Warkentin

Preface to the First Edition

Physical, chemical, and biological changes proceed ceaselessly in the soil. These changes make the existence of life in the soil possible; plants and animals of the field cannot live without these changes. Soil water is not only indispensable for life, but plays a most important role in the changes through its movement. Ion exchange and transfer of heat and solutes are brought about in the processes of wetting and drying of soil. In spite of the importance of soil water, there are yet many problems that must be solved in scientific fields related to soil water. The clarification of these problems is of great importance for the existence of humankind.

During the past three decades, significant developments have occurred in science fields connected with soil water. Mechanisms of flocculation and dispersion of soil particles in soil solution, soil freezing, and soil swelling have been clarified with concepts from physical chemistry. Analyses of water uptake by plant roots, water movement due to temperature differences, water movement under a near-saturated condition, and flow of solution through clay layers have developed to a notable extent. It is regrettable, however, that a systematic account of the results obtained from this research has yet to be written. In Japan, the phenomena related to water movement in paddy fields, such as unsaturated percolation, capillary flow, and outflow of fertilizers from paddy fields, have formed the subjects of investigation for many researchers over a number of years. Some of their achievements have attracted a great deal of attention; however, since their research papers have been reported in Japanese, soil scientists in other countries have been less able to benefit from this new knowledge.

We hope that this book will help to satisfy both the need for a clear and systematic description of the phenomena in question and a basic introduction to pioneering achievements that have recently come to the fore in science fields connected with soil water.

The eager and generous support of Dr. Benno P. Warkentin of Oregon State University has encouraged the authors to write this book in a way that will interest and stimulate researchers from many fields to open new directions in the rapidly expanding theoretical and empirical notions of soil physics. Thus, the appearance of this book owes much to Dr. Warkentin.

We wish also to acknowledge the late Dr. A. D. McLaren of the University of California, Berkeley, and Dr. T. Hattori of Tohoku University, who provided the opportunity to write this book. Finally, we recall with gratitude the early influence of our teacher, Dr. F. Yamazaki of Tokyo University.

Shingo Iwata
Toshio Tabuchi

Contents

1

ENERGY CONCEPT AND THERMODYNAMICS OF WATER IN SOIL

1.1 ENERGY CONCEPT OF WATER IN SOIL

1.1.1 Historical Review of the Energy Concept

Capillary Potential

Buckingham (1907) introduced the concept that flow of water results from a difference in capillary potential between two points in the soil. He considered that the soil exerts an attraction sufficient to hold water against the action of gravity, and that this attraction decreases as the amount of water held by the soil increases. He proposed the term "capillary potential," ψ, to describe this attraction. $\psi(\theta)$ was defined as the work required to pull a unit mass of water away from the large mass of soil whose water content is θ, the volume fraction of water. He introduced the equation

$$\psi = g\,x \tag{1.1}$$

where x is the height above a water surface in a column of soil standing in water, at equilibrium.

By measuring water content at different heights in these columns, he could obtain the relation between capillary potential and water content for various soils. He obtained curves of the type shown in Fig. 1.1, which have proved to be of enormous value in the study of soil water. Before that time, only "water content" had been used to express the state of water in soil. But if the water contents of two soils are equal, the state of water in each soil need not be the same. The concept of capillary potential enabled scientists to compare quantitatively, with the same scale, the state of water in various moist soils. Buckingham established the foundation for scientific studies on soil water. It should be noted that the sign of g in Eq. (1.1) should be negative, as pointed out

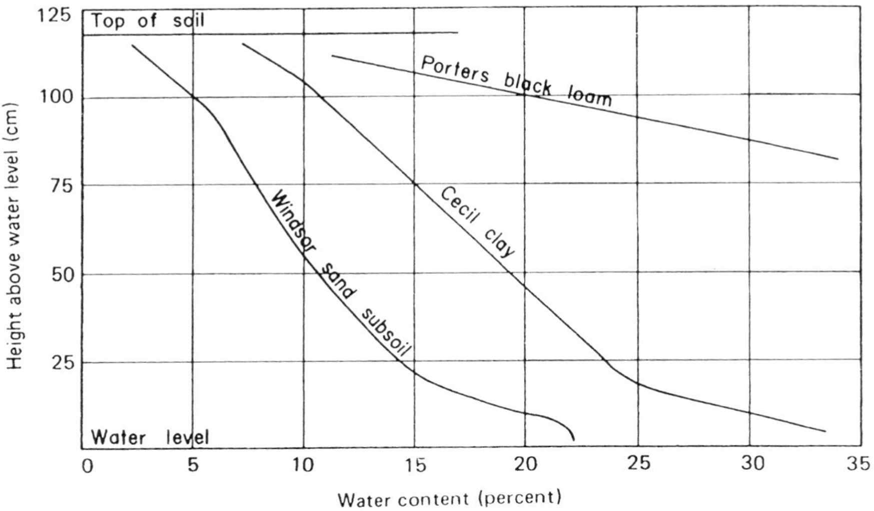

Fig. 1.1. Distribution of water in 48-in. columns of soil after 53 to 68 days. (From Buckingham, 1907)

by Gardner et al. (1922). Consequently, Eq. (1.1) should be written

$$\psi = -g\ x \tag{1.2}$$

This equation always shows a negative capillary potential.

The capillary potential as defined by Buckingham has weaknesses from both theoretical and practical points of view. He was aware of some of these deficiencies. Consider the system shown in Fig. 1.2, with the soil column in contact with a free water surface. Assuming that the system comes to equilibrium with pure water in the soil, the combined force acting on an infinitesimal mass of water in the soil is zero. One of the forces is gravity, whose direction is perpendicular downward. The force, equal to the magnitude of gravity and opposite in direction, may be the capillary force. As gravity can be expressed by gravitational potential, so the capillary force could also have a potential that is numerically equal to the gravitational potential and opposite in sign. This capillary force, however, does not have a potential in the mechanical sense. The capillary force's field, or capillary potential considered by Buckingham, can appear only under conditions where the system reaches an equilibrium state in the gravitational field. Then the work required to move an infinitesimal mass of

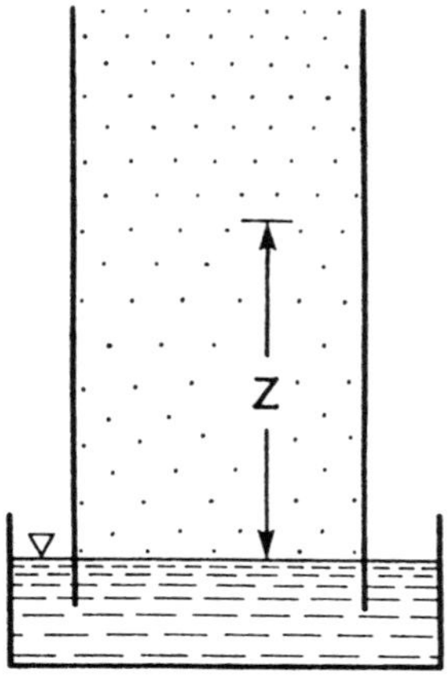

Fig. 1.2. Soil column at moisture equilibrium with a free water surface.

water from one point to another is zero, as pointed out correctly by Buckingham. On the other hand, when a soil column of infinite height is brought into contact with a free water surface, and if the gravitational field does not exist, water continues to rise by capillary force. The capillary force field or capillary potential does not exist in this system.

The same conclusion is recognized in a system where a soil column with a free water surface is placed in a centrifugal force field. A capillary potential appears to exist, which is numerically equal to the potential of the centrifugal force and opposite in sign.

Capillary potential, as used by Buckingham, should have been related to a physical quantity other than a potential in the mechanical sense. A potential whose existence is possible only under conditions where the system comes to equilibrium in an external force field should be related to the change of state of the substance in the system. Buckingham did not recognize this point. The theoretical analysis of the system shown in Fig. 1.2 can be performed exactly by use of a chemical potential, as discussed by Russell (1942). That is,

$$\Delta \mu_w = g\, x = 0$$

Therefore,

$$\Delta \mu_w = -g\, x$$

where $\Delta\mu_w$ is the chemical potential of water in soil, based on that of bulk water.

This inability to relate capillary potential to a state quantity of water in soil prevented effective application of the concept of capillary potential to analysis of various phenomena connected with soil solutions in the field and the laboratory. In most cases, water in soil is not in contact with a free water surface. One cannot measure the value of capillary potential of such water with the concept of capillary potential proposed by Buckingham.

This problem was solved to some degree by Gardner et al. (1922), who proposed an equation that shows the relationship between the pressure of water in soil (a state quantity of water) and its capillary potential.

$$\psi = \int \frac{dP}{\rho} \tag{1.3}$$

where ρ is the density of water in soil and P is the pressure of water in soil relative to atmospheric pressure. In the general case, ρ is a function of pressure; however, assuming that water is incompressible, ψ becomes equal to P/ρ. The introduction of this relationship between ψ and P led to the development of tensiometer and suction plate apparatus for measuring capillary potential of water in soil which was not in contact with a free water surface.

Establishment of the Energy Concept

The further development of studies on soil water required a more general concept to express the state of water in soil. Studies of the relationship between plant growth and soil water could not ignore the existence of solutes in soil solution. Moreover, research into the physical and mechanical properties of soil needed a concept to compare the state of water over the whole water content range from saturation to oven-dry.

Various concepts were proposed in the decade after 1935 to satisfy this demand. They include pF by Schofield (1935), water potential by Veihmeyer and Edlefsen (1937), osmotic potential and pressure potential by Day (1947), and soil moisture stress, the sum of water suction and osmotic pressure, by Wadleigh and Ayers (1945). Although these concepts have different names, they are all expressions of chemical potential (total potential) of water in soil relative to free water at the same temperature. pF is the base 10 logarithm of the absolute value of chemical potential expressed in a gravitational system of units, assuming that solutes are negligible. Water potential is the chemical potential of water. Pressure potential and osmotic potential are equivalent to the decrement of chemical potential due to surface tension effects and/or electric and van der Waals force fields, and due to solutes. Water suction and osmotic pressure are, respectively, equal to the value of pressure potential and osmotic potential divided by the partial volume of water in soil.

Schofield defined pF as the logarithm of the specific Gibbs free energy of water in soil expressed in gravitational units. He indicated that the use of pF allows removal of the word "capillary" from the energy concept. This is most advisable because the capillary force is only one of the forces influencing the state of water in soil. Also, the logarithmic scale permits the expression on one graph of the energy over the entire range of water content.

Schofield apparently thought of soil water as pure water, as Buckingham had done, and did not develop the idea that the energy was actually a function of many variables of the state of water in soil, such as temperature and concentration of solutes. This is brought out by the fact that he applied Gibbs specific free energy instead of chemical potential in his energy concept. In spite of this weakness, high value should be set on the work done by Schofield. By applying Gibbs free energy, it became possible to use new methods of measurement, such as vapor pressure or freezing-point depression, in measuring the relationship between water content and the free energy of water in soil.

In 1943, Edlefsen and Anderson published a highly valuable report entitled "Thermodynamics of Soil Moisture." It was the first systematic study of thermodynamics of water in soil, and it emphasized the importance of the use of chemical potential. The authors pointed out the advantages of using the chemical potential:

1. If a heterogeneous system arrives at equilibrium, the chemical potential of each substance involved in the system shows the same value through all phases. Therefore, one can measure the value of the chemical potential of water in soil under a certain water content by allowing the soil to arrive at equilibrium with ice or water vapor whose chemical potential is already known.
2. If the chemical potential of any substance is greater in one part of the system than in anther, that substance will pass from the former to the latter place. Therefore, the direction of water movement between two points in a soil or between a soil and a plant can be determined. This point plays an important role in understanding water absorption by roots.
3. It becomes possible to estimate the change of the state quantity of water in soil resulting from a temperature change.

The state of water in soil can be described exactly on the basis of thermodynamic concepts such as chemical potential. Buckingham (1907) wrote: "It is obvious that a rigorous treatment of the subject, with no restrictions imposed on either the water content or the soluble salt content of soil, would have to use thermodynamic reasoning. But the simple conception will suffice for the present purpose, though it is not impossible that with more comprehensive experimental data available we should have to use thermodynamic potential or free energy."

However, a long period of 36 years was required to establish this general energy concept.

In the early 1980s, a noteworthy thermodynamic analysis of unsaturated soil was done by Sposito and Chu (1981, 1982). They attempted to describe the state of an unsaturated soil with a set of independent variables: the masses of the three components of soil—solids, water, and air—together with the applied pressure and absolute temperature. At the present stage the analysis is not sufficiently complete to be used for practical purposes such as a thermodynamic study of the total soil system, including not only the water but the solids and air. Such an analysis will be important in future research on soil water. In addition, the balance of internal energy for water in soil has been studied (Hassanizadeh and Gray, 1979a,b, 1980; Sposito and Chu, 1981), and the fundamental Richards equation was deduced as a special case of internal energy balance for water in an unsaturated soil (Sposito and Chu, 1982).

1.1.2 Potential of Water in Soil

Total Potential and Its Components

In 1974, the second Terminology Committee of Commission I of the International Society of Soil Science (ISSS) (1976) adopted the total potential as an energy concept to express the state of water in soil. The total potential was defined as follows:

> The total potential, ψ_t, of the constituent water in soil at temperature T_0, is the amount of useful work per unit mass of pure water that must be done by means of externally applied forces to transfer reversibly and isothermally an infinitesimal amount of water from the state S_0 to the soil liquid phase at the point under consideration. It is convenient to divide the transfer process referred to above into several steps, by introducing substandards.
>
> S_1: a pool of pure, free water as in S_0, but situated at the same height as the soil liquid phase under consideration, h_x, i.e., S_1 is at T_0, h_x, P_0.
>
> S_2: a pool of free solution identical in composition with the soil liquid phase at the point under consideration, having an osmotic pressure, π, but otherwise identical with S_1, i.e., S_2 is at T_0, h_x, P_0.

The total potential is composed of three potentials: the gravitational potential ψ_g, the osmotic potential ψ_o, and the tensiometer-pressure potential (the pressure potential) ψ_p, which are all defined by the concept of useful work in the same manner as the total potential. The relation between the total potential and the three component potentials is

$$\psi_t = \psi_g + \psi_o + \psi_p = g\ \Delta h - \int_0^{\pi} \bar{V}_w dP + \int_0^{P} \bar{V}_w dP \qquad (1.4)$$

where Δh is the difference in height between S_0 and S_1, π the value of the experimentally accessible osmotic pressure of the soil solution, P the value of the experimentally accessible tensiometer pressure of the soil liquid phase *in situ*, and $\bar{V}_w$ the partial specific volume of the constituent water in the soil solution.

At present there is consensus on these concepts of soil water potential: The total potential of water in soil, referenced to the chemical potential of pure liquid, is equivalent to the chemical potential of the soil water at a chosen temperature and pressure (Taylor and Ashcroft, 1972; Sposito, 1981). The osmotic potential of water in soil is equal to a component of the chemical potential related to solutes existing in the water, and the pressure potential to another component of the chemical potential related to surface tension effects and/or electric and van der Waals force fields. Consequently, readers who are more familiar with the concept of total potential can understand that total potential is equivalent to chemical potential.

Physical Meaning of Tensiometer Pressure

A tensiometer is a useful device for measuring the total potential or the pressure potential (the tensiometer pressure potential). If the wall material of a tensiometer cup completely restricts passage of solute ions, the pressure measured by the tensiometer is equal to the value of total potential of the water (the sum of osmotic potential and the pressure potential) divided by the partial volume of the soil water. The pressure measured by a tensiometer whose wall material allows passage of both water molecules and solute ions equals the value of the pressure potential divided by the partial volume of soil water.

However, there is possible confusion about the physical meaning of the tensiometer pressure. Water pressure in a soil does not have the same value at all points within the soil water. In water-filled pores, or in wedges of water around points of contact of soil particles, water exists only under the influence of surface tension forces, and the water has a constant pressure. Where the water is close to the surfaces of soil particles, it is under electric and van der Waals force fields, and the water pressure changes with thickness of the water film. Therefore, as water content of the soil decreases, the amount of water in which the pressure shows the same value decreases.

It may then be concluded that the pressure measured by a tensiometer actually represents only the pressure in water under the influence of surface tension effects, even when the pressure measured by the tensiometer is above -1 bar. If it would be possible to measure a pressure below -1 bar with an ideal tensiometer, the value of the pressure would reflect the energy status of the soil water, but the measured pressure would not be a real pressure in the soil water because a pressure below -1 bar does not have physical significance.

Difficulties with the Total Potential

The definition of total potential, which is based on classical mechanics, belongs to the school of Buckingham. Most technicians and applied scientists interested in soil physics are more familiar with mechanical than with thermodynamic terminology. Second, mechanical concepts are more suitable for studies on water movement in soil. Although the mechanical potential has these advantages, there remains a theoretical problem relating to the definition.

The problem is whether or not the amount of useful work in the definition is dissipated to increase the potential energy or the free energy of the transferred water. If work was done reversibly on the transferred water, the amount of work must be equal to the increase in potential energy or free energy of water. However, the amount of useful work does not satisfy this condition except for the gravitational potential (Iwata, 1972b).

A simple mechanical model to discuss a few points about the physical meaning of useful work will confirm the statement above. Consider the pressure potential of water in the soil at point A (Fig. 1.3) submerged at depth h cm below the water table. Then the water pressure at A is $\int_0^h \rho g \, dh$, where ρ is the density of the water. Accordingly, the pressure of water at A, ψ_P^A, is given by

$$\psi_P^A = \int_0^{\int_0^h \rho g dh} \bar{V}_w \, dP \tag{1.5}$$

Assuming that water is incompressible and ρ is 1.0, Eq. 1.5 becomes

$$\psi_P^A = gh \tag{1.6}$$

We shall now derive this result on the basis of the definition of the pressure potential. Suppose that we have a system consisting of M grams of water in a

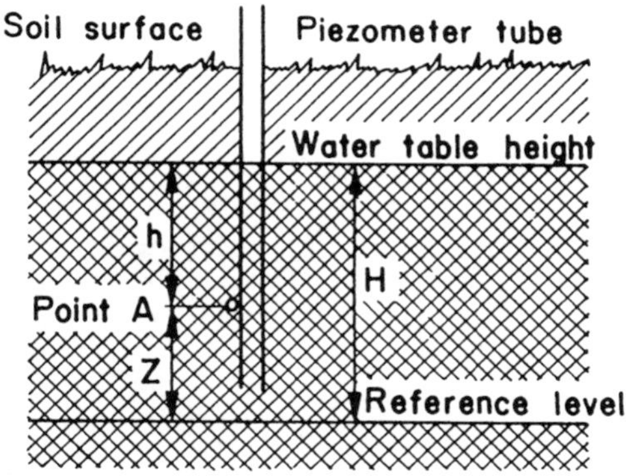

Fig. 1.3. Schematic diagram illustrating the definition of submergence potential at point A.

container and ΔM grams of water in a tube attached to the container (Fig. 1.4). The cross-sectional area of the tube is dA (cm^2).

1. Initially, there is a partition wall between the container and the tube, and atmospheric pressure P_0 is applied to the water in the tube through a piston. The level of point A is taken as a reference level for the gravitational potential energy. Then the water in the tube is in the standard state S_0, and the potential energy of the water in the container is given by the equation

$$\psi(1) = Mg\left(h - \frac{H}{2}\right) \tag{1.7}$$

Therefore, the total potential energy of the system $\psi_t(1)$, is

$$\psi_t(1) = Mg\left(h - \frac{H}{2}\right) \tag{1.8}$$

2. The pressure exerted on the water in the tube by the piston is increased from P_0 to (P_0 + P), which is equal to the pressure at point A, where P is an external force and its value is gh. Since the work done on the water in the tube is only the compression work, then, if water is incompressible, the work done by the external force is zero.

3. When the partition wall is removed, the energy of the system remains constant.

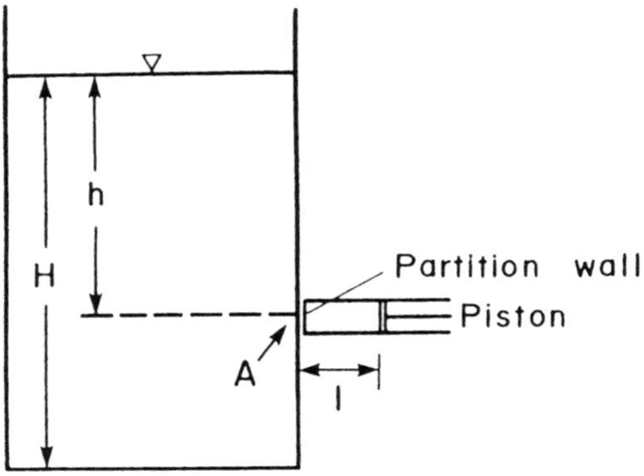

Fig. 1.4. Schematic representation explaining the physical meaning of pressure potential.

4. The water in the tube is transferred into the water in the container by increasing the pressure of the piston by an infinitesimal amount. When the transfer is completed, the amount of water in the container is increased by ΔM grams of water, whose potential energy is equal to ΔMgh. Therefore, $\psi_t(4)$ can be expressed by the equation

$$\psi_t(4) = Mg\left(h - \frac{H}{2}\right) + \Delta Mgh \tag{1.9}$$

Subtracting Eq. (1.9) from Eq. (1.8), one has

$$\psi_t(4) - \psi_t(1) = \Delta Mgh$$

On the other hand, the work W done by the external force is

$$W = P\ell\, dA = P\, dV = P\, \Delta M = \Delta Mgh = \psi_t(4) - \psi(1) \tag{1.10}$$

The result leads to the following conclusions. By procedure (2) the state of water in the tube becomes identical to the state of water at point A. It is possible to consider that the work W is expended to lift ΔM grams of water from the level of point A to the level of the water surface. That is, the work W that is done by the external force when the water in the tube is transferred to point A does not mean the work done on the water in the tube. The same conclusion is reached for the osmotic potential by applying the same reasoning.

Envelope Pressure Potential or Overburden Potential

The envelope pressure potential has now been accepted as a component potential, although it was introduced some years ago by Coleman and Croney (1952). In 1969, Philip (1969a,b,c) applied the concept of overburden potential to hydrostatics and hydrodynamics of swelling soils. Groenevelt and Bolt (1972) further clarified the meaning of this potential as a component of the thermodynamic potential of water.

The overburden potential was defined in the 1976 terminology report of ISSS as follows: "The envelope-pressure potential (overburden potential), ψ_P^e, is the increment of ψ_P following the application of an envelope pressure P_e to a soil sample with wetness W and excess gas pressure, $\Delta Pa = 0$, and originally under zero envelope pressure," according to

$$\psi_P^e = \int_0^{Pe} \left(\frac{\partial \psi_P}{\partial Pe}\right)_{W,\Delta Pa=0} dP \tag{1.11}$$

in which ψ_P is the pressure potential, W the water content on a mass basis relative to the total solid content, and ΔPa the pressure in the gas phase relative to atmospheric pressure, P_0.

From the boundary condition, where $\psi_P^e = 0$ at $P_e = 0$, Eq. (1.11) becomes

$$\psi_P^e = \psi_P(Pe,W) - \psi_P(0,W) \tag{1.12}$$

When a soil is saturated with water and the system is at equilibrium, ψ_P^e is generally zero. Therefore, it can be defined only under unsaturated water conditions. It is usually obtained under the condition where the sample is constrained to change volume one-dimensionally. The pressure of water in the sample is measured with a tensiometer inserted into the sample under initial unloaded conditions. The tensiometer reading corresponds to $\psi_p(0,W)$ if $\Delta Pa = 0$. An external pressure P_1 is then brought to act on the sample by applying a load to the upper surface. After equilibrium is achieved, that is, when the sample has been compressed, the water pressure is again measured with a tensiometer to give the pressure reading $\psi_p(P_1,W)$. By Eq. (1.12), the envelope pressure potential at that time is $\psi_P^{e_1} = \psi_P(P_1,W) - \psi_P(0,W)$. We can get $\psi_P(P_2,W)$, $\psi_P(P_3,W)$, ..., $\psi_P(P_n,W)$ and $\psi_P^{e_2}$, $\psi_P^{e_3}$, ..., $\psi_P^{e_n}$ by the same procedure, where $P_1 < P_2 < P_3 \ldots < P_n$. The relationship between ψ_p^e and P_e is shown in Figure 1.5.

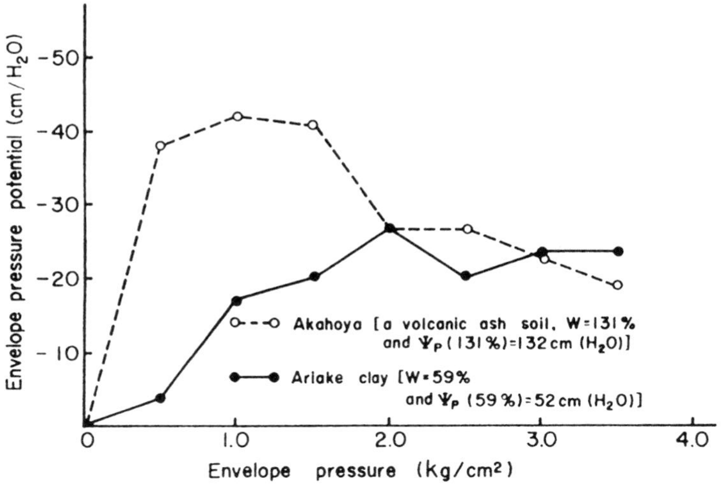

Fig. 1.5. Relationship between envelope pressure and envelope pressure potential. W denotes the water content on a weight basis. (From Miyauchi, 1978)

It is clear from the description above that the envelope pressure corresponds to the pore pressure in the terminology used in soil mechanics for a partially saturated soil. The envelope pressure potential will play an important role in research on the behavior of soils with a nonrigid matrix, such as swelling soils. Equation (1.12) has also been expressed

$$\psi_P^e = \alpha P_e \tag{1.12'}$$

where α is a compressibility (Coleman and Croney, 1952) or load factor (Groenevelt and Bolt, 1972). Talsma (1977) measured the contribution of the overburden potential to total potential in swelling field soils and found that, within the limits of accuracy of his methods, α did not depend on P_e, and average values of α measured at high moisture content on a number of swelling soils were all <0.25. In addition, Towner (1981) published a short review to clarify the role of the overburden pressure in the interpretation of tensiometer readings for swelling soils.

1.1.3 Weakness of the Energy Concept

Neither the pore geometry nor the mechanism by which water is retained needs to be considered in using the energy concept. This constitutes both its strength and its weakness. Use of the energy concept has produced a tendency to ignore pore space geometry and the mechanisms retaining water. The idea that only one law of water movement is applicable for the whole water range from saturation to air-dry is an example of this tendency. Water can be held mainly by surface tension effects or by van der Waals forces between water molecules and particles; while total potential is the same in each case, the physical properties of the water differ significantly. The laws of motion governing water that flows and fills up pores should be different from those governing water that flows along the walls of particles in a water film, even when the total potential is the same. It is therefore important to know how water is held in order to understand the properties of water retention and movement. We need to tie together the energy concept with the pore space geometry and with the mechanisms for retaining water.

1.2 THERMODYNAMICS OF WATER IN SOILS

The soil solution is a very heterogeneous system. It is generally influenced by a van der Waals force field, an electric field, and by surface tension effects. In addition, ions contained in the solution are distributed nonuniformly in the electric field. The solution consists of an infinite number of apparently homogeneous systems (phases), and its thermodynamic quantities, apart from chemical potential, have different values depending on distance from the surface of a soil particle.

The following assumptions have been made in this section in working out the thermodynamic description of soil water: (1) surfaces of soil particles are either planar or spherical; (2) electric charges possessed by soil particles are distributed uniformly over their surfaces; (3) potential fields acting on water in soil do not affect the inner function of the water molecules; and (4) if factors having effects on thermodynamic quantities of water in soil were removed, the state of water in soil would become the same as that of bulk water. This means that the potential energy of the interaction between the water molecule under consideration and the residual water molecules in soil water is equal to that of bulk water.

The general background in thermodynamics has not been developed here. For discussions of concepts of the first and second laws of thermodynamics, entropy, free energy, and chemical potential, the reader is referred to textbooks on thermodynamics, such as that by Kirkwood and Oppenheim (1962).

1.2.1 Thermodynamic Terminology Relating to the Solution Phase in Soil

A system to be treated in thermodynamics has to attain its equilibrium state, and thermodynamics deals with phenomena relating to mutual transformation of heat and work from a macroscopic standpoint. Thermodynamics requires that the state of a system is determined by its properties (its variables of state), such as temperature, volume, and pressure. It has been confirmed by countless experiments that this requirement is correct and necessary.

Types of Work Relating to Water in Soil

In general, internal energy as used in thermodynamics does not include the potential and kinetic energy of the system as used in mechanics. But when soil solution is treated thermodynamically, potential energies should be included in the internal energy because water in soil is under force fields such as the van der Waals force field. The five types of work that are used in thermodynamics of water in soil are as follows:

1. Compression work, $-P\,dV$, where P is pressure and V is the volume of the system. Within the pressure range in which water can be considered incompressible, this work is zero.
2. Work due to surface tension, σdA, where σ is the surface tension (surface free energy per unit area) of water and A is the interfacial area between air and water in soil or between soil particles and water. Note that σ is not a body force but a surface force.

   The surface tension of pure water at 20°C is 72.75 dyne/cm. But the surface tension of soil solution in a surface soil that is rich in organic matter content is considered to be less than that of pure water. Tschapek et al. (1978) found values of surface tension of soil solutions to be 63 to 64 dyne/cm in surface soils.
3. Electrical work, $-P_t\,dD = V/4\pi[(D/\bar{\varepsilon}) - D]\,dD$ (Iwata, 1972a), where D is the electric displacement, $\bar{\varepsilon}$ the partial dielectric constant, and P_t the total dielectric polarization. This work is obtained by subtracting the energy stored in a vacuum space by true electric charges from the energy stored when a dielectric substance occupies the space.

A dielectric substance changes its state when placed in an electric field, and this gives rise to a change in energy. For a system involving charges on a parallel-plate condenser, the following equation for a unit mass of dielectric substance can be used:

$$vd\left(f_1 + f_{12} + f_2^{/} + f_2^{//} + f_2^{///}\right) = \frac{v}{4\pi} E\, dD \tag{1.13}$$

where

f_1 = electrostatic energy per unit volume due to true charges of the plate, equal to $D^2/8\pi$
f_{12} = energy of interaction per unit volume between dipoles in the dielectric substance and charges
$f_2^{/}$ = energy of interaction per unit volume between dipoles and polarized charges appearing at the plate-dielectric substance boundary
$f_2^{//}$ = energy of dipole-dipole interaction per unit volume in the dielectric substance
$f_2^{///}$ = elastic energy per unit volume caused by the electric field
E = intensity of the electric field
v = volume per unit mass of dielectric substance

The work done on the dielectric substance, which was adapted by Iwata (1972a), is evidently equal to

$$vd\left(f_1 + f_{12} + f_2^{/} + f_2^{//} + (f_2^{///}\right) - v\, df_1 = \frac{v}{4\pi} E\, dD - \frac{v}{4\pi} D\, dD$$

$$= \frac{v}{4\pi}\left(\frac{D}{\overline{\epsilon}} - D\right) dD \tag{1.14}$$

$$= -P_t\, dD$$

4. Work due to the gravitational field, M dh, where M is the mass of the system under consideration and equals the sum of masses of every component, and h is the distance from a given reference height in the field. In general, this work is omitted in the thermodynamic study of water in soil, with the exception of studies on water movement and on the equilibrium water distribution in a vertical column of soil.
5. Work due to the conservative force field acting on water in soil, $M(\partial\psi/\partial z)$ dz, where ψ is the potential energy of intermolecular interaction between a clay particle surface and water molecules of unit mass, and z is the distance from the surface.

In general, a potential energy of intermolecular interaction is expressed in the form

$$\psi = -\frac{A}{z^n} \tag{1.15}$$

Accordingly, Eq. (1.15) will be adopted as the equation that gives the potential energy of interaction between an uncharged clay surface and a water molecule.

General Differential Forms of Gibbs Free Energy and Chemical Potential (Total Potential)

For an open system that can exchange both energy and matter with its surroundings, and for the types of work discussed above, the general differential form for Gibbs free energy is given by

$$\begin{aligned} dG = &\left(\frac{\partial G}{\partial T}\right)_{P,A,D,\ldots} dT + \left(\frac{\partial G}{\partial P}\right)_{T,A,D,\ldots} dP + \left(\frac{\partial G}{\partial A}\right)_{P,T,D,\ldots} dA \\ &+ \left(\frac{\partial G}{\partial D}\right)_{P,T,A,\ldots} dD + \left(\frac{\partial G}{\partial z}\right)_{P,T,A,\ldots} dz + \left(\frac{\partial G}{\partial h}\right)_{P,T,A,\ldots} dh \\ &+ \sum \left(\frac{\partial G}{\partial m_i}\right)_{P,T,A,\ldots} dm_i = -S\,dT + V\,dP + \sigma\,dA - P_t\,dD \\ &+ M\frac{\partial \psi}{\partial z}\,dz + Mg\,dh + \Sigma \mu_i\,dm_i \end{aligned} \tag{1.16}$$

where G is the Gibbs energy, S the entropy, V the volume, T the temperature, μ_i the chemical potential (total potential) defined by $(\partial G/\partial m_i)_{T,P,A,\ldots}$, and m_i the mass of the ith component.

This equation is valid for a system having curved surfaces, provided that the surface tension does not depend on the magnitude of the curvature. Strictly speaking, the effect of the curvature dependency of surface tension cannot be neglected when the principal radius of curvature at the air-water interface is very small. According to the results of Kim and Chang (1979), it should be taken into consideration where the principal radius of curvature is smaller than about 100 Å.

To simplify the explanation, we shall consider a spherical drop whose radius is r (Pitzer and Brewer, 1961). In such a case, the surface area is no longer an independent variable but is dependent on the volume (i.e. on the amount of material in the drop). The volume change on addition of increments dm_i grams of the various components is

$$dV = \Sigma \bar{v}_i \, dm_i \tag{1.17}$$

where $\bar{v}_i$ is the partial volume of the ith component. In a sphere, the relationship between the surface area and the volume is

$$dA = \frac{2}{r} dv \tag{1.18}$$

Combining Eqs. (1.16) and (1.18), we arrive at

$$dG = -S\,dT + V\,dP - P_t\,dD + M\frac{\partial \psi}{\partial z}dz + Mg\,dh + \Sigma \mu_i\,dm_i + \sum \frac{2\sigma}{r}\bar{v}_i\,dm_i \tag{1.19}$$

Since chemical potential is also a thermodynamic quantity, we have

$$d\mu_w = \left(\frac{\partial \mu_w}{\partial T}\right)dT + \left(\frac{\partial \mu_w}{\partial P}\right)dP + \left(\frac{\partial \mu_w}{\partial D}\right)dD + \left(\frac{\partial \mu_w}{\partial z}\right)dz + \left(\frac{\partial \mu_w}{\partial h}\right)dh + \sum_{j=1}^{k-1}\left(\frac{\partial \mu_w}{\partial Cj}\right)dCj + \left(\frac{\partial \mu_w}{\partial m_w}\right)dm_w \tag{1.20}$$

where μ_w is the chemical potential of water, C_j the concentration of the jth component, k the number of components, and m_w the mass of water. It should be noted that μ_w is not a function of the mass of every component, but of the concentration of every component, as μ_w is a physical amount per one gram of water. Because $\Sigma_{j=1}^{k} Cj$ = constant, the number of independent variables is (k - 1).

The last term on the right-hand side in Eq. (1.20) represents the change in chemical potential of water due to the change of r, the principal radius of curvature at the air-water interface, accompanying the increment dm_w. Evidently

$$\left(\frac{\partial \mu_w}{\partial D}\right) = \left[\frac{\partial}{\partial D}\left(\frac{\partial G}{\partial m_w}\right)\right] = \left[\frac{\partial}{\partial m_w}\left(\frac{\partial G}{\partial D}\right)\right] = -\left(\frac{\partial P_t}{\partial m_w}\right) = -\bar{P}_t$$
$$= \frac{\bar{v}_w}{4\pi}\left(\frac{1}{\bar{\bar{\epsilon}}} - 1\right) D \tag{1.20'}$$

Applying the same procedure to $(\partial\mu_w/\partial z)$, $(\partial\mu_w/\partial h)$, and $(\partial\mu_w/\partial m_w)$, we have

$$d\mu_w = -\bar{S}_w\, dT + \bar{v}_w\, dP + \frac{\bar{v}_w}{4\pi}\left(\frac{1}{\bar{\bar{\epsilon}}} - 1\right) dD + \left(\frac{\partial \psi}{\partial z}\right) dz + g\, dh$$
$$+ \sum_{j=1}^{k-1}\left(\frac{\partial \mu_w}{\partial Cj}\right) dCj + \frac{\partial}{\partial m_w}\left(\frac{2\sigma}{r}\,\bar{v}_w\right) dm_w \tag{1.21}$$

where $\bar{S}_w$ is the partial entropy of water and $\bar{v}_w$ the partial volume of water. Because m = f(r) and dm = (dm/dr)dr, we get

$$\frac{\partial}{\partial m_w}\left(\frac{2\sigma}{r}\,\bar{v}_w\right) dm_w = \frac{\partial}{\partial r}\left(\frac{2\sigma}{r}\,\bar{v}_w\right) dr \tag{1.22}$$

Substituting Eq. (1.22) into Eq. (1.21), we have

$$d\mu_w = -\bar{S}_w\, dT + \bar{v}_w\, dP + \frac{\bar{v}_w}{4\pi}\left(\frac{1}{\bar{\bar{\epsilon}}} - 1\right) dD + \left(\frac{\partial \psi}{\partial z}\right) dz$$
$$+ g\, dh + \sum_{j=1}^{k-1}\left(\frac{\partial \mu_w}{\partial Cj}\right) dCj + \frac{\partial}{\partial r}\left(\frac{2\sigma}{r}\,\bar{v}_w\right) dr \tag{1.23}$$

This result is applicable to a system having concave curved surfaces, such as the solution in soil. P is the sum of the external pressure, P_{ex}, and the internal pressure, P_{in}, as described later.

1.2.2 Chemical Potential Water in Soil

At equilibrium, the chemical potential of water in soil has a constant value at any point, and can be measured. Therefore, it is the most useful energy concept by which to understand the state of water in soil. The chemical potential of water in soil is commonly expressed by the decrement in chemical potential, taking bulk water at the same temperature and external pressure (generally atmospheric pressure) as standard. In other words, the chemical potential of water in soil is equivalent to the total potential of water in soil.

Factors Affecting Chemical Potential

Bolt and Miller (1958) have discussed the calculation of total and component water potentials. The chemical potential of water in soil, μ_w, is given approximately by the sum of the decrement in chemical potential due to each factor described below.

1) Surface Tension Effects. Water menisci are formed in soil pores in the unsaturated condition. The curved surfaces decrease the chemical potential of the water. The decrement μ_c, as represented in Eq. (1.23), is

$$\mu_c = \int_{\infty}^{r} \frac{\partial}{\partial r}\left(\frac{2\sigma}{r}\,\bar{v}_w\right) dr = \frac{2\sigma}{r}\,\bar{v}_w \tag{1.24}$$

where $\bar{v}_w$ is the partial volume of water. Equation (1.24) is equivalent to $\Delta P = \frac{2\sigma}{r}$ where ΔP is the difference between the water pressure and the air pressure when a hemispherical air-water interface exists. This is the well-known Kelvin or Laplace equation.

Radius of curvature r depends on the arrangement of soil particles and on water content, θ. In soils with a rigid matrix, r is a function only of θ. μ_c causes the condensation of water vapor from soil air when the chemical potential of water vapor relative to that of saturated vapor is larger than μ_c. The chemical potential of water vapor, μ_v, whose pressure is P, taking that of saturated vapor as standard, is given by $\mu_v = \frac{RT}{M} \ln \frac{P}{P_0}$ where P_0 is saturated vapor pressure at T. This phenomenon is called capillary condensation. It induces hysteresis in an isothermal adsorption curve for a soil, as described later. In addition, it should be noted that μ_c is a physical quantity different from work because σ is a surface force, not a body force.

2) Solutes. Chemical potential of water in soil is affected by the existence of various solutes in the soil solution. When the variables except C_j are constant, Eq. (1.23) becomes

$$[d\mu_w]_{P,T,...} = \sum \left(\frac{\partial \mu_w}{\partial Cj} \right)_{P,T,...,C_k} dC_j \tag{1.25}$$

Assuming an ideal solution, Eq. (1.25) becomes

$$[d\mu_w]_{P,T,...} = d\left[\frac{RT}{M} \ln x_w\right] = d\left[\frac{RT}{M} \ln\left(1 - \Sigma x_j\right)\right] \tag{1.26}$$

where x_w and x_j are mole fractions of water and solute j, respectively. R ln x_w in Eq. (1.26) is the entropy of mixing. Consequently, the decrement of chemical potential due to solutes, μ_0, is

$$\mu_0 = \int_1^{x_w} d\left[\frac{RT}{M} \ln x_w\right] = \frac{RT}{M} \ln\left(1 - \Sigma x_j\right) \tag{1.27}$$

If $\Sigma x_j << 1$, Eq. (1.27) becomes

$$\mu_0 = -\frac{RT}{M} \Sigma x_j = -\frac{RT}{1000} \Sigma n_j = -\Pi \bar{v}_w \tag{1.28}$$

where n_j is the molality of solute j and Π is the osmotic pressure. Consequently, the osmotic pressure is due to the entropy of mixing and is not related to work. Furthermore, under a condition where the soil solution is not ideal, μ_0 is given by the equation

$$\mu_0 = -\frac{RT}{1000} \Sigma \pi_j n_j = -\Pi \bar{v}_w \tag{1.29}$$

where π_j is the osmotic coefficient. μ_0 is equivalent to the osmotic potential.

3) Electric Field. Clay minerals and organic matter possess electric charges, and the soil solution is influenced by electric fields resulting from those charges. From Eq. (1.23), the change of chemical potential of water due to electric displacement is

$$[d\mu]_{T,P,...} = -\bar{p}_t \, dD \tag{1.30}$$

where $-\bar{p}_t$ is the partial dielectric polarization of water. Therefore, the decrement in chemical potential of water in soil due to an electric field, μ_e, is

$$\mu_e = -\int_0^D \bar{p}_t \, dD = \int_0^D \frac{D\bar{v}_w}{4\pi}\left(\frac{1}{\bar{\epsilon}} - 1\right) dD \tag{1.31}$$

where $\bar{\varepsilon}$ is the partial dielectric constant of water and D is the value of electric displacement at the point at which the water exists. The value of D depends on the shape and size of the soil particle or the organic matter that adsorbs the water, the surface density of electric charge, and the distance from the surface to the water under consideration.

4) van der Waals Force Field. Water in soil is under the influence of $\psi(z)$ expressed in Eq. (1.15). The change in chemical potential of water when we transfer 1 g of water from z to z + dz is

$$[d\mu]_{T,P,D,\ldots} = \left(\frac{\partial\psi}{\partial z}\right) dz \tag{1.32}$$

Thus, if we take ψ at $z = \infty$ as standard, the decrement of chemical potential of water in soil produced by the van der Waals force field, μ_f, is

$$\mu_f = \int_\infty^z \left(\frac{\partial\psi}{\partial z}\right) dz = [\psi(z)]_\infty^z = \frac{-A}{z^n} \tag{1.33}$$

The magnitude of $\psi(z)$ at z also depends on the shape and size of the clay or soil particle holding the water.

5) Internal Pressure. Internal pressure is a state quantity developed when water is placed in a nonuniform potential field such as the van der Waals force field or the gravitational field, and is based on external pressure as the standard. The drop in chemical potential of water near a particle surface due to the van der Waals force field, an electric field, and a concentration gradient of solutes is compensated for by increasing internal pressure from the gas-liquid interface toward the inside. In other words, the requirement of equilibrium, where the chemical potential of each component should be equal at all points in the system, is satisfied by introducing the concept of internal pressure.

We will take concrete examples as follows: Suppose that we have a system consisting of water in a container under a gravitational field as shown in Fig. 1.6. Then the chemical potentials of water at points A and B, μ_A and μ_B, are expressed by

$$\mu_A = \mu_{0A} + gH \tag{1.34}$$

$$\mu_B = \mu_{0B} + gz_B \tag{1.35}$$

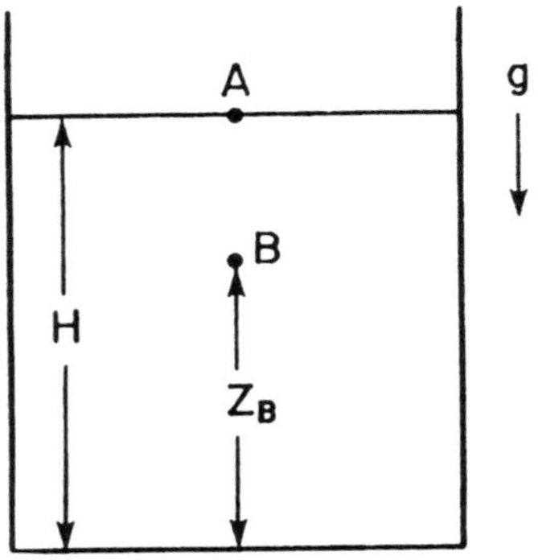

Fig. 1.6. Concept of internal pressure due to the gravitational field.

where μ_{0A} and μ_{0B} are the chemical potentials of water at points A and B, respectively, assuming that the gravitational field vanishes, but that each pressure at points A and B remains unaltered. Because of the requirement for equilibrium,

$$\mu_A = \mu_B \tag{1.36}$$

Hence

$$\mu_{0B} - \mu_{0A} = g(H - z_B) \tag{1.37}$$

How is the difference between μ_{0B} and μ_{0A} produced? We shall consider each pressure at points A and B. If atmospheric pressure (the external pressure) is taken as standard, the pressure at point A, P_A, is evidently equal to zero. The pressure at point B, P_B, is

$$P_B = \int_{z_B}^{H} \rho g \, dz = \rho g(H - z_B) \tag{1.38}$$

P_B, produced by the gravitational field, is the internal pressure.

On the other hand, when the variables except pressure are constant, Eq. (1.23) becomes

$$[d\mu]_{T,D,\ldots} = \bar{v}_w \, dP \tag{1.39}$$

Applying Eq. (1.39) to the calculation of μ_{0B}, we get

$$\mu_{0B} - \mu_{0A} = \int_0^{P_B} \bar{v}_w \, dP = \rho g(H - z_B)\bar{v}_w = g(H - z_B) \tag{1.40}$$

where ρ is the density of water. Equation (1.40) is the same as Eq. (1.38). This means that at point B the increment of chemical potential due to the internal pressure produced by the gravitational field, $\int_0^{P_B} \bar{v}_w \, dP$, is counterbalanced by the decrease of the gravitational potential, $g(H - z_B)$.

In much the same manner we can obtain the same conclusion for a system consisting of water in a container under a centrifugal force field. In Fig. 1.7, O indicates the axis of rotation, R is the radius of rotation at A, and ω is the number of revolutions per minute (rpm). The difference between the decrement of chemical potential of water due to the centrifugal potential at point B, μ_B, and that at A, μ_A, is

$$\mu_B - \mu_A = -\int_R^{R+h} x\omega^2 \, dx \tag{1.41}$$

The pressure at A is equal to atmospheric pressure (internal pressure is zero) and the internal pressure at B, P_B, is

$$P_B = \int_R^{R+h} \rho x\omega^2 \, dx \tag{1.42}$$

The increment of chemical potential due to P_B, $(\mu_B)_p$, is

$$(\mu_B)_p = \int_R^{R+h} \bar{v}_w \, \rho x\omega^2 \, dx = \int_R^{R+h} x\omega^2 \, dx \tag{1.43}$$

As the change in chemical potential due to the internal pressure at A, P_A, is zero, Eq. (1.43) becomes

$$(\mu_B)_p - (\mu_A)_p = \int_R^{R+h} x\omega^2 \, dx \tag{1.44}$$

Evidently, at point B, the increment of chemical potential due to the internal pressure is compensated by the decrement due to the centrifugal potential.

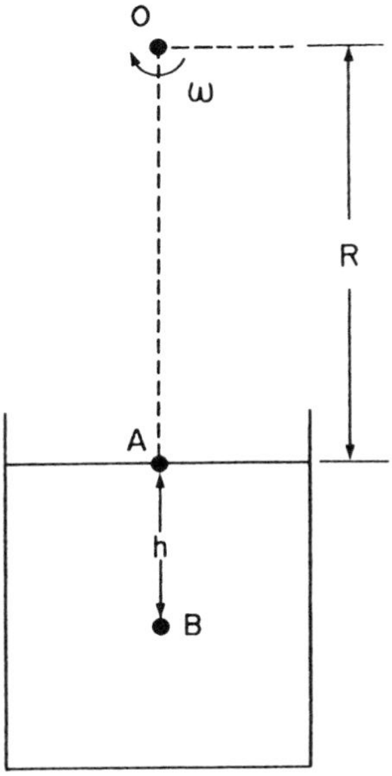

Fig. 1.7. Diagram to introduce the concept of internal pressure due to the centrifugal force field.

Internal pressure is produced only by placing a system in a nonuniform potential field. Hence if the potential energy as a function of z and the boundary conditions are given, we can calculate the value of the internal energy which depends on z. In addition, the requirement of equilibrium in a solution, that the chemical potential of each component should be equal at all points in the system, is satisfied not only by the appearance of internal pressure but also by the change in the distribution of components. Taking internal pressure as P_{in}, the increment of chemical potential due to internal pressure, $\mu_{P_{in}}$, is

$$\mu_{P_{in}} = \int_0^{P_{in}} \bar{v}_w \, dP \tag{1.45}$$

Both Gilpin (1980) and Iwata (1980) introduced a driving force for water migration in frozen soil based on the concept of internal pressure (see Section 2.4.3).

6) External Pressure. The change of external pressure also affects the value of chemical potential of water in soil. The change in chemical potential, μ_{Pex}, due to the external pressure change from P_1 to P_2 is

$$\mu_{Pex} = \int_{P_1}^{P_2} \bar{v}_w \, dP \tag{1.46}$$

If $\bar{v}_w$ is constant, Eq. (1.46) becomes

$$\mu_{Pex} = \bar{v}_w (P_2 - P_1) \tag{1.47}$$

Where water in a soil is in contact with the atmosphere, we may neglect the effect of the external pressure change on the chemical potential of water in soil because the change of atmospheric pressure is usually small. However, when measuring the water retention curve of a soil with a pressure plate apparatus, the external pressure in which the soil sample is placed plays an important role in determining the water energy status. At equilibrium we have $\mu_{Pex} + \mu_w = \mu_0 = 0$; therefore,

$$\mu_w = -\mu_{Pex} = -\int_{P_0}^{Pex} \bar{v}_w \, dP = (P_0 - P_{ex}) \bar{v}_w \tag{1.48}$$

where μ_w is the chemical potential of water in the soil sample (the pressure potential); μ_0 the chemical potential of free water at atmospheric pressure, P_0, which is taken as being zero by definition; and P_{ex} is the additional external pressure.

7) Gravitational Field. The gravitational field is considered only in studying the equilibrium state of water through a soil profile, such as the equilibrium water distribution in a vertical column of soil in contact with a free water surface. From Eq. (1.23), the change of the chemical potential due to the gravitational potential $d\mu_g$ is

$$d\mu_g = g \, dh \tag{1.49}$$

Hence

$$\mu_g = \int_{h_1}^{h_2} g \, dh \tag{1.50}$$

If the height at which water exists in the standard state is taken as zero, Eq. (1.50) becomes

$$\mu_g = gh \tag{1.51}$$

8) Water Content. The chemical potential of water in a soil changes markedly with water content. Precisely speaking, this change is due to the changes of radius of curvature of the air-water interface, and the distributions of ions and D in water in the soil arising from the increase or decrease of water content, as discussed in Sec. 1.2.2.

9) Temperature. An increase in temperature results in an increase in the chemical potential of bulk water. For example, the chemical potential of bulk water at 30°C is larger than that at 20°C by 3.85×10^9 ergs/g. This is large compared to the chemical potential of water at the permanent wilting percentage, 1.6×10^7 ergs/g.

Variables of State for Water in Soil

In a thermodynamic analysis of a system, one of the most important things is to determine those state variables that can define absolutely the state of the system.

Since the introduction by Schofield (1935) and Edlefsen and Anderson (1943) of specific free energy or chemical potential as an energy concept of water in soil, many researchers have studied the state variables required to define the chemical potential of water in soil (Gardner and Chatelain, 1946; Chatelain, 1949; Gardner et al., 1951; Babcock and Overstreet, 1955; Takagi, 1954; Slatyer and Taylor, 1960; Bolt and Frissel, 1960; Babcock, 1963; Slatyer, 1967; Sposito, 1981). The thermodynamic treatments by these authors have two common characteristics: water in soil is considered as a homogeneous system, and temperature T, external pressure P, and the volumetric water content are adopted as state variables. Specifically, the differential form of the chemical potential of water in soil with a rigid matrix has been expressed by the equation

$$d\mu_w = -\bar{S}_w\, dT + \bar{v}_w\, dP + \left(\frac{\partial \mu_w}{\partial \theta}\right)_{n_j,P,T} d\theta \tag{1.52}$$

where μ_w, $\bar{S}_w$, and $\bar{v}_w$ represent the chemical potential, the partial entropy, and the partial volume of water in soil, respectively, and n_j is the number of moles of the jth solute.

The first term of the right-hand side of Eq. (1.52) shows the dependence of the chemical potential on temperature, and the second term describes the effect of external pressure. The last term represents the contribution of the sorptive forces, that is, the electric force due to the surface charges of clay, the intermolecular force between a water molecule and clay, and the capillary force.

Equation (1.52) has been introduced on the basis of the assumption that thermodynamic quantities such as $\overline{S}_w$ and n_j would show the same values at any point in the soil water. For example, it is assumed in Eq. (1.52) that the value of $\overline{S}_w$ of water molecules in direct contact with a clay surface would be equal to that at an air-liquid interface. Is this true? The water molecules in direct contact with the clay receive strong forces from the electric field due to the surface charges of the clay and the intermolecular force field produced by interaction between a water molecule and the clay surface, in addition to the large internal pressure. On the other hand, the water molecules at the air-liquid interface receive only a very weak force from capillary action. It is more reasonable to consider $\overline{S}_w$ of the former as considerably different from that of the latter. Consequently, the introduction of Eq. (1.52) seems to be based on an incorrect assumption. This weakness in the Eq. (1.52) arises because water in soil has been dealt with as a homogeneous system.

Soil solution must be a heterogeneous system, being influenced by an electric field and a van der Waals force field. At equilibrium, although the chemical potential of water in soil shows a constant value at any point in the water, other thermodynamic quantities, such as the partial internal energy and entropy, would take different values, corresponding to the distance from the clay surface.

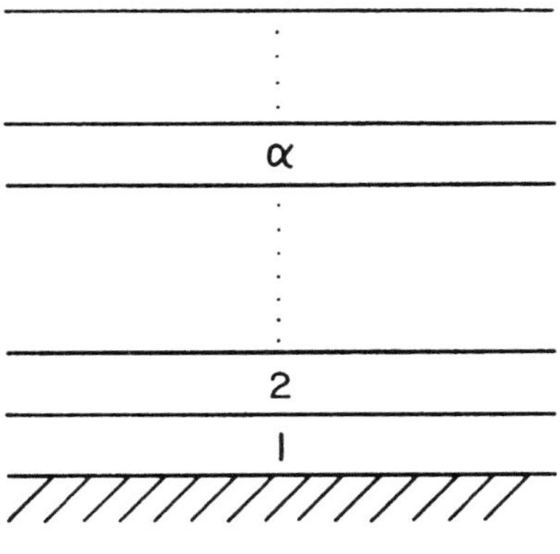

Fig. 1.8. Division of soil water into an infinite number of apparently homogeneous systems.

Therefore, we should divide the soil solution into an infinite number of apparently homogeneous systems having a uniform thickness of dz and take each of those homogeneous systems into consideration in a thermodynamic analysis of water in soil (Fig. 1.8). The variables determining the state of water in soil may be listed as follows: (1) pore radius R(x,y,z), a quantity governed by the arrangement of soil particles (x, y, and z are rectangular coordinates); (2) species, j, and the total molality of solute in the system, N_j; (3) surface density of electric charge, Γ; (4) distance from particle surface, z; (5) temperature T; (6) external pressure P_{ex}; and (7) water content θ. But in order to calculate theoretically the chemical potential of water in soil, these variables should be transformed into the following variables: (1) the principal radius of curvature at the air-water interface, r(R,θ); (2) species and molality of solute at z, $n_j(\theta,\Gamma,z,N_j)$; (3) electric displacement at z, $D(\theta,\Gamma,z,N_j)$; (4) z; (5) T; (6) P_{ex}; and (7) internal pressure $P_{in}(\theta,\Gamma,z,N_j)$.

For phase α in the solution, from Eqs. (1.24), (1.28), (1.31), and (1.33), we can obtain the following equations:

$$\begin{aligned}\mu^{(\alpha)} &= \mu^{(\alpha)\prime} + \frac{2\sigma}{r}\bar{v}_w - \int_0^{D^{(\alpha)}} \bar{P}_t \, dD + \psi\left(z^{(\alpha)}\right) \\ &= \mu^{(\alpha)\prime} + \mu_c^{(\alpha)} + \mu_e^{(\alpha)} + \mu_f^{(\alpha)} \\ &= \mu\left(T, P_{in} + P_{ex}, n_{j=0}\right) + \mu_0^{(\alpha)} + \mu_c^{(\alpha)} + \mu_e^{(\alpha)} + \mu_f^{(\alpha)}\end{aligned} \tag{1.53}$$

where $\mu^{(\alpha)\prime}$ is the chemical potential of water, assuming that the force fields vanish, but the pressure and concentration of phase α remain unaltered.

$$
\begin{aligned}
d\mu^{(\alpha)} &= \left(\frac{\partial \mu^{(\alpha)}}{\partial T}\right)_{P_{ex},D^{(\alpha)},\ldots} dT + \left(\frac{\partial \mu^{(\alpha)}}{\partial r}\right)_{P_{ex},D^{(\alpha)},\ldots} dr \\
&+ \left(\frac{\partial \mu^{(\alpha)}}{\partial D}\right)_{P_{ex},T,\ldots} dD^{(\alpha)} + \left(\frac{\partial \mu^{(\alpha)}}{\partial z^{(\alpha)}}\right)_{P_{ex},T,\ldots} dz^{(\alpha)} \\
&+ \left(\frac{\partial \mu^{(\alpha)}}{\partial P_{ex}}\right)_{T,D,\ldots} dP_{ex} + \left(\frac{\partial \mu^{(\alpha)}}{\partial P_{in}^{(\alpha)}}\right)_{P_{ex},T,\ldots} dP_{in}^{(\alpha)} \\
&+ \sum \left(\frac{\partial \mu^{(\alpha)}}{\partial n_j}\right)_{P_{ex},T,\ldots} dn_j^{(\alpha)} \\
&= -\bar{S}_w^{(\alpha)}\, dT - \frac{2\sigma}{r^2}\,\bar{v}_w\, dr + \frac{D^{(\alpha)}\,\bar{v}_w}{4\pi}\left(\frac{1}{\bar{\epsilon} - 1}\right) dD^{(\alpha)} \\
&+ \bar{v}_w\left(dP_{ex} + dP_{in}^{(\alpha)}\right) - \sum \frac{\pi_j RT}{1000}\, dn_j^{(\alpha)}
\end{aligned}
\tag{1.54}
$$

It is important to remember that the internal pressure is a function of Γ, N_j, z, and θ. As described in the preceding paragraph, the accurate values of internal pressure P_{in} can be calculated easily as a function of z, at least in a simple system such as water placed in a centrifugal field, although it has not been possible up to the present time to obtain P_{in} accurately as a function of Γ, N_j, z, and θ.

In general, since no change in $z^{(\alpha)}$ may take place, the term $dz^{(\alpha)}$ becomes zero. Moreover, for a soil with a rigid matrix, r becomes a function of θ only. In addition, as Γ, and N_j are usually constant, $D^{(\alpha)}$, $n_j^{(\alpha)}$, and $P_{in}^{(\alpha)}$ also become functions of θ only. Then we have

$$\left(\frac{\partial\mu^{(\alpha)}}{\partial r}\right)_{P_{ex},T,\ldots} dr + \left(\frac{\partial\mu^{(\alpha)}}{\partial D}\right)_{P_{ex},T,\ldots} dD^{(\alpha)}$$

$$+ \sum \left(\frac{\partial\mu^{(\alpha)}}{\partial n_j^{(\alpha)}}\right)_{P_{ex},T,\ldots} dn_j^{(\alpha)} + \left(\frac{\partial\mu^{(\alpha)}}{\partial P_{in}}\right)_{P_{ex},T,\ldots} dP_{in}^{(\alpha)} \tag{1.55}$$

$$= \left(\frac{\partial\mu^{(\alpha)}}{\partial\theta}\right)_{P_{ex},T,\Gamma,N_j,z^{(\alpha)}} d\theta$$

Thus Eq. (1.54) becomes

$$d\mu^{(\alpha)} = -\bar{S}^{(\alpha)}\, dT + \bar{v}_w\, dP_{ex} + \left(\frac{\partial\mu_w}{\partial\theta}\right) d\theta \tag{1.56}$$

This is equivalent to the equation that has been used so far. If phase α is in proximity to the charged surface and not on the outermost boundary, we may assume that

$$d\mu^{(\alpha)} = -\bar{S}^{(\alpha)}\, dT - \bar{P}_t^{(\alpha)}\, dD^{(\alpha)} + \bar{v}_w\, dP_{in}^{(\alpha)} - \sum \frac{\pi_j RT}{1000}\, dn_j^{(\alpha)} \tag{1.57}$$

Equation (1.57) has considerable significance. Generally speaking, it is useful in the case where the charges on a clay surface show changes, or where solutes move in and out of the system. In addition, the equation is applicable if the values of T are changed, where the changes in $n_i^{(\alpha)}$, $D^{(\alpha)}$, and $P_{in}^{(\alpha)}$, which occur concomitantly in phase α, cannot be controlled.

Dependence of Chemical Potential on Position in the Soil Solution

Bolt and Miller (1958), assuming that the total potential of water in moist soil was the sum of gravitational, pressure, osmotic, and adsorption components, have discussed the dependence of component potentials on position in the soil solution. A high value should be set on their research, although the component potential due to the intermolecular force field was neglected. We shall explain, referring to their work, the dependence of decrements of chemical potential due to various factors such as the electric field in an ideal clay-water system, in a coarse-grained system, and in a mixed system.

Figure 1.9 has been drawn to illustrate the essential features of the chemical potential of water only under the influence of a clay particle in an ideal clay-water system. Line e indicates the borderline beyond which the influence of the electric field may be neglected; that is, d is the thickness of the diffuse ion layer. Line f denotes the borderline beyond which the influence of the van der Waals force field may be neglected. It is assumed that the gravitational potential is neglected, as the scale of the system is very small. Then, if the external pressure is taken as standard, the chemical potentials at points A, B, C, and D are

$$\mu_A = \mu_e^{(A)} + \mu_f^{(A)} + \mu_0^{(A)} + \int_0^{P_{in}^{(A)}} \bar{v}_w \, dP \tag{1.58}$$

$$\mu_B = \mu_e^{(B)} + \mu_0^{(B)} + \int_0^{P_{in}^{(B)}} \bar{v}_w \, dP \tag{1.59}$$

$$\mu_C = \mu_0^{(C)} \tag{1.60}$$

$$\mu_D = \mu_0^{(D)} \tag{1.61}$$

and

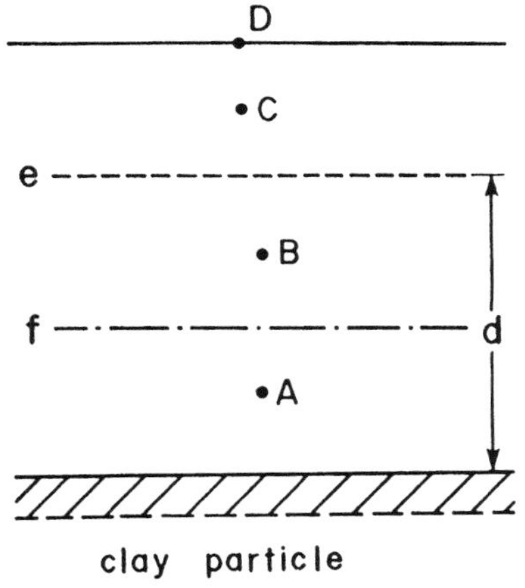

Fig. 1.9. Schematic diagram of water adsorbed by a clay particle.

$$\mu_A = \mu_B = \mu_C = \mu_D = \frac{RT}{M} \ln \frac{P}{P_0} \tag{1.62}$$

where the subscripts A, B, C, and D represent the thermodynamic quantities at points A, B, C, and D, respectively, P is the water vapor pressure in equilibrium with this water film, and P_0 is the saturated vapor pressure.

In addition, when the air-liquid interface of the water film exists on the line drawn through A or B, and point A is not too close to the periphery, μ_A and μ_B become

$$\mu_A = \mu_f^{(A)} + \mu_0^{(A)} \tag{1.63}$$

$$\mu_B = \mu_f^{(B)} + \mu_0^{(B)} \tag{1.64}$$

The term μ_e does not exist in Eqs. (1.63) and (1.64) because the strength of the electric field is considered zero at the interface when the distance of the interface from the clay surface is larger than the average distance between adjacent electrons on the clay. This will be discussed later.

Figure 1.10 shows the chemical potential of water between parallel interacting plates in an ideal clay-water system. Point E is located somewhere between the plates, and point F exists at the midplane. The values of chemical potential at E and F are

$$\mu_E = \mu_f^{(E)} + \mu_e^{(E)} + \mu_0^{(E)} + \int_0^{P_{in}^{(E)}+P_{ex}} \bar{v}_w \, dP + z^{(E)}g \tag{1.65}$$

$$\mu_F = \mu_f^{(F)} + \mu_0^{(F)} + \int_0^{P_{in}^{(F)}+P_{ex}} \bar{v}_w \, dP + z^{(F)}g \tag{1.66}$$

Of course, if points E and F are located where the intermolecular force does not act, the term μ_f in Eq. (1.65) or (1.66) disappears. It should be noted that the pressure depends on the position in the soil solution; this is easily understood by comparing Eqs. (1.58), (1.59), and (1.60) or Eq. (1.63) and Eq. (1.64). In addition, if boundary conditions such as the surface charge density and the potential of intermolecular forces are given, we can calculate the approximate values of μ_e, μ_f, and μ_0 at each point. This will be discussed later. An example of the calculation is shown in Fig. 2.29.

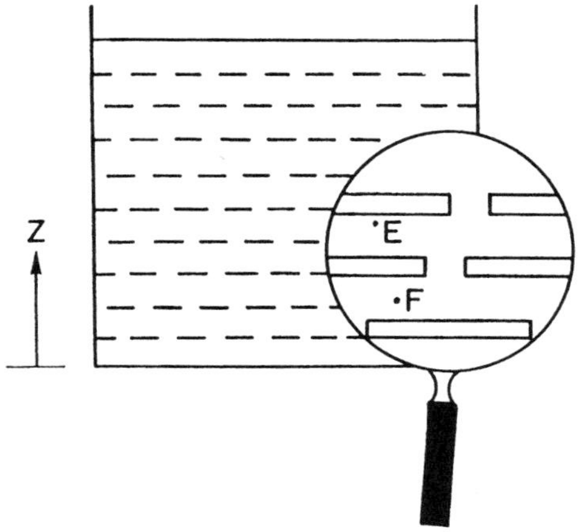

Fig. 1.10. Schematic representation explaining chemical potential of water between parallel, interacting clay plates. (From Bolt and Miller, 1958)

In the description above, the influence of the curved air-water interface on the chemical potential of water inside the interface is neglected by assuming that the interface is a plane. When the soil solution has a curved surface, the analysis of chemical potential due to position becomes very difficult, because it is not clear how far the influence of the curved surface extends. Therefore, the reader should not accept an analysis of the curved surface without careful consideration. Let us take an example. In Fig. 1.11a, water in a tube is evidently under the influence of the curved surface. The correctness of this conclusion is easily confirmed by the following equations:

$$\mu_A = \frac{2\sigma}{r}\bar{v}_w + gh \tag{1.67}$$

$$\mu_B = \frac{2\sigma}{r}\bar{v}_w + gz_B + \int_{z_B}^{h} \rho g \bar{v}_w \, dz = \frac{2\sigma}{r}\bar{v}_w + gh \tag{1.68}$$

$$\mu_C = \frac{2\sigma}{r}\bar{v}_w + \int_{0}^{h} \rho g \bar{v}_w \, dz = \frac{2\sigma}{r}\bar{v}_w + gh \tag{1.69}$$

Therefore,

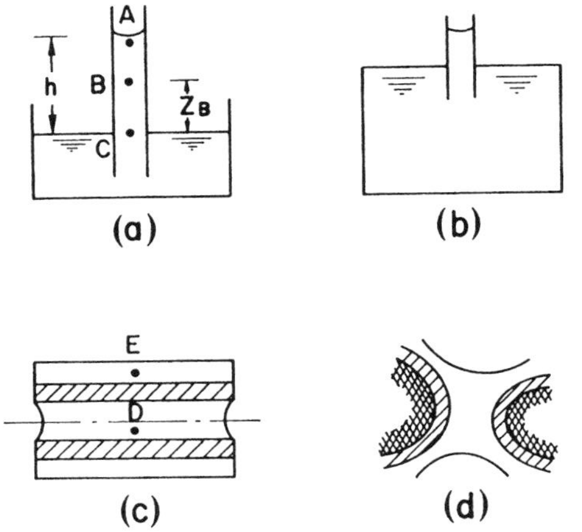

Fig. 1.11. Schematic diagram illustrating the influence of a curved surface on the chemical potential of water.

$$\mu_A = \mu_B = \mu_C \tag{1.70}$$

where r is the radius of curvature of the meniscus.

In Fig. 1.11b, it can be confirmed, in much the same way, that the existence of the meniscus extends in influence to all the water in the tube and the container. Figure 1.11c is the schematic representation of a water droplet containing parallel clay plates, proposed by Bolt and Miller (1958). In this case, it may be considered that the influence of menisci extends to the water between the plates but not to the water adsorbed at the outer surfaces. That is,

$$\mu_D = \frac{2\sigma}{r}\bar{v}_w + \mu_e^{(D)} + \mu_f^{(D)} + \mu_P^{(D)} + \mu_0^{(D)} \tag{1.71}$$

$$\mu_E = \mu_e^{(E)} + \mu_f^{(E)} + \mu_P^{(E)} + \mu_0^{(E)} \tag{1.72}$$

where μ_p is the increment of chemical potential due to pressure. In Fig. 1.11d, most of the unshaded portion is under the influence of the curved air-water interface, but it is difficult to draw a clear line between the affected and unaffected areas, because the degree of influence changes continuously from 1 to 0 between them.

For a coarse-grained system, it is reasonable to assume that an electric field is not produced. Therefore, if the gravitational potential is neglected because the scale of the system is very small, the values of chemical potential at each point in Fig. 1.12 are given by

$$\mu_A = \frac{2\sigma}{r}\bar{v}_w + \mu_0^{(A)} \tag{1.73}$$

$$\mu_B = \frac{2\sigma}{r}\bar{v}_w + \mu_0^{(B)} = \frac{2\sigma}{r}\bar{v}_w + \mu_0^{(A)} \tag{1.74}$$

$$\mu_C = \frac{2\sigma}{r}\bar{v}_w + \mu_0^{(C)} + \mu_f^{(C)} + \mu_P^{(C)} \tag{1.75}$$

$$\mu_D = \mu_0^{(D)} + \mu_f^{(D)} + \mu_P^{(D)} \tag{1.76}$$

where the dashed line in Fig. 1.12 indicates the distance to which the influence of the van der Waals force field extends. The concentration of solutes is uniform outside the dashed line.

In a mixed system, small fragments of an ideal clay-water system are scattered through a coarse-grained system. From our earlier statements it is easy to understand the dependence of decrements in chemical potential due to the various factors on position in the system. In addition, as is evident from the description above, the pressure potential is equivalent either to μ_c or to the sum of μ_e, μ_f, $\mu_{P_{in}}$, and μ_c of the water under consideration, where μ_c, μ_e, μ_f, and $\mu_{P_{in}}$ are the decrements of the chemical potential of water in soil due to surface tension effects, the electric field, the van der Waals force field, and the internal pressure, respectively.

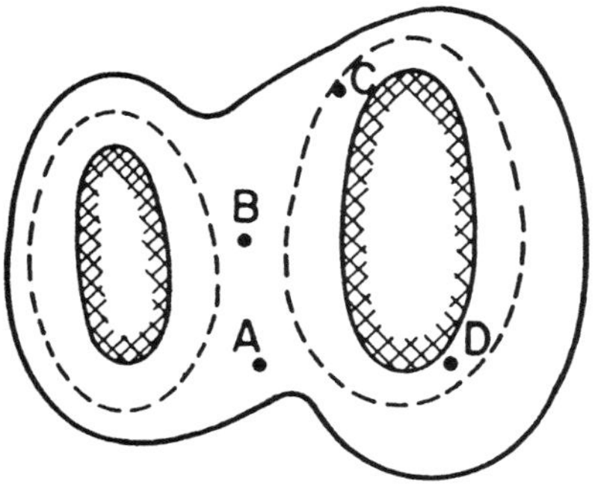

Fig. 1.12. Schematic representation for values of chemical potential of water in a coarse-grained soil system.

1.2.3 Chemical Potential as a Driving Force for Liquid Water Flow Under Nonisothermal Conditions

Soil temperature fluctuates in time and space. This fluctuation produces both vapor and liquid water flow, generally occurring simultaneously. Water flow induced by temperature is the result of interaction among various forces. Therefore, it is very difficult to analyze the phenomenon by the usual mechanical techniques. At present, researchers are making strenuous efforts to introduce new techniques to analyze water flow arising from temperature differences.

Temperature Dependence of the Tensiometer Pressure-Water Content Relationship

Temperature changes are accompanied by changes in specific energy of solutions and changes in the distribution of ions in the diffuse layers around soil particles. The specific surface energy of the solution, surface tension, decreases with temperature. The distribution of ions obeys Boltzmann's law of energy distribution [Eq. (2.10)] and the dielectric constant of the solution depends on temperature. The distributions of electric potential and internal pressure also change with changes in the distribution of ions. At constant water content, when the temperature T increases to T + dT, the change of chemical potential of water in soil is expressed by the following equation [see Eq. (1.54)]:

$$(d\mu)_T = -\bar{S}_w^{(\alpha)}\, dT + \bar{v}_w^{(\alpha)} \left(dP_{in}^{(\alpha)}\right)_T + \frac{2\bar{v}_w}{r} (d\sigma)_T + \frac{\bar{v}_w (D^{(\alpha)})^2}{8\pi} \left(d\frac{1}{\epsilon}\right)_T + \frac{\bar{v}_w D^{(\alpha)}}{4\pi} \left(\frac{1}{\epsilon} - 1\right) \left(dD^{(\alpha)}\right)_T - \sum \frac{\pi_j RT}{1000} \left(dn_j^{(\alpha)}\right)_T \tag{1.77}$$

where the subscript T means the change of a physical quantity produced by temperature change. Consequently, the change of chemical potential, $\Delta\mu$, corresponding to a temperature change from T_1 to T_2, is given by

$$(\Delta\mu)_T = \int_{T_1}^{T_2} -\overline{S}_w^{(\alpha)}\, dT + \int_{T_1}^{T_2} \overline{v}_w^{(\alpha)} \left(dP_{in}^{(\alpha)}\right)_T + \frac{2\overline{v}_w}{r} \int_{T_1}^{T_2} \left(\frac{d\sigma}{\partial T}\right) dT$$
$$+ \int_{T_1}^{T_2} \frac{D^{(\alpha)}\overline{v}_w}{4\pi} \left(\frac{1}{\epsilon} - 1\right) \left(dD^{(\alpha)}\right)_T - \sum \frac{\pi_j RT}{1000} \int_{T_1}^{T_2} \left(dn_j\right)_T \tag{1.78}$$

If a system consisting of a tensiometer and a soil is in thermodynamic equilibrium at temperature T, the tensiometer pressure is equivalent to $2\sigma/r$. Consequently, the change of tensiometer pressure P due to the change of temperature is given by

$$\Delta P_T = \frac{2}{r}\left[\sigma(T') - \sigma(T)\right] = P_T\left[\frac{\sigma(T')}{\sigma(T)} - 1\right] \tag{1.79}$$

where P_T is the tensiometer pressure at T.

Some researchers have studied the temperature dependence of the tensiometer pressure-water content relationship (Richards et al., 1937; Gardner, 1955; Peck, 1960; Wilkinson and Klute, 1962; Chahal, 1964; Haridasan and Jensen, 1972). Their results are summarized as follows. The tensiometer pressure at the same water content decreases as temperature increases, and the decrement of the tensiometer pressure is larger than the value predicted by the change of surface tension due to the temperature change [Eq. (1.79)]. Peck (1960) and Chahal (1964) explained this difference as the effect of air in soil water. However, Haridasan and Jensen (1972) reported that a difference still remained even when enclosed air was excluded. Constantz (1991) measured isothermal water retention over a range in matric potential from 0 to -100 kPa and isobaric water retention experiments at 20° and 80°C on core samples of a sandy soil and a nonwelded tuff. The temperature dependence of matric potential at constant water content has a mean magnitude several times greater than would be predicted from the temperature sensitivity of the surface tension of water. Nimmo and Miller (1986) obtained experimental results using glass beads, a sand and a silt loam, and concluded that the difference is mainly due to the increase of concentration and effectiveness of dissolved surfactants such as fulvic and humic acid with temperature. The cause of this disagreement has not yet been clarified.

Driving Forces

Under isothermal conditions, a component moves from phase α toward phase β when the chemical potential of the component in phase α is larger than that in phase β. The gradient of chemical potential is the driving force in transport phenomena. However, if the temperature in a system is not uniform, the statement above is incorrect. Let us show an example. The chemical potential of water vapor decreases with the increase of temperature. Then water vapor in soil should move from an area of lower temperature to one of higher temperature. We have not recognized such a phenomenon.

According to the thermodynamics of irreversible processes, the driving force X_i related to transport of the component i in a system when an isothermal condition is not satisfied, is given by

$$X_i = \nabla_T \mu_i = \nabla\mu_i + \bar{S}\,\nabla T = \nabla\mu_i - \left(\frac{\partial\mu_i}{\partial T}\right)\nabla T \tag{1.80}$$

The subtraction of $(\partial\mu_i/\partial T)\,\nabla T$ performs an operation that converts a real system into an imaginary isothermal system. Consequently, X_i signifies the gradient of chemical potential of i in the imaginary isothermal system.

At constant water content, if a temperature gradient exists, the driving force for liquid water in soil is obtained from Eq. (1.77)

$$\begin{aligned} X_i = \nabla_T\mu &= \bar{v}_w\,\nabla\left(dP_{in}^{(\alpha)}\right)_T + \frac{\bar{v}_w\left(D^{(\alpha)}\right)}{4\pi}\left(\frac{1}{\epsilon} - 1\right)\nabla\left(D^{(\alpha)}\right)_T \\ &+ \frac{2\bar{v}_w}{r}\left(\frac{\partial\sigma}{\partial T}\right)\nabla T - \sum \frac{\pi_j RT}{1000}\,\nabla\left(n_i^{\alpha}\right)_T \end{aligned} \tag{1.81}$$

When both water content θ and temperature change, the driving force becomes

$$\begin{aligned} X_i = \nabla_T\mu &= \bar{v}_w\,\nabla\left[\left(P_{in}^{(\alpha)}\right)_T + \left(P_{in}^{(\alpha)}\right)_\theta\right] + \frac{\bar{v}_w D^{(\alpha)}}{4\pi}\left(\frac{1}{\epsilon} - 1\right) \\ &\times \nabla\left[\left(D^{\alpha}\right)_T + \left[D^{\alpha}\right)_\theta\right] + \frac{2\bar{v}_w}{r}\left(\frac{\partial\sigma}{\partial T}\right)\nabla T + \frac{\partial}{\partial\theta}\left(\frac{1}{r}\right)2\bar{v}_w\sigma \\ &- \sum \frac{\pi_i RT}{1000}\,\nabla\left[\left(n_j^{\alpha}\right)_T + \left(n_j^{\alpha}\right)_\theta\right] \end{aligned} \tag{1.82}$$

where the subscript θ means the change of a physical quantity produced by water content. It should be noted that the magnitude of each driving force in Eq. (1.82) is considered to be dependent on the distance from the surface of a particle.

1.3 HYSTERESIS IN SOIL WATER PHENOMENA

Suppose that a system is taken from state A to state B along a given path by slowly changing an independent external variable of the system, such as the vapor pressure. Then a dependent variable (e.g. water content) will pass through a certain set of values. If the independent variable is returned along the same path from B to A at an exceedingly slow rate, the dependent variable may pass through the same set of values as those found in the forward change. This is a reversible or equilibrium process. However, we also find a class of processes for which the path taken by a dependent variable during the change from B to A is different from that during the change from A to B, while all points on both paths correspond to stable and reproducible values of the dependent variable. Then the change AB is said to exhibit hysteresis (Everett and Whitton, 1952). Hysteresis complicates the physical analysis of a process. Unfortunately, hysteresis is common in soil water phenomena, so it must be considered.

1.3.1 Hysteresis

We call curve ABC (Fig. 1.13), which starts from saturation and reaches a limiting value of pressure potential ψ_{min}, the first drainage curve. The two branches, CDE and EFC, of the main hysteresis loop CDEFC are named the drying and wetting boundary curves or the main drying and wetting curves. The region enclosed within CDEFC is called the hysteresis region. Depending on the location of the reversal point in the process of successive wetting and drying, we have various "scanning" curves. Curves DIC and FGE are primary wetting and drying (scanning) curves, and curves GH and IJ are called the secondary drying and wetting (scanning) curves, respectively.

Hysteresis in the relationship between the pressure potential ψ and the saturation percent of water in soil S was first discussed by Haines (1930). Later, hysteresis was observed by many researchers in the relationship between hydraulic conductivity K, and ψ or S. Hysteresis is found in the high water content range, and is due mainly to complicated shapes of capillary pores in soil. Mechanisms of hysteresis are described in Sec. 1.3.2.

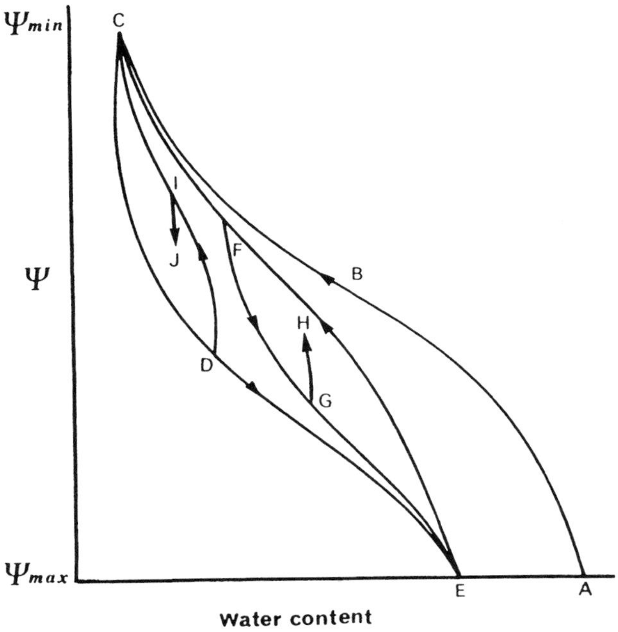

Fig. 1.13. Schematic representation of hysteresis in soil water potential, Ψ, versus water content curves.

Figure 1.14a shows $\psi(S)$ hysteresis curves for a monodispersed glass bead sample. The experimental values were obtained by Topp and Miller (1966). The solid lines and dashed lines in the figure represent the curves predicted by using the dependent domain theory, model III (Mualem and Dagan, 1975), and the independent theory, model II (Mualem, 1973, 1974), respectively. Figure 1.14b and Fig. 1.15 give the $K(\psi)$ and $K(S)$ hysteresis curves for the same monodispersed glass bead sample. The theoretical curves calculated by Mualem (1976b), shown by the solid lines, are compared with the values measured by Topp and Miller (1966). The $\psi(S)$ and the $K(\psi)$ relationships show considerable hysteresis, while the $K(S)$ relationship is almost a unique curve.

Hysteresis in the $\psi(\theta)$ relationship has been observed widely for many soil samples in the laboratory. The magnitude of the measured hysteresis varies with the sample. It has been clearly shown that $\psi(\theta)$ hysteresis is also important in the field and cannot be neglected (Royer and Vachaud, 1975; Watson et al., 1975; Beese and van der Ploeg, 1976; Tzimas, 1979) (Fig. 1.16). It is also commonly recognized among soil physicists that $K(\psi)$ hysteresis is significant.

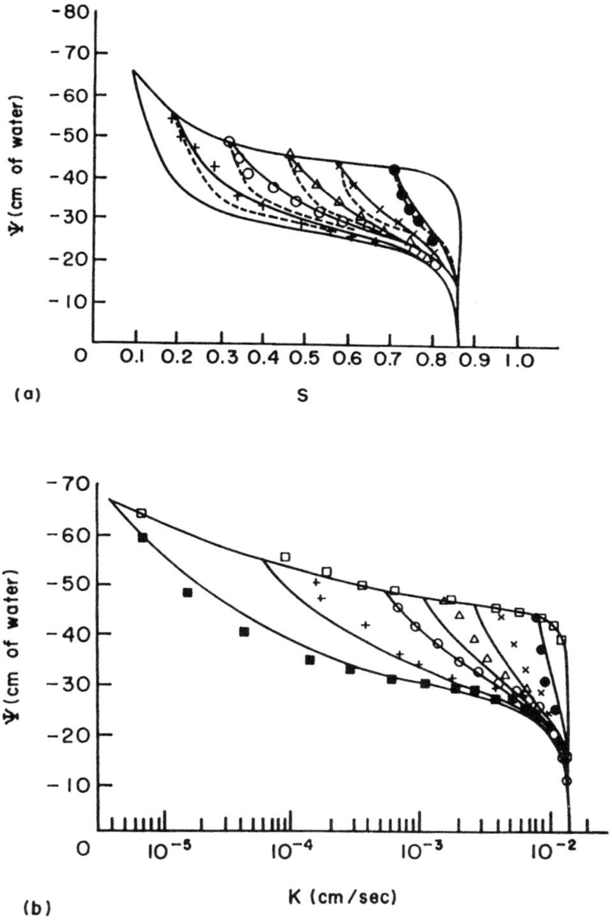

Fig. 1.14. S-ψ and K-ψ curves for monodispersed glass bead samples. (From Topp and Miller, 1966)

K(S) hysteresis was negligible in data reported by Nielsen and Biggar (1961) for a silt loam, Narr et al. (1962) for a consolidated sandstone, Green et al. (1964) for a silt loam, Elrick and Bowman (1964) for a loam, Topp and Miller (1966) for two glass bead samples and an admixture of different grades of glass beads, Topp (1969) for a sandy loam, Talsma (1970) for sands, Topp (1971a) for a silt loam and a clay loam soil, and Tzimas (1979) for a soil monolith. However, hysteresis was found by Narr et al. (1962) for a sand, Youngs (1964) for porous materials, Staple (1965, 1966) for loam soils, Collis-George and Rosenthal (1966) for a sand, Poulovassilis (1969, 1970) for sands, Poulovassilis

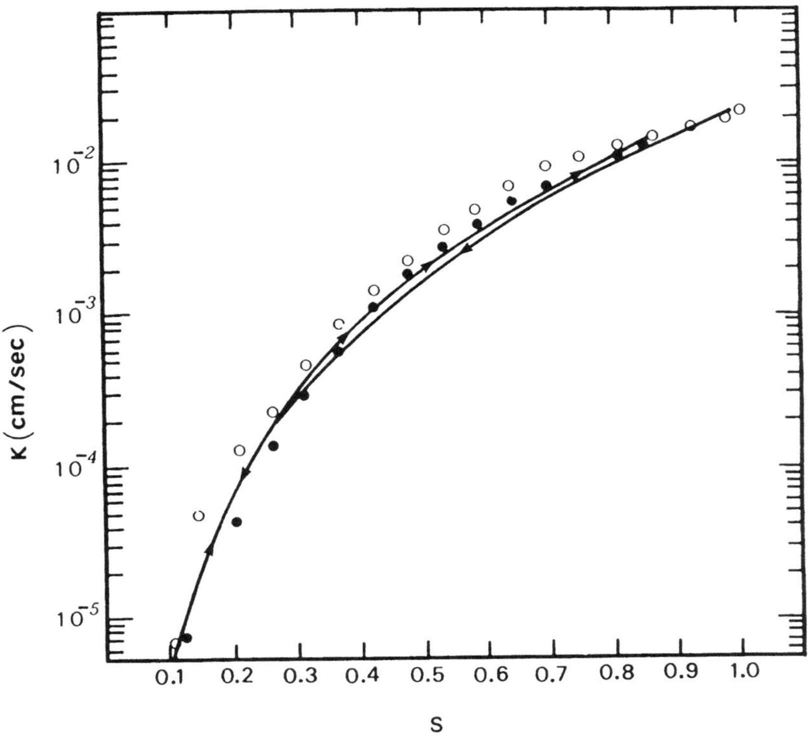

Fig. 1.15. Predicted K(S) for first drainage and rewetting (solid lines) compared with experimental results in drainage (open circles) and wetting (solid circles). (From Topp and Miller, 1966; Mualem, 1976b)

and Tzimas (1974, 1975) for a sand and an admixture of glass beads, Pavlakis and Barden (1972) for clay soils, and Dane and Wierenga (1975) for a sand and a clay loam. Two examples showing a large hysteresis in K(θ) and K(S) are shown in Figs. 1.17 and 1.18 for a sand and a clay.

Although further research is needed, it seems better to use the K(S) function rather that a K(ψ) function, as pointed out by Mualem and Morel-Seytoux (1979). It is interesting that in spite of large hysteresis in the ψ(S) relationship, the K(S) hysteresis is often small. We can suppose two water conditions, (ψ_1,S_1) and (ψ_2,S_2), where ψ_1 is different from ψ_2, while $S_1 = S_2$. Then the spatial distribution of water corresponding to ψ_1 must be different from that corresponding to ψ_2. Why can the different distributions result in identical values of hydraulic conductivity?

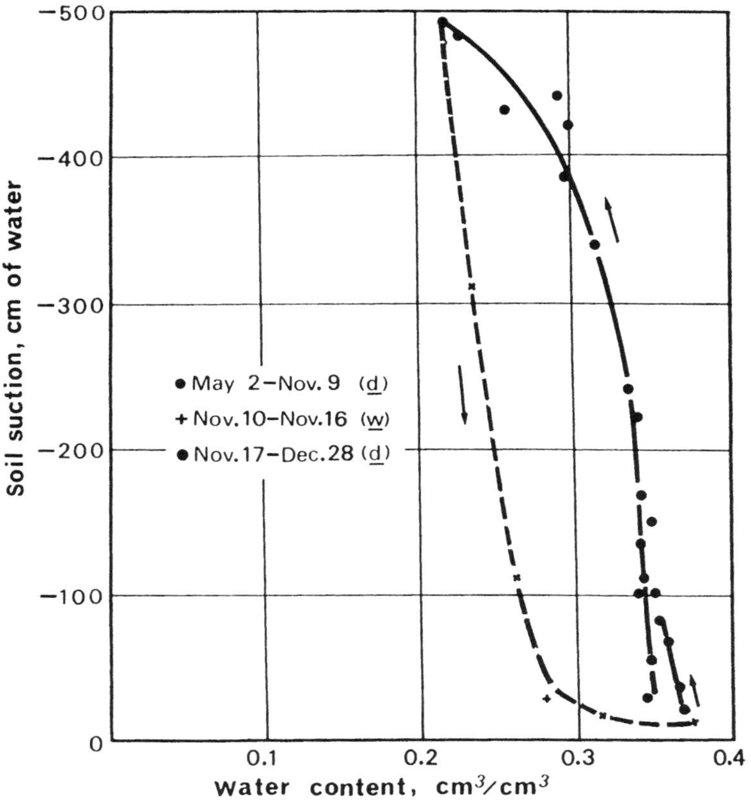

Fig. 1.16. Soil suction-water content relationships for a heavy clay soil in the field. (From Royer and Vachaud, 1975)

Hysteresis has also been observed in the relationship between vapor pressure P (or chemical potential of water μ) and the amount of adsorbed water, n, at the low water range. Results obtained by van Olphen (1965) are shown in Fig. 1.19 as an example. Mason (1982) proposed a method for determining the connectivity of pore space of an adsorbent, using hysteresis of the adsorption-desorption isotherm. Hysteresis also exists in the relationship between the conductivity of air (or liquid), K_v, and μ; or K_v and the volumetric water content, θ. However, experimental data on K_v-μ and K_v-θ hysteresis seem to be very limited (Jackson, 1964; Rose, 1971). Figure 1.20 shows the result for Highfield clay loam obtained by Rose.

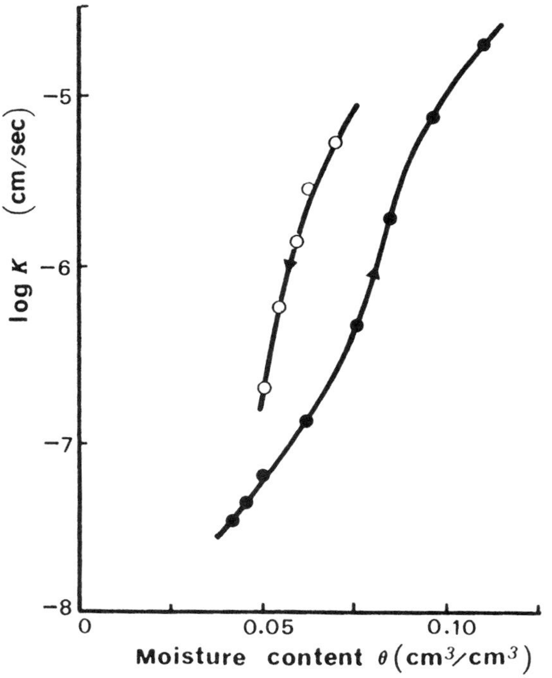

Fig. 1.17. Hysteresis in the relationship between measured hydraulic conductivity K and water content θ for 0.1 to 0.2 mm sand; open circle and solid circles indicate drying and wetting, respectively. (From Collis-George and Rosenthal, 1966)

1.3.2 Mechanisms for Hysteresis

Hysteresis in the relationship between water content and pressure potential or chemical potential may be considered to result from three specific mechanisms.

Different Pore Sizes

Assume two capillary tubes at right angles to a free water surface (Fig. 1.21), where d and D are capillary diameters, h_D a distance from the free water surface to the section with the larger diameter, P_0 the atmospheric pressure, and P the pressure on the free water surface. P can take any value because the gas in contact with the surface is isolated from air. We assume that equilibrium is

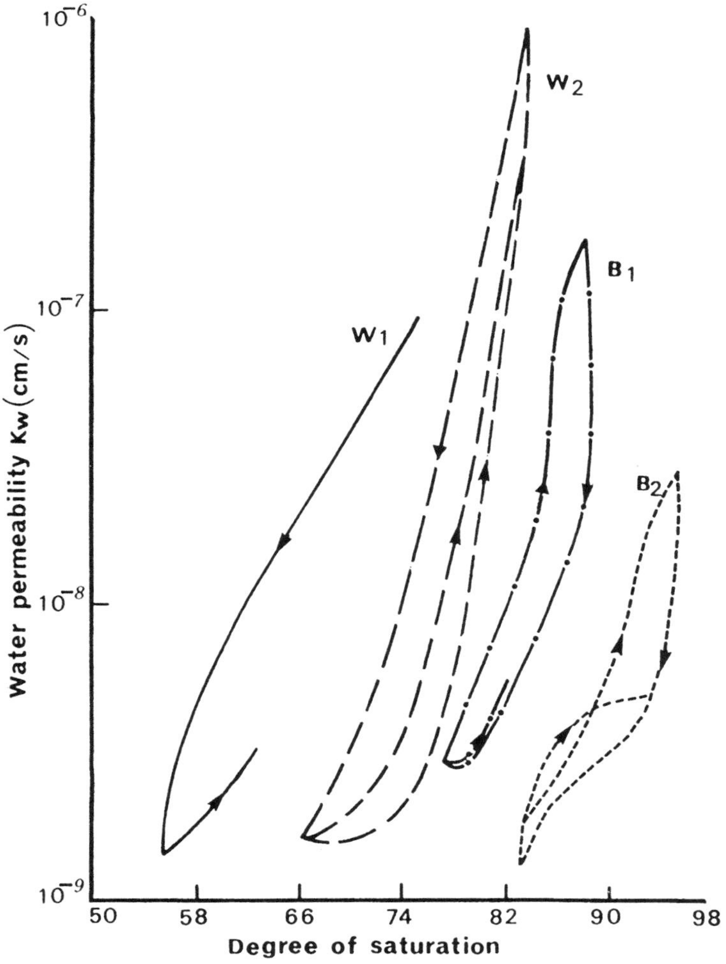

Fig. 1.18. Hysteresis curves in the relationship between degree of saturation and water permeability for two clays, B and W. Subscripts 1 and 2 represent different initial properties of test samples. (From Pavlakis and Barden, 1972)

reached with the meniscus positions as shown in Fig. 1.21 when P equals P_1. Then, if P is decreased, the air-water interfaces in capillaries A and B are lowered, and when P is reduced to P_0 - $[4\sigma/d]$ (σ is the surface tension), the interfaces reach the level of the free water surfaces. If P is then increased to P_1, both air-water interfaces return again to the positions shown in Fig. 1.21, where

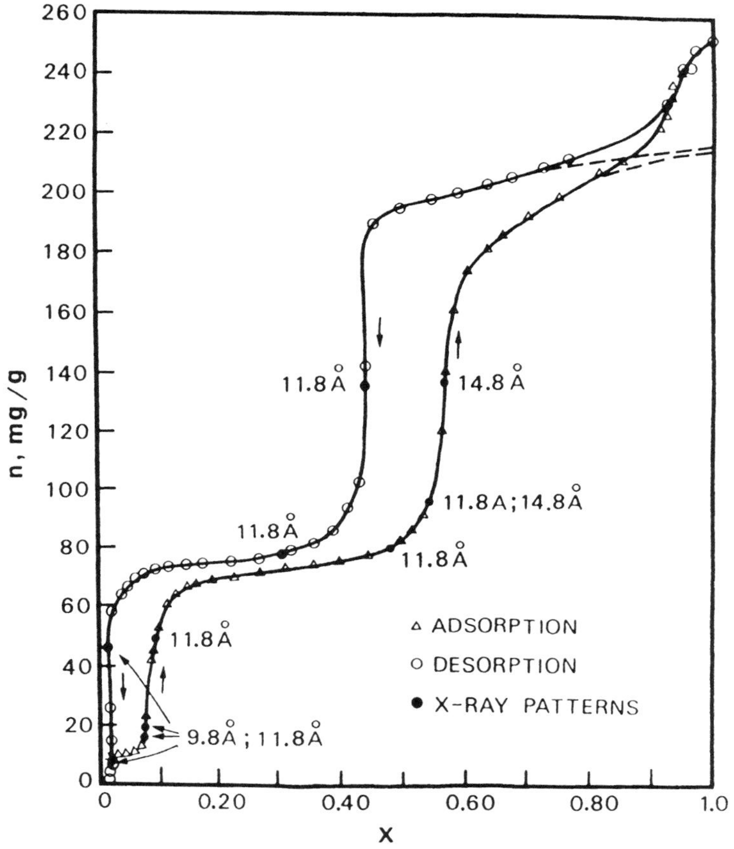

Fig. 1.19. Adsorption and desorption isotherms for water vapor on sodium vermiculite at 25°C. (From van Olphen, 1965)

a condition, $P_o + \rho g h_D - [4\sigma/D] < P_1$, is satisfied ($\rho$ is the density of water and g is the acceleration of gravity).

Figures 1.22a and b show the relationships between pressure and the quantity of water retained in capillary A and capillary B of Fig. 1.21. The changes indicated by the solid lines in Fig. 1.22 are stable and reversible; the changes given by the dashed lines are spontaneous and irreversible. Curves I and II represent the P_B water content relationship in the cases where the condition ($P_o + \rho g h_D - [4\sigma/D] < P_1$) is satisfied and where the condition is not satisfied, respectively. Evidently, the relationship including curve I exhibits hysteresis. If the condition is not satisfied, the air-water interface in capillary B cannot go up to h_D with increasing P. In this case, the relationship between P and water content does not exhibit hysteresis. Water in soil is retained in a great number

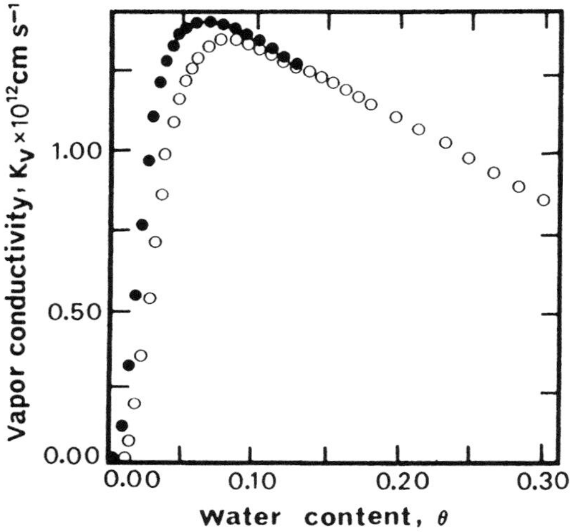

Fig. 1.20. Hysteresis in vapor conductivity. ● sorption; o desorption. (From Rose, 1971)

of pores with very different sizes and more complicated shapes than the capillaries shown in Fig. 1.21. Consequently, water content-pressure potential hysteresis can be found over a considerable water content range.

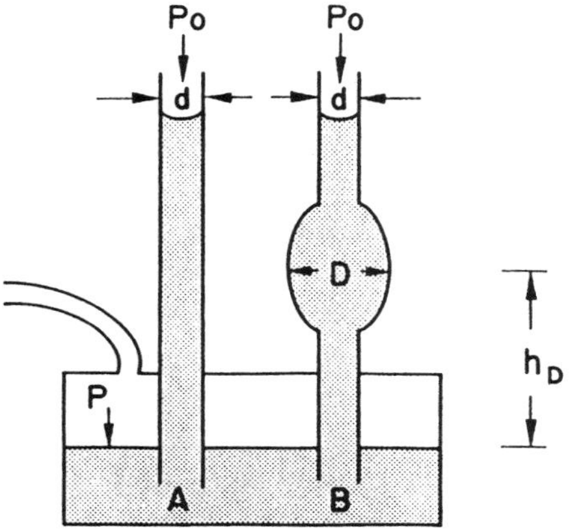

Fig. 1.21. Schematic representation to explain capillary hysteresis.

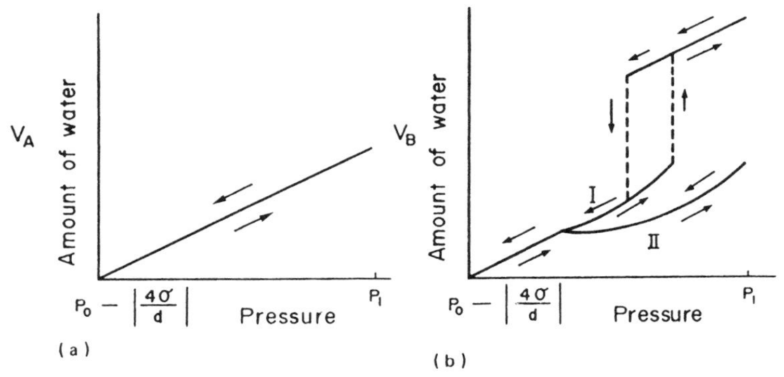

Fig. 1.22. Pressure-water content relationships in the models shown in Fig. 1.21: (a) relationship for model A and (b) relationship for model B.

Adsorbed Water on Clay Surfaces

In Fig. 1.23 water exists between two flat, parallel clay surfaces; the distance between surfaces is 2h, and the thickness of adsorbed water is h′. Water in state A is adsorbed by van der Waals forces and that in state B is retained by surface tension effects. Generally, h is recognized to be larger than h′ when water under the two states is in equilibrium with the same water vapor pressure. For example, if h is 50 Å, h′ is less than 10 Å. The chemical potential of water in state A, equal to that in state B, is 2.9 x 10^8 ergs/g. State A corresponds to a wetting process, and state B corresponds to a drying process.

We will assume the vapor pressure p(h) to be in equilibrium with water of thickness h adsorbed by a flat clay surface. If vapor pressure is increased sufficiently slowly from 0 to p(h) and afterward decreased to 0 again, the relationship between water vapor pressure p and quantity of adsorbed water V is given by the curves shown in Fig. 1.24. Evidently, the p-V relationship exhibits hysteresis. The mechanism described above may be regarded as capillary condensation, which explains adsorption hysteresis. Other capillary condensation theories are described in detail by Everett (1967).

Hysteresis induced by different pore sizes or different water layers is due to irreversible changes accompanying changes of the external pressure or the vapor pressure. Consequently, the water retention curve in the range where hysteresis can be found should be considered to consist of discontinuous curves on a micro- [illegible] as shown in Fig. 1.25. The discontinuities are not experimentally [illegible] because they are so close together.

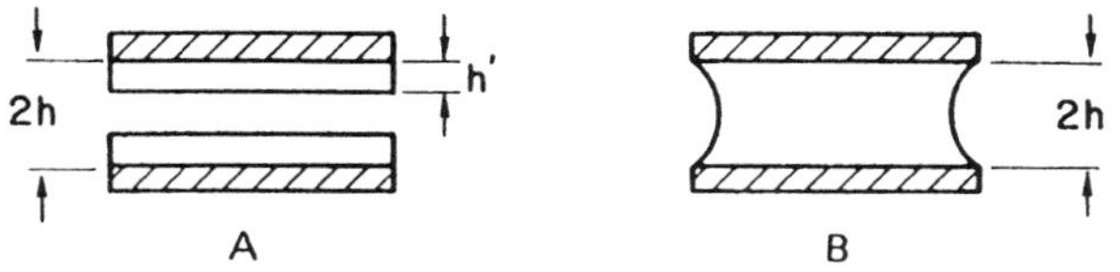

Fig. 1.23. Schematic representation to explain hysteresis due to capillary condensation.

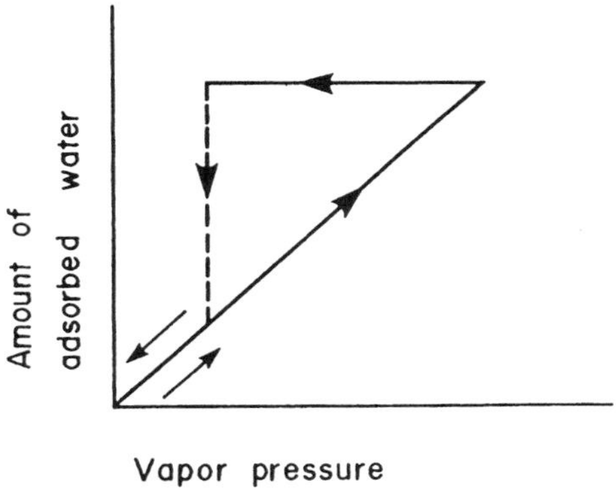

Fig. 1.24. Relationship between vapor pressure (chemical potential) and amount of adsorbed water in the model shown in Fig. 1.23.

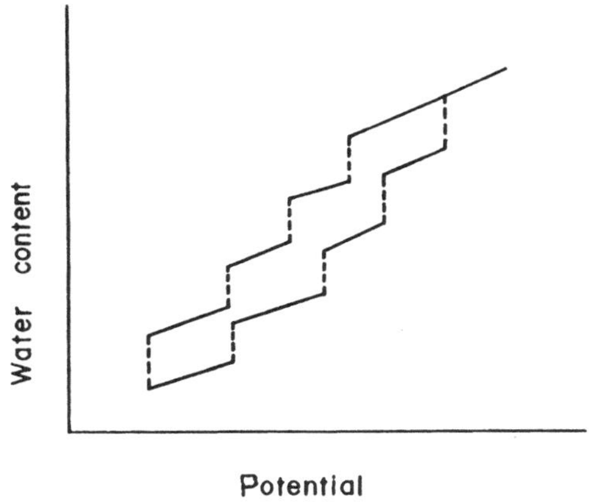

Fig. 1.25. Schematic representation of the relationship between chemical potential, or pressure potential, and water content in a soil exhibiting hysteresis.

Contact Angle

The contact angle between an air-water interface and a solid surface exhibits hysteresis. The angle reaches its maximum value when water moves toward a dry surface and takes its minimum value when water recedes. Letey et al. (1962) measured contact angles for various soils and solutions in wetting (e.g. 66.9° for Hanford soil and 65.2° for Aiken soil for water at 20°C). Laroussi and De Backer (1979) obtained 66° for wetting and 41° for drying in glass beads. In addition, they analyzed the influence of the contact angles and of the geometric properties of the porous media on the hysteresis loop. As a result of this contact angle hysteresis, a row consisting alternatively of air bubbles and of water drops can be maintained against a significant pressure drop between the two ends of a tube (Mualem and Morel-Seytoux, 1979).

1.3.3 Theory of Hysteresis Based on the Concept of Domains

Fundamental Independent Domain Theory

The fundamental theory of hysteresis based on the independent domain model was first proposed by Preisach (1935) and Neel (1942, 1943), and developed by Everett and his co-workers (1952-1954) and Enderby (1955). They divided a system exhibiting hysteresis into domains, where each domain exists in one of two states, with $\lambda_{12}^{(\alpha)} > \lambda_{21}^{(\alpha)}$, where α indicates the number of the domain (α = 1,2,...,α,...). The values of $\lambda_{12}^{(\alpha)}$ and $\lambda_{21}^{(\alpha)}$ are not influenced by the neighboring domains. The conversion from one state to the other is brought about by increasing or decreasing an external independent variable, λ, such as chemical potential or temperature. Further, λ_{12} is not equal to λ_{21} in a domain, where λ_{12} and λ_{21} indicate the values of λ when state I converts into state II and state II changes into state I. They investigated the path of state changes induced in each domain by the change of λ and concluded that hysteresis can be found only where at least one of the changes (I $\rightarrow$ II and II $\rightarrow$ I) occurs irreversibly, in a thermodynamic sense. In other words, if a system exhibits hysteresis, there must exist at least one discontinuous part (jumping part) in the curve describing the relationship between the external variable and the dependent variable in each domain. The irreversible changes need not occur in all domains. Even if irreversible changes are found only in some of the domains, hysteresis will be observed.

On the basis of these assumptions, we can represent each domain by a representative point in rectangular coordinates related to λ_{12} and λ_{21}. We may choose as coordinates λ_{12} and λ_{21} (Neel, 1942, 1943), λ_{12} and (λ_{12} - λ_{21}) (Everett, 1954), $1/2(\lambda_{12} + \lambda_{21})$ and $(1/2)(\lambda_{12} - \lambda_{21})$ (Preisach, 1935), or two quantities a

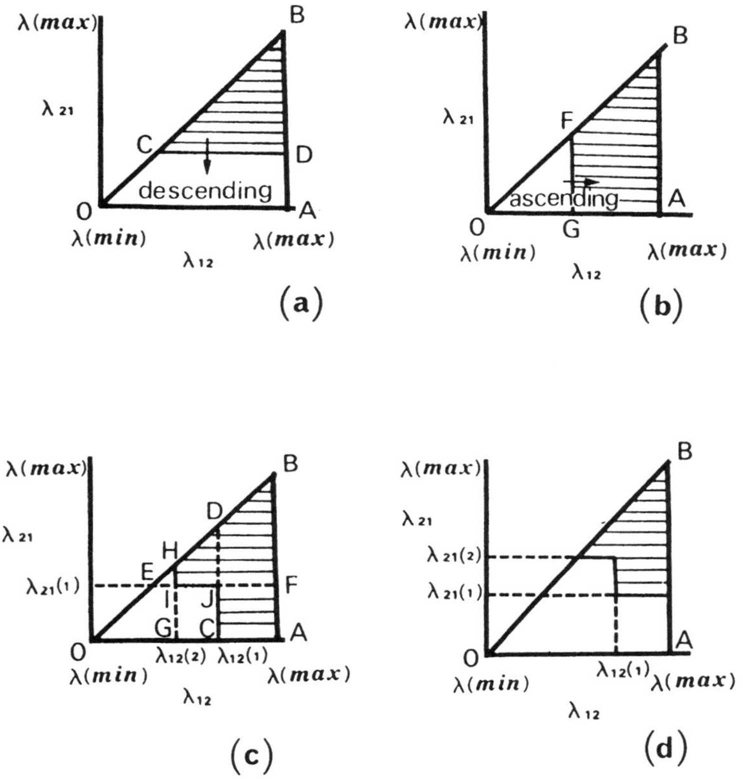

Fig. 1.26. Neel diagrams with shaded regions showing domains in state I: (a) the main descending process; (b) the main ascending process; (c) the process of $\lambda_{(min)} \rightarrow \lambda_{12}(1) \rightarrow \lambda_{21}(1) \rightarrow \lambda_{12}(2)$; (d) the process of $\lambda_{(min)} \rightarrow \lambda_{(max)} \rightarrow \lambda_{21}(1) \rightarrow \lambda_{12}(1) \rightarrow \lambda_{21}(2)$.

and b related to λ_{12} and λ_{21} by parametric equations (Enderby, 1955). Here we will take the simplest Neel diagram. All representative points must lie in the triangle OAB, since $\lambda_{12}^{(\alpha)} > \lambda_{21}^{(\alpha)}$ (Fig. 1.26). Domains that behave reversibly lie on line OB. The main ascending and descending processes are given by the downward movement of line CD parallel to the abscissa (Fig. 1.26a) and the movement of FG parallel to the ordinate from left to right (Fig. 1.26b) across the triangle, respectively. All domains in the shadowed regions in Fig. 1.26 are in state I and those in the empty regions are in state II. $\lambda_{(min)}$ and $\lambda_{(max)}$ are the maximum and minimum values of λ in the range where the system exhibits hysteresis with a change of λ. Figure 1.26c shows the state of the domains after λ is changed following the order

$$\lambda_{(min)} \xrightarrow{\text{ascending process}} \lambda_{12(1)} \xrightarrow{\text{descending process}}$$

$$\lambda_{21(1)} \xrightarrow{\text{ascending process}} \lambda_{12(2)}$$

$$\left(\lambda_{12(1)} > \lambda_{12(2)} > \lambda_{21(1)} > \lambda_{(min)}\right)$$

We will take ϕ as the fraction of the number of molecules contained in the domains in state II to the total number of molecules in the system. This definition is valid only when the total number of molecules is constant; that is, it is not changed with the change of λ. Then $\phi(\lambda_{12},\lambda_{21})$ represents the change of ϕ corresponding to the change of λ from λ_{12} to λ_{21}. In addition, we will adopt the convention that $\phi(\lambda_{12},\lambda_{21}) = -\phi(\lambda_{21},\lambda_{12})$ $[\phi(\lambda_{21},\lambda_{12}) > 0]$. Then ϕ is expressed by

$$\begin{aligned} \phi &= \phi\left(\lambda_{(min)}\ {}^{\lambda_{12(1)}}\ \lambda_{21(1)}\ {}^{\lambda_{12(2)}}\right) = \phi\left(\lambda_{min}, \lambda_{12(1)}\right) \\ &\quad + \phi\left(\lambda_{12(1)}, \lambda_{21(1)}\right) + \phi\left(\lambda_{21(1)}, \lambda_{12(2)}\right) \end{aligned} \tag{1.83}$$

When λ is changed by the order $(\lambda_{(min)} \rightarrow \lambda_{(max)} \rightarrow \lambda_{21(1)} \rightarrow \lambda_{12(1)} \rightarrow \lambda_{21(2)})$ (Fig. 1.26d), ϕ is given by

$$\begin{aligned} \phi &= \left(\lambda_{(min)}\ {}^{\lambda_{(max)}}\ \lambda_{21(1)}\ {}^{\lambda_{12(1)}}\ \lambda_{21(2)}\right) \\ &= \phi\left(\lambda_{(min)}, \lambda_{(max)}\right) + \phi\left(\lambda_{(max)}, \lambda_{21(1)}\right) + \phi\left(\lambda_{21(1)}, \lambda_{12(1)}\right) \\ &\quad + \phi\left(\lambda_{12(1)}, \lambda_{21(2)}\right) \end{aligned} \tag{1.84}$$

Next, we will consider a surface $F(\lambda_{12},\lambda_{21})$ to be erected over the triangle OAB in the Neel diagram, where F is defined such that $F(\lambda_{12},\lambda_{21})\, d\lambda_{12}\, d\lambda_{21}$ is the number of molecules in domains having transition points in $d\lambda_{12}\, d\lambda_{21}$ near $(\lambda_{12},\lambda_{21})$. This surface determines completely the distribution characteristics of the domains. The changes of ϕ due to processes starting from one of the hysteresis regions can always be expressed as the sums and differences of integrals of F over triangles, all of whose hypotenuse lie on line OB. The integral of F over the triangle whose hypotenuse is between λ_{12} and λ_{21} on OB, $\phi(\lambda_{21(1)},\lambda_{12(1)})$, is given by

$$\phi(\lambda_{21(1)}\lambda_{12(1)}) = \int_{\lambda_{21(1)}}^{\lambda_{12(1)}} \int_{\lambda_{21(1)}}^{\lambda_{12(1)}} F(\lambda_{21}, \lambda_{12}) d\lambda_{12} d\lambda_{21} \tag{1.85}$$

Using Eq. (1.85), we get

$$\begin{aligned} \phi &= \left(\lambda_{(min)} \quad {}^{\lambda_{12(1)}} \quad \lambda_{21(1)} \quad {}^{\lambda_{12(2)}}\right) \\ &= \phi(\lambda_{(min)}, \lambda_{12(1)}) + \phi(\lambda_{12(1)}, \lambda_{21(1)}) + \phi(\lambda_{21(1)}, \lambda_{12(2)}) \\ &= (\text{integral of F over } \Delta\text{OCD}) - (\text{integral of F over } \Delta\text{EJD}) \\ &+ (\text{integral of F over } \Delta\text{EIH}) \end{aligned} \tag{1.86}$$

According to the convention, $\phi(\lambda_{12},\lambda_{21})$ is negative. Thus, if F is known, the values of ϕ corresponding to any process can be obtained using the Neel diagram and F. From Eq. (1.85), we obtain

$$F = -\frac{\partial^2\phi}{\partial\lambda_{12}\,\partial\lambda_{21}} = -\frac{\partial}{\partial\lambda_{21}}\left(\frac{\partial\phi}{\partial\lambda_{12}}\right) \tag{1.87}$$

Consequently, if we have complete knowledge of either the primary ascending scanning curves or the primary descending ones for a system exhibiting hysteresis, we can calculate F.

Suppose that (1) we have a family of primary ascending scanning curves which have reversal points, respectively, at $\lambda_{21(1)}, \lambda_{21(2)},\ldots, \lambda_{21(\alpha)},\ldots, \lambda_{21(n)}$ on the boundary descending curve (Fig. 1.27a), (2) $\lambda_{21(1)} > \lambda_{21(2)} > \ldots > \lambda_{21(\alpha)} \ldots > \lambda_{21(n)}$, and (3) the difference $\lambda_{21(\alpha+1)}$ and $\lambda_{21(\alpha)}$ is small. Then we can obtain F by the following procedures:

1. The slope at a given value of λ, for example $\lambda_{12(\beta)}$, on each of the primary ascending scanning curves, $(d\phi/d\lambda)_{\lambda=\lambda_{12(\beta)}}$, is determined.
2. The value of $(d\phi/d\lambda)_{\lambda=\lambda_{12(\beta)}}$ on each curve is plotted against the value of the reversal point to obtain the λ_{21} - $(d\phi/d\lambda)$ curve having $\lambda_{12(\beta)}$ as a parameter (Fig. 1.27b).
3. Then the slope of the curve at $\lambda_{21(\alpha)}$ gives $-F[\lambda_{12(\beta)},\lambda_{21(\alpha)}]$.
4. Performing the same procedure for various values of λ_{12} - $\lambda_{12(1)}, \lambda_{12(2)},\ldots, \lambda_{12(\beta)},\ldots, \lambda_{12(n)}$—a family of λ_{21} - $(d\phi/d\lambda)_{\lambda=\lambda_{12(\beta)}}$ curves having $\lambda_{12(1)}, \lambda_{12(2)},\ldots, \lambda_{12(\beta)},\ldots$, respectively, as a parameter is obtained.
5. $F(\lambda_{12},\lambda_{21})$ for a pair of arbitrary λ_{12} and λ_{21} within the hysteresis region can be obtained using the family of λ_{21} - $(d\phi/d\lambda)$ curves.

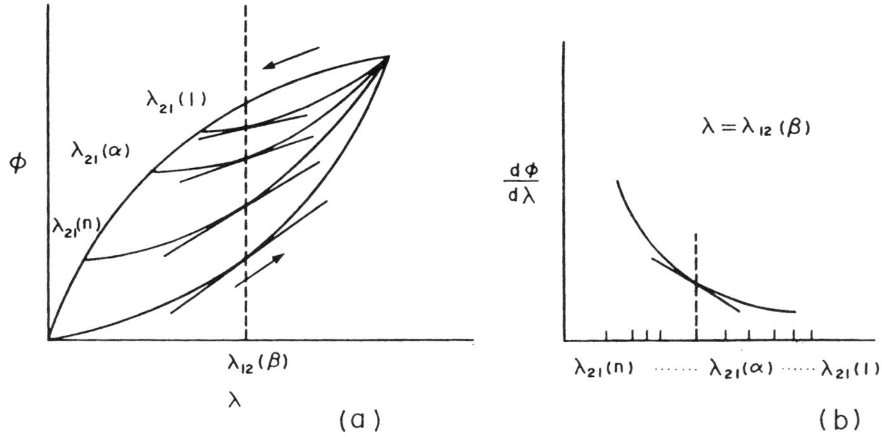

Fig. 1.27. Schematic representation to explain the procedure for obtaining the distribution function, F, from a family of primary ascending scanning curves.

The independent domain theory can be tested either by comparing the distribution function, F, derived from primary ascending scanning curves with that derived from the primary descending scanning curves or by calculating one set of primary scanning curves from knowledge of the other set and then comparing the calculated with the experimental curves.

Application of the Domain Theory to Hysteresis in Soil Water Phenomena

Several researchers have investigated the validity of the domain theory for capillary hysteresis of water in soil. The soil is viewed as a system consisting of independent domains having only two states, either empty or full. Each domain is characterized by two pressure potential values, ψ_w and ψ_d, for pores filled with water or for empty pores.

Poulovassilis (1962, 1970) reported the theory to be valid for glass beads and sands, and Lai et al. (1981) found the theory applicable for consolidated material. However, other investigators observed considerable discrepancies between predicted and observed scanning curves. Morrow and Harris (1965), Topp and Miller (1966), and Bomba and Miller (1967) found discrepancies for glass beads; Talsma (1970), Vachaud and Thony (1971), and Poulovassilis and Childs (1971) observed them for sands; and Topp (1969, 1971a) found them for various loam soils.

With those facts as background, a new approach was taken by Mualem (1973, 1974). The characteristics of his theory can be summarized in the following points:

1. $\bar{\rho}$, $\bar{r}$, and $\bar{R}$ are defined, respectively, by

$$\bar{r} = \frac{r - R_{min}}{r_{max} - R_{min}} \qquad \bar{\rho} = \frac{\rho - R_{min}}{R_{max} - R_{min}} \qquad \bar{R} = \frac{R - R_{min}}{R_{max} - R_{min}}$$

where ρ and r are the radius of a pore and the radius of its connection with other pores, R is related to pressure potential ψ using the Laplace equation $R = 2\sigma/|\psi|$, and ρ, r, and R vary in the range between R_{min} and R_{max} which correspond to ψ_{min} and ψ_{max}.

2. In a wetting process under a change from $\psi(\bar{R})$ to $\psi(\bar{R} + d\bar{R})$, pores characterized by $\bar{R} \leq \bar{\rho} \leq \bar{R} + d\bar{R}$ are filled, while in a drying process when ψ decreases from $\psi(\bar{R})$ to $\psi(\bar{R} - d\bar{R})$, only the pores characterized by $\bar{R} - d\bar{R} \leq \bar{\rho} \leq \bar{R}$ and $\bar{R} \leq \bar{r} < 1$ are drained. The domain diagrams in this case are represented in Fig. 1.28. It should be noted that in this model the condition assumed by Everett and his co-workers (1952-1954) and Poulovassilis (1962, 1970), $\rho \geq r(\psi_d \leq \psi_w)$, is omitted in the wetting process. As a result, the model includes a reversible part. The area above triangle OBC covered equally in drainage and wetting represents a reversible contribution, whereas triangle OAB represents the hysteretic contribution. Under the condition where $\bar{\rho} \geq \bar{r}$, the domains behaving reversibly lie on the diagonal OB.

3. According to the similarity approach suggested by Philip (1964), the pore water distribution function, $F(\bar{r},\bar{\rho})$, which corresponds to $F(\lambda_{12},\lambda_{21})$, is assumed to be represented as a product of two independent distribution functions $h(\bar{r})$ and $\ell(\bar{\rho})$, namely,

$$F(\bar{r}, \bar{\rho}) = h(\bar{r})\, \ell(\bar{\rho}) \tag{1.88}$$

This hypothesis simplifies the computational procedures. In addition, only the measured boundary curves are required in calibration of the model, and the scanning curves are expressed analytically as described below.

First, the two integral distribution functions are defined as

$$L(\bar{R}) = \int_0^{\bar{R}} \ell(\bar{\rho})\, d\bar{\rho} \quad \text{and} \quad H(\bar{R}) = \int_0^{\bar{R}} h(\bar{r})\, d\bar{r} \tag{1.89}$$

Assuming that the effective water content θ is

$$\theta = \Theta - \Theta_{min}$$

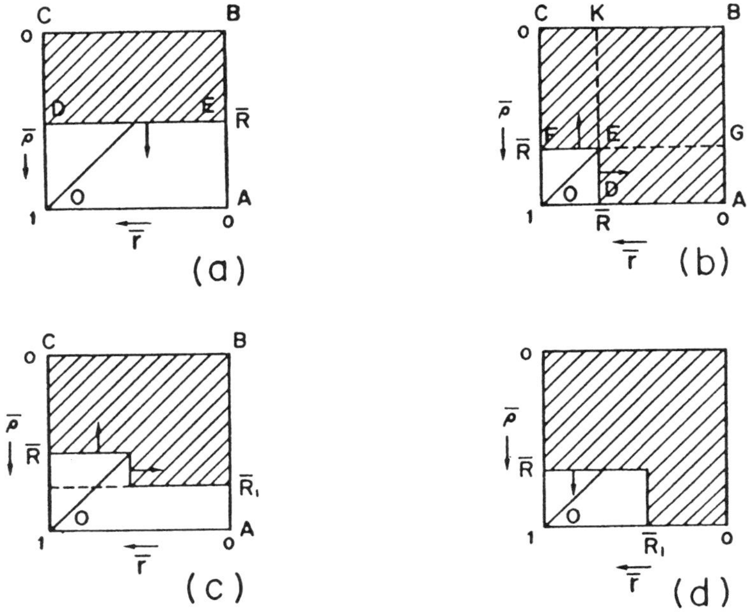

Fig. 1.28. Filled-pore diagrams in the $\bar{r}$-$\bar{\rho}$ plane (the shaded domains) for (a) the main wetting process, $\psi(R_{min}) \rightarrow \psi(\bar{R})$; (b) the main drying process, $\psi(R_{max}) \rightarrow \psi(\bar{R})$; (c) the primary drying process, $\psi(R_{min}) \rightarrow \psi(\bar{R}_1)$; $\rightarrow \psi(R)$; (d) the primary wetting process; $\psi(R_{max}) \rightarrow \psi(\bar{R}_1) \rightarrow \psi(\bar{R})$. (From Mualem, 1974)

where Θ and Θ_{min} are the actual and the residual water content, $\theta_w(\bar{R})$ in the main wetting process (Fig. 1.28a) is determined by

$$\theta_w(\bar{R}) = \int_0^1 \int_0^{\bar{R}} F(\bar{r}, \bar{\rho})\, d\bar{\rho}\, d\bar{r} \text{ (over path BCDE)}$$
$$= \int_0^{\bar{R}} \ell(\bar{\rho})\, d\bar{\rho} \int_0^1 h(\bar{r})\, d\bar{r} = L(\bar{R})H(1) \tag{1.90}$$

Similarly, in the main drying process (Fig. 1.28b),

$$\theta_d(\bar{R}) = \int_0^1 \int_0^{\bar{R}} F(\bar{r}, \bar{\rho})\, d\bar{\rho}\, d\bar{r} \text{ (over path BCFG)}$$
$$+ \int_0^{\bar{R}} \int_0^1 F(\bar{r}, \bar{p})\, d\bar{\rho}\, d\bar{r} \text{ (over path ABKD)} \tag{1.91}$$
$$- \int_0^{\bar{R}} \int_0^{\bar{R}} F(\bar{r}, \bar{p})\, d\bar{\rho}\, d\bar{r} \text{ (over path BKEG)}$$
$$= L(\bar{R})H(1) + \left[L(1) - L(\bar{R})\right]H(\bar{R})$$

Choosing H(1) = 1 and considering that there is a unique relationship between $\bar{R}$ and ψ, we may express H and L as functions depending directly on ψ by

$$L(\psi) = \theta_w(\psi) \tag{1.92}$$

$$H(\psi) = \theta_d(\psi) - \frac{\theta_d(\psi) - \theta_w(\psi)}{\theta_m - \theta_w(\psi)} \tag{1.93}$$

where θ_m is the maximum effective water content at the point at which the wetting boundary curve joins the drying one and equals L(I). Consequently, if the wetting and drying boundary curves for a soil are given, we can calculate the values of $L(\psi)$ and $H(\psi)$ at an arbitrary value of ψ using Eqs. (1.92) and (1.93). Once $L(\psi)$ and $H(\psi)$ are determined, any hysteretic path is described in a simple way. For example, we can obtain the water content along a primary drying scanning curve (Fig. 1.28c) from the equation

$$\theta\left(\psi_{\min} \overset{\psi_1}{} \psi\right) = \theta_m(\psi) + H(\psi)\left[L(\psi_1) - L(\psi)\right] \tag{1.94}$$

The scanning curves predicted by this theory are in better agreement with the observations than those derived using the original independent domain theory. This theory has the merits of including a reversible part of the process, simplifying the computational procedures, decreasing the number of experimental data necessary to predict the scanning curves, and better agreement with observations.

Parlange (1976) proposed a model that corresponds to a special case of Mualem's similarity hypothesis $F(\psi_w,\psi_d) = h(\psi_w)\, \ell(\psi_d)$, in which $h(\psi)$ is set at unity for all ψ values. Mualem and Morel-Seytoux (1978) extensively tested the applicability of the model for soil water hysteresis for different types of porous media in addition to analyzing it theoretically, and showed that the model contradicts well-known properties of the water retention curves.

The K-S (hydraulic conductivity-water saturation) hysteresis has also been studied on the basis of domain theory. Poulovassilis (1969) considered, in a qualitative way, the influence of capillary hysteresis on hydraulic conductivity using the basic concept of the independent domain theory. Poulovassilis and Tzimas (1974, 1975) obtained the distribution diagrams from an analysis of the primary depletion K-θ (water content) curves and calculated the primary accumulation curves from the distribution diagrams. Mualem (1976a) established a new model, similar to that of Childs and Collis-George (1950), for predicting the K-θ or K-ψ curve using a ψ-θ curve and a measured value of K at saturation. Mualem (1976b) then proposed a theory to predict K-θ or K-ψ relationships quantitatively using the new model and the models adapted to calculate capillary hysteresis by Mualem (1974) and Mualem and Dagan (1975). The theory requires the same amount of experimental data as the previous models for capillary hysteresis, in addition to a measured value of K at a given point. Good agreement is found between the theory and observations. It is very interesting that the predicted K-S hysteresis loops almost shrink to a unique curve (Fig. 1.15), whereas the loops in the ψ-S and K-ψ relationships are of considerable magnitude (Figs. 1.14a and b).

Further Developments of Domain Theories

Several researchers proposed dependent domain theories, in which the influence of the neighboring domains on the values of λ_{12} and λ_{21} in a domain is taken into account. Everett (1967) and Topp (1971b) suggested a theory that accounts for pore blockage against air entry near saturation and air blockage against water entry at low water content. Mualem and Dagan (1975) and Mualem (1976b) generalized Mualem's model described above to incorporate the fundamental ideas of Everett and Topp, and established a simpler computational procedure. These theories brought about better agreement with observations, especially where the effect of pore blockage against air entry is large. Afterward, Mualem and Miller (1979) proposed a quantitative model of hysteresis, assuming the pore-water blockage against air entry to be a direct function of the volume of pore domains located in the completely filled region of the diagram. Mualem (1984a,b) also improved his models.

Poulovassilis and Childs (1971), Poulovassilis (1973), and Poulovassilis and El-Ghamry (1978) presented another dependent domain theory by extending the concept of domains. This theory is not easy to understand and requires more experimental data. Laroussi and De Backer (1979) presented a universal hysteresis loop showing the relation between nondimensional variables of water potential and water content for porous media composed of spherical particles whose diameters have a small range of variation. The hysteresis was assumed to be due only to hysteresis of the contact angle at the solid-liquid interface.

Controversial Points in the Application of the Domain Theory to Hysteresis in Soil Water Phenomena

At present, the application of domain theory is regarded as the most useful method to obtain a quantitative representation of hysteresis in soil water phenomena. However, the application has two weak points:

1. In the domain theory, each domain is required to exist in either of two states. Consequently, soil is assumed to be a system made up of pores, each of which empties or refills in jump transition. Strictly speaking, this assumption is generally not satisfied in soil pores. The assumption is correct only for exceptional pores such as those in Fig. 1.29a and b. Most pores in soil have shapes similar to that shown in Fig. 1.30. Such pores go through partially full states in the wetting process, and the pore is about half full of water when the jump occurs. This does not satisfy the requirement of the domain theory. The partially full states show reversible wetting and drying.
2. Complicated sequences of interconnected pores (Fig. 1.29c), each of which is regarded to be an independent domain, do not behave according to the independent domain model (Poulovassilis, 1962; Mualem, 1974). Such sequences could exist in soils. A theory considering this possibility has not yet been established.

These may be the reasons why most values predicted by the independent domain theory based on the Neel-Everett-Poulovassilis model did not agree well with observed values. Values predicted from the Mualem (1974) model often agree better with measured results, but this is a conceptual rather than a physical model, and it is difficult to visualize what is happening. It is not clear how this conceptual model removes the weak points of the usual model.

Model d shown in Fig. 1.29 represents a pore for which drying and wetting processes are reversible, and part of the wetting process for the pore model shown in Fig. 1.30 is also reversible. It is very useful for the study of mechanisms of water retention and water movement in a soil to evaluate the proportion of the reversible to the irreversible part in the drying and wetting. If all pores in a soil showing hysteresis empty and refill in jump transition, the distribution function, F, which is calculated from the scanning curves on the Neel diagram, is related only to the irreversible part because the procedure shown in Fig. 1.27b eliminates the reversible part. In this case, we can evaluate the ratio of the reversible to the irreversible part. However, as only a small fraction of the pores in soil is considered to have only two states, either empty or full, we

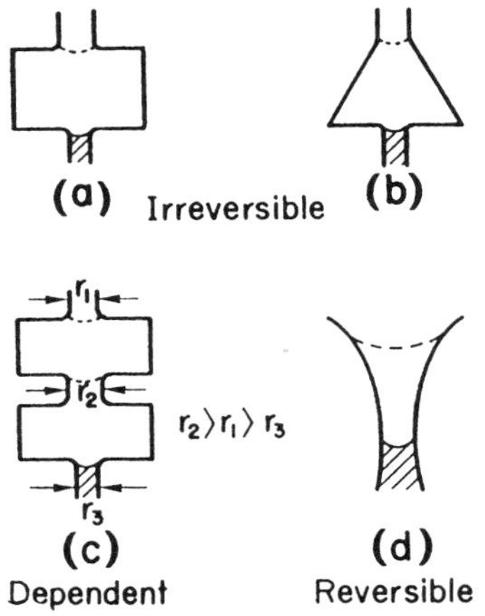

Fig. 1.29. Various models of soil pores to explain reversible, irreversible, and dependent domains.

have not yet been able to calculate the proportion of the reversible to the irreversible part.

1.3.4 Other Theories

Hogarth et al. (1988) derived a simple set of equations to describe the main, primary and secondary wetting and drying curves. They used the Brooks and Corey (1964) equations relating soil water content and matric potential, and the hysteresis model of Parlange (1976). They proposed a method which can predict

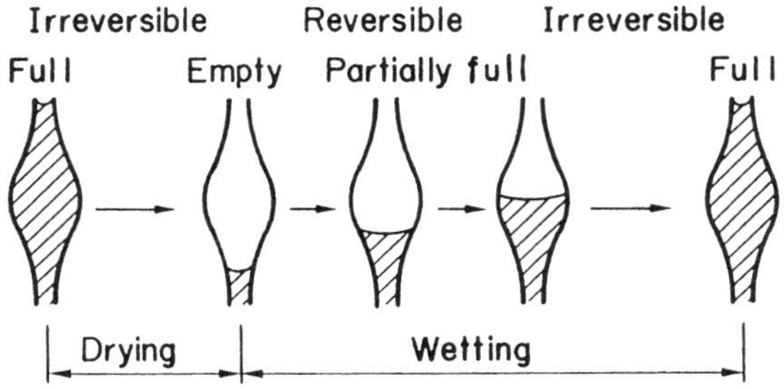

Fig. 1.30. Schematic representation illustrating reversible and irreversible filling and emptying of a pore.

the wetting boundary and all scanning curves from knowledge of a boundary or primary drying curve. Parker and Lenhard (1987), and Lenhard and Parker (1987) proposed a theoretical model for hysteresis in relative permeability (k) - fluid saturation (S) - pressure (P) relations of air-water and air-nonaqueous phase liquid-water systems in rigid porous media subjected to complex saturation histories. The model considers hysteresis due to nonwetting fluid entrapment in addition to hysteresis caused by irregular pore geometry and contact angle. The results suggested that 1) fluid entrapment effects lead to greater hysteresis in the three-phase S-P relation than are evident in the two phase data, and 2) hysteresis in k-S-P relations of three phase systems, including fluid entrapment effects, may be substantial. Lenhard et al. (1991) also compared simulated and experimental hysteretic two-phase transient fluid flow phenomena and concluded that consideration should be given to effects of hysteresis in k-S-P relations to accurately predict fluid distributions.

Jones and Watson (1987) simulated the effect of soil water hysteresis on solute movement during intermittent leaching. Their results indicated that when hysteresis data is included in the analysis, the solute moves through the soil profile in a series of well-defined peaks. When hysteresis is neglected, the peaks disappear after the first infiltration event and the solute becomes fairly uniformly distributed much deeper in the profile.

1.3.5 Thermodynamics of Systems Exhibiting Hysteresis

Systems exhibiting hysteresis are subject to some limitations in a thermodynamic analysis, as pointed out by Enderby (1955). We can measure the derivative of an extensive quantity of a hysteretic system such as chemical potential or partial entropy. If the measurement of the chemical potential of a system not exhibiting hysteresis is performed under a condition where the temperature and pressure are kept constant, the area under the chemical potential against composition curve corresponding to a process gives the work done on the system in the process. However, since the change of state of a hysteretic system has irreversible changes, this description is not correct according to the second law of thermodynamics, i.e. the area is not equal to the external work done on the system. A similar conclusion applies to the relationships between the partial entropy and the total entropy change, and between the differential heat of adsorption obtained by the adsorption isoteres and the total heat measured directly. On the other hand, since the internal energy and total heat require only the first law of thermodynamics for their experimental definition, they can be measured directly in the usual ways. Research on the thermodynamic quantities of a hysteretic system must keep this distinction in mind.

ACKNOWLEDGEMENTS

Figures 1.15 and 1.16 are from *Soil Science Society of America Proceedings* by permission of the Soil Science Society of America. Figures 1.14, 1.15, and 1.28 are from *Water Resources Research* (© the American Geophysical Union). Figure 1.17 is reproduced from *Australian Journal of Soil Research* (© CSIRO Australia). Figures 1.18 and 1.20 are reprinted from *Journal of Soil Science* (© Blackwell Scientific Publications). Figure 1.19 is reproduced from *Journal of Colloid Science* (© Academic Press).

REFERENCES

Babcock, K. L., Theory of chemical properties of soil colloidal systems at equilibrium, *Hilgardia* 34:417-542 (1963).

Babcock, K. L., and R. Overstreet, Thermodynamics of soil moisture, *Soil Sci.* 80:257-263 (1955).

Beese, F., and R. R. van der Ploeg, Influence of hysteresis on moisture flow in an undisturbed soil monolith, *Soil Sci. Soc. Am. J.* 40:480-484 (1976).

Bolt, G. H., and M. J. Frissel, Thermodynamics of soil moisture, *Neth. J. Agric. Sci.* 8:57-78 (1960).

Bolt, G. H., and R. D. Miller, Calculation of total and component potentials of water in soil, *Trans. Am. Geophys. Union* 39:917-928 (1958).

Bomba, S. J., and E. E. Miller, Secondary-scan hysteresis in glass-bead media, Mimeo for discussion at Soil Sci. Soc. Am., Meeting (1967).

Brooks, R. H., and A. J. Corey, Hydraulic properties of porous media, *Hydrol. Pap. 3*, Colorado State Univ., Ft. Collins, CO (1964).

Buckingham, E., Studies on the movement of soil moisture, *U.S. Dept. Agric. Bur. Soils Bull. 38* (1907).

Chahal, R. S., Effect of temperature and trapped air on the energy status of water in porous media, *Soil Sci.* 98:107-112 (1964).

Chatelain, J., Thermodynamics of soil moisture, *Soil Sci.* 67:305-309 (1949).

Childs, E. C., and N. Collis-George, The permeability of porous materials, *Proc. Roy. Soc. London* A201:392-405 (1950).

Coleman, J. D., and D. Croney, The estimation of the vertical moisture distribution with depth in unsaturated cohesive soils, *Road Res. Lab. U.K., Note* RN/1709/JDC.DC (1952).

Collis-George, N., and M. J. Rosenthal, Proposed outflow method for the determination of the hydraulic conductivity of unsaturated porous materials, *Aust. J. Soil Res.* 4:165-180 (1966).

Constantz, J., Comparison of isothermal and isobaric water retention paths in nonswelling porous materials, *Water Resour. Res.* 27:3165-3170 (1991).

Dane, J. H., and P. J. Wierenga, Effect of hysteresis on the prediction of infiltration, redistribution and drainage of water in a layered soil, *J. Hydrol.* 25:229-242 (1975).

Day, P. R., The moisture potentials of soils, *Soil Sci.* 38:391-400 (1947).

Edlefsen, N. E., and B. C. Anderson, Thermodynamics of soil moisture, *Hilgardia* 15:31-298 (1943).

Elrick, D. E. and D. H. Bowman, Note on an improved apparatus for soil moisture flow measurements, *Soil Sci. Soc. Am. Proc.* 28:450-453 (1964).

Enderby, J. A., The domain model of hysteresis: 1. Independent domains, *Trans. Faraday Soc.* 51:835-848 (1955).

Everett, D. H., A general approach to hysteresis: 3. A formal treatment of the independent domain model of hysteresis, *Trans. Faraday Soc.* 50:1077-1096 (1954).

Everett, D. H., Adsorption hysteresis, in *Solid-Gas Interface*, E. A. Flood, ed., Marcel Dekker, New York, pp. 1055-1113 (1967).

Everett, D. H., and F. W. Smith, A general approach to hysteresis: 2. Development of the domain theory, *Trans. Faraday Soc.* 50:187-197 (1954).

Everett, D. H., and W. I. Whitton, A general approach to hysteresis: 1, *Trans. Faraday Soc.* 48:749-752 (1952).

Gardner, W. R., Relation of temperature to moisture tension of soil, *Soil Sci.* 79:257-265 (1955).

Gardner, W., and J. Chatelain, Thermodynamic potential and soil moisture, *Soil Sci. Soc. Am. Proc.* 11:100-102 (1946).

Gardner, W., O. W. Israelsen, N. E. Edlefsen, and H. Contad, The capillary potential function and its relation to irrigation practice, *Phys. Rev.* 20:196 (1922).

Gardner, W. H., W. H. Gardner, and W. Gardner, Thermodynamics and soil moisture, *Soil Sci.* 72:101-105 (1951).

Gilpin, R. R., A model for the prediction of ice lensing and frost heave in soils, *Water Resour. Res.* 16:918-930 (1980).

Green, R. E., R. J. Hanks, and W. E. Larson, Estimates of field infiltration by numerical solution of the moisture flow equation, *Soil Sci. Soc. Am. Proc.* 28:15-19 (1964).

Groenevelt, P. H., and G. H. Bolt, Water retention in soil, *Soil Sci.* 113:238-245 (1972).

Haines, W. B., Studies in the physical properties of soils: 5, *J. Agric. Sci.* 20:97-116 (1930).

Haridasan, M., and R. D. Jensen, Effect of temperature on pressure head-water content relationship and conductivity of two soils, *Soil Sci. Soc. Am. Proc.* 36:703-708 (1972).

Hassanizadeh, M., and W. G. Gray, General conservation equations for multiphase systems: 1. Averaging procedure, *Adv. Water Resour.* 2:131-144 (1979a).

Hassanizadeh, M., and W. G. Gray, General conservation equations for multiphase systems: 2. Mass, momenta, energy and entropy equations, *Adv. Water Resour.* 2:191-203 (1979b).

Hassanizadeh, M., and W. G. Gray, General conservation equations for multiphase systems: 3. Constitutive theory for porous media flow, *Adv. Water Resour.* 3:25-40 (1980).

Hogarth, W. L., J. Hopmans, J.-Y. Parlange, and R. Haverkamp, Application of a simple soil-water hysteresis model, *J. Hydrol.* 98:21-29 (1988).

Iwata, S., Thermodynamics of soil water: 1. *Soil Sci.* 113:162-166 (1972a).

Iwata, S., On the definition of soil water potential as proposed by the ISSS in 1963, *Soil Sci.* 114:88-92 (1972b).

Iwata, S., Driving force for water migration in frozen clayey soil, *Soil Sci. Plant Nutr. Tokyo* 26:215-227 (1980).

Jackson, R. D., Water vapor diffusion in relatively dry soil: II. Desorption experiments, *Soil Sci. Soc. Am. Proc.* 28:464-466 (1964).

Jones, M. J., and K. K. Watson, Effect of water hysteresis on solute movement during intermittent leaching, *Water Resour. Res.* 23:1251-1256 (1987).

Kim, M., and S. Chang, Effect of the curvature dependency of surface tension on pendular ring condensation calculations, *J. Colloid Interface Sci.* 69:4-44 (1979).

Kirkwood, J. G., and I. Oppenheim, *Chemical Thermodynamics*, McGraw-Hill, New York (1962).

Lai, F. S. Y., I. F. Macdonald, F. A. L. Dullien, and I. Catzis, A study of the applicability of the independent domain model of hysteresis to capillary pressure hysteresis in sandstone samples, *J. Colloid Interface Sci.* 84:362-378 (1981).

Laroussi, C., and L. W. De Backer, Relations between geometrical properties of glass bead media and their main hysteresis loops, *Soil Sci. Soc. Am. J.* 43:646-650 (1979).

Lenhard, R. J., and J. C. Parker, A model for constitutive relations governing multiphase flow, 2. Permeability-saturation relations, *Water Resour. Res.* 23:2197-2206 (1987).

Lenhard, R. J., J. C. Parker, and J. J. Kaluarachchi, Comparing simulated and experimental hysteretic two-phase transient fluid flow phenomena, *Water Resour. Res.* 27:2113-2124 (1991).

Letey, J., J. Osborn, and R. E. Pelishek, Measurement of liquid-solid contact angles in soil and sand, *Soil Sci.* 93: 149-153 (1962).

Mason, G., The effect of pore space connectivity on the hysteresis of capillary condensation in adsorption-desorption isotherms, *J. Colloid Interface Sci.* 88:36-46 (1982).

Miyauchi, S., Unpublished private communication (1978).

Morrow, N. R., and C. C. Harris, Capillary equilibrium in porous materials, *Soc. Pet. Eng. J.* 5:12-14 (1965).

Mualem, Y., Modified approach to capillary hysteresis based on a similarity hypothesis, *Water Resour. Res.* 9:1324-1331 (1973).

Mualem, Y., A conceptual model of hysteresis, *Water Resour. Res.* 10:514-520 (1974).

Mualem, Y., A new model for predicting the hydraulic conductivity of unsaturated porous media, *Water Resour. Res.* 12:513-521 (1976a).

Mualem, Y., Hysteretical models for prediction of the hydraulic conductivity of unsaturated porous media, *Water Resour. Res.* 12:1248-1254 (1976b).

Mualem, Y., A model dependent-domain theory of hysteresis, *Soil Sci.* 137:283-289 (1984a).

Mualem, Y., Prediction of the soil boundary wetting curve, *Soil Sci.* 137:379-390 (1984b).

Mualem, Y., and G. Dagan, A dependent domain model of capillary hysteresis, *Water Resour. Res.* 11:452-460 (1975).

Mualem, Y., and E. E. Miller, A hysteresis model based on an explicit domain-dependence function, *Soil Sci. Soc. Am. J.* 43:1067-1073 (1979).

Mualem, Y., and H. J. Morel-Seytoux, Analysis of a capillary hysteresis model based on a one-variable distribution function, *Water Resour. Res.* 16:605-610 (1978).

Mualem, Y., and H. J. Morel-Seytoux, *Capillary Pressure, The Encyclopedia of Soil Science*, Part 1, Dowden, Hutchinson & Ross, Stroudsburg, PA, pp. 49-60 (1979).

Narr, J., R. J. Wygal, and J. H. Henderson, Imbibition and relative permeability in unconsolidated porous media, *Am. Inst. Min. Met. Eng. Trans. (Pet. Div.)* 225:12-17 (1962).

Neel, L., Théories des lois d'aimantation de Lord Rayleigh: 1, *Cah. Phys.* 12:1-20 (1942).

Neel, L., Théories des lois d'aimantation de Lord Rayleigh: 2, *Cah. Phys.* 13:19-30 (1943).

Nielsen, D. R., and J. W. Biggar, Measuring capillary conductivity, *Soil Sci.* 92:192-193 (1961).

Nimmo, J. R., and E. E. Miller, The temperature dependence of isothermal moisture vs. potential characteristics of soils, *Soil Sci. Soc. Am. J.* 50:1105-1113 (1986).

Parker, J. C., and R. J. Lenhard, A model for hysteretic constitutive relations governing multiphase flow, I. Saturation-pressure relations, *Water Resour. Res.* 23:2187-2196 (1987).

Parlange, J.-Y., Capillary hysteresis and the relationship between drying and wetting curves, *Water Resour. Res.* 12:224-228 (1976).

Pavlakis, G., and L. Barden, Hysteresis in the moisture characteristics of clay soil, *J. Soil Sci.* 23:350-361 (1972).

Peck, A. J., Changes of moisture tension with temperature and air pressure, *Soil Sci.* 89:303-310 (1960).

Philip, J. R., Similarity hypothesis for capillary hysteresis in porous materials, *J. Geophys. Res.* 69:1553-1562 (1964).

Philip, J. R., Hydrostatics and hydrodynamics in swelling soils, *Water Resour. Res.* 5:1070-1077 (1969a).

Philip, J. R., Moisture equilibrium in the vertical in swelling soils, I. Basic theory, *Aust. J. Soil Res.* 7:99-120 (1969b).

Philip, J. R., Moisture equilibrium in the vertical in swelling soils, II. Application, *Aust. J. Soil Res.* 7:121-141 (1969c).

Pitzer, K. S., and L. Brewer, *Thermodynamics*, McGraw-Hill, New York (1961).

Poulovassilis, A., Hysteresis of pore water, an application of the concept of independent domains, *Soil Sci.* 93:405-412 (1962).

Poulovassilis, A., The effect of hysteresis of pore water on the hydraulic conductivity, *J. Soil Sci.* 20:52-56 (1969).

Poulovassilis, A., The hysteresis of pore water in granular porous bodies, *Soil Sci.* 109:5-12 (1970).

Poulovassilis, A., The hysteresis of pore water in the presence of nonindependent water elements, in *Ecological Studies*, Vol. 4, *Physical Aspects of Soil Water and Salts in Ecosystems*, A. Hadas, D. Swartzendruber, P. E. Rijtema, M. Fuchs, and B. Yaron, eds., Springer-Verlag, Berlin, pp. 161-180 (1973).

Poulovassilis, A., and E. C. Childs, The hysteresis of pore water: the nonindependence of domains, *Soil Sci.* 112:301-312 (1971).

Poulovassilis, A., and W. M. El-Ghamry, The dependent domain theory applied to scanning curves of any order in hysteresis soil water relationships, *Soil Sci.* 126:1-8 (1978).

Poulovassilis, A., and E. Tzimas, The hysteresis in the relationship between hydraulic conductivity and suction, *Soil Sci.* 117:250-256 (1974).

Poulovassilis, A., and E. Tzimas, The hysteresis in the relationship between hydraulic conductivity and soil water content, *Soil Sci.* 120:327-331 (1975).

Preisach, F., Uber die magnetische Nachwirkung, *Z. Phys.* 94:277-302 (1935).

Richards, L. A., M. B. Russell, and O. R. Neal, Further developments on apparatus for field moisture studies, *Soil Sci. Soc. Am. Proc.* 1:55-63 (1937).

Rose, D. A., Water movement in dry soils: II. An analysis of hysteresis, *J. Soil Sci.* 22:490-507 (1971).

Royer, J. M., and G. Vachaud, Field determination of hysteresis in soil water characteristics, *Soil Sci. Soc. Am. Proc.* 39:221-223 (1975).

Russell, M. B., The utility of the energy concept of soil moisture, *Soil Sci. Soc. Am. Proc.* 7:90-94 (1942).

Schofield, R. K., The pF of the water in soil, *Trans. 3rd Int. Cong. Soil Sci.* 2:37-48 (1935).

Slatyer, R. O., *Plant-Water Relationships*, Academic Press, New York, pp. 78-82 (1967).

Slatyer, R. O., and S. A. Taylor, Terminology in plant and soil water relations, *Nature* 187:922-924 (1960).

Sposito, G., *The Thermodynamics of Soil Solutions* Clarendon Press, Oxford (1981).

Sposito, G., and S.-Y. Chu, The statistical mechanical theory of groundwater flow, *Water Resour. Res.* 17:885-892 (1981).

Sposito, G., and S.-Y. Chu, Internal energy and the Richards equation, *Soil Sci. Soc. Am. J.* 46:889-893 (1982).

Staple, W. J., Moisture tension, diffusivity, and conductivity of a loam soil during wetting and drying, *Can. J. Soil Sci.* 45:78-86 (1965).

Staple, W. J., Infiltration and redistribution in vertical columns of loam soil, *Soil Sci. Soc. Am. Proc.* 30:553-558 (1966).

Takagi, S., Criticism of a viewpoint on "Thermodynamics and soil moisture," *Soil Sci.* 77:303-312 (1954).

Talsma, T., Hysteresis in two sands and the independent domain model, *Water Resour. Res.* 6:964-970 (1970).

Talsma, T., Measurement of the overburden component of total potential in swelling field soils, *Aust. J. Soil Res.* 15:95-102 (1977).

Taylor, S. A., and G. L. Ashcroft, *Physical Edaphology*, W. H. Freeman, San Francisco (1972).

Terminology Committee, Commission I, ISSS, Terminology in Soil Physics, *ISSS Bull.* 49:26-36 (1976).

Topp, G. C., Soil water hysteresis measured in a sandy loam compared with the hysteresis domain model, *Soil Sci. Soc. Am. Proc.* 33:645-651 (1969).

Topp, G. C., Soil water hysteresis in silt loam and clay loam soils, *Water Resour. Res.* 7:914-920 (1971a).

Topp, G. C., Soil water hysteresis: the domain model theory extended to pore interaction conditions, *Soil Sci. Soc. Am. Proc.* 35:219-225 (1971b).

Topp, G. C., and E. E. Miller, Hysteresis moisture characteristics and hydraulic conductivities for glass-bead media, *Soil Sci. Soc. Am. Proc.* 30:156-162 (1966).

Towner, G. D., The correction of in situ tensiometer readings for overburden pressure in swelling soils, *J. Soil Sci.* 32:499-504 (1981).

Tschapek, M., C. O. Scoppa, and C. Wasowski, The surface tension of soil water, *J. Soil Sci.* 29:17-21 (1978).

Tzimas, E., The measurement of soil water hysteretic relationships on a soil monolith, *J. Soil Sci.* 30:529-534 (1979).

Vachaud, G., and J. L. Thony, Hysteresis during infiltration and redistribution in a soil column at different initial water contents, *Water Resour. Res.* 7:111-127 (1971).

van Olphen, H., Thermodynamics of interlayer adsorption of water in clays: I. Sodium vermiculite, *J. Colloid Sci.* 20:822-837 (1965).

Veihmeyer, F. J., and N. E. Edlefsen, Interpretation of soil moisture problems by means of energy changes, *Trans. Am. Geophys. Union, 18th Ann. Meet., Hydrol.* :302-318 (1937).

Wadleigh, C. H., and A. D. Ayers, Growth and biochemical composition of bean plants as conditioned by soil moisture tension and salt concentration, *Plant Physiol.* 20:107-132 (1945).

Watson, K. K., R. J. Reginato, and R. D. Jackson, Soil water hysteresis in a field soil, *Soil Sci. Soc. Am. Proc.* 39:242-246 (1975).

Wilkinson, G. E., and A. Klute, The temperature effect on equilibrium energy status of water held by porous media, *Soil Sci. Soc. Am. Proc.* 26:326-329 (1962).

Youngs, E. G., An infiltration method of measuring the hydraulic conductivity of unsaturated porous materials, *Soil Sci.* 97:307-311 (1964).

2

INTERACTION BETWEEN SOIL PARTICLES AND SOIL SOLUTION

Intermolecular and electrostatic interactions exist between soil solution and soil particles. Unfrozen water in a soil below 0°C is an example of the results of these interactions. In this chapter, fundamental knowledge necessary to analyze these phenomena is introduced and various phenomena discussed on the basis of this knowledge.

2.1 PHYSICAL PROPERTIES OF SUBSTANCES TAKING PART IN INTERACTIONS BETWEEN SOIL SOLUTION AND SOIL PARTICLES

Water, ions, and soil particles all take part in the interactions between soil solution and soil particles. The physical properties of these substances will be reviewed briefly. To achieve generality, clay minerals which possess electric charges, and which are the most active part of soils, are selected as typical soil particles.

2.1.1 Water

Structure of the Water Molecule

The bond length and bond angle of the water molecule found by Benedict et al. (1956), which are generally recognized as the most reliable, are shown in Fig. 2.1. A water molecule is regarded as a sphere with a radius of 1.38 Å, the center lying at the oxygen nucleus. This structure gives the water molecule an electric dipole. A simple model adopted by many scientists to illustrate the electric properties of the water molecule is shown in Fig. 2.2. The electric charge from the outer electrons is distributed to provide localized concentrations around the four apexes of a regular tetrahedron inscribed in the sphere. The two apexes in the direction from the center to two hydrogen nuclei are charged positively and the remaining two negatively. The resulting electric dipole moment of the water molecule is 1.84 D (Moelwyne-Hughes, 1964).

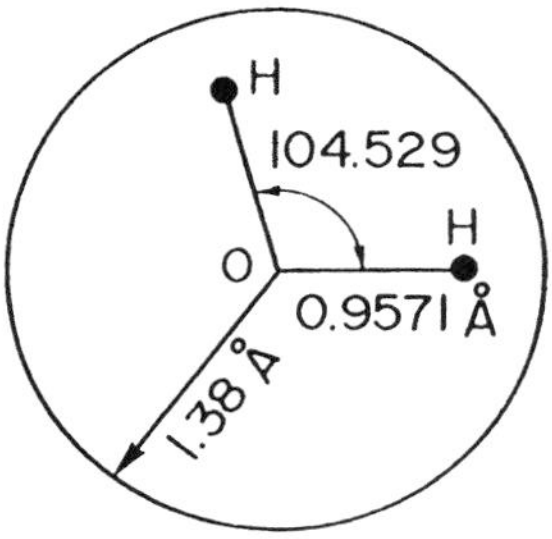

Fig. 2.1. Molecular dimensions of the H_2O molecule. (From Benedict et al., 1956)

Hydrogen Bond

When a large electronegative atom such as N, O, F, Cl, and B, approaches another large electronegative atom through a hydrogen atom attached to it, a relatively weak secondary bond is produced, and the system is stabilized. This is the hydrogen bond. It is generally shown by notations such as X-H···Y, where X and Y represent large electronegative atoms and H is the hydrogen atom. Because of the large electronegativity of X, the H atom is slightly positively charged. The atom Y is charged negatively by its large electron affinity. The electrostatic force between them was considered, until 1950, to be the main source of the hydrogen bond. Since then, the stabilization by resonance between two structures of X^--H^+···Y^- and X^-···H^+-Y^-, in other words, the occurrence of the covalent bond between H and Y, has also been regarded as a source of the hydrogen bond.

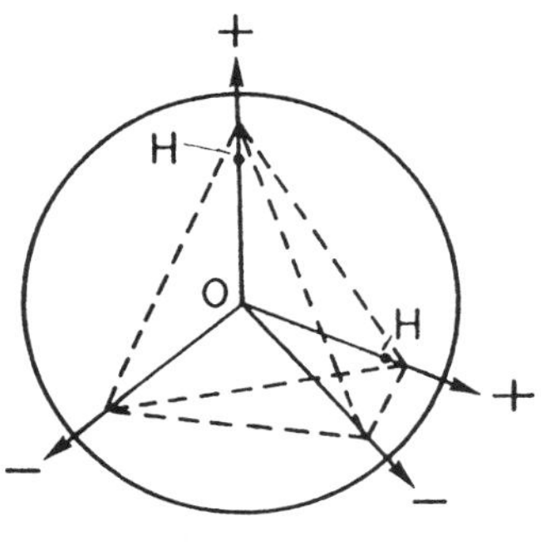

Fig. 2.2. Model for charge distribution in a water molecule.

Although its bond energy is generally as small as 2 to 8 kcal/mol, the hydrogen bond has a significant influence on physical properties such as heat of vaporization, dielectric constant, infrared and ultraviolet absorption, and nuclear magnetic resonance absorption. Water has a simple molecular structure, but because of the hydrogen bonds between water molecules, it has a high boiling point and high heat of vaporization. Ice is a crystal that is solidified by the hydrogen bond, whose energy is 6.7 kcal/mol.

The hydrogen bond length in ice is 1.01 Å, which is slightly larger than the O-H distance of an isolated water molecule. The H···O distance is 1.75 Å, which is much smaller than the sum of the van der Waals radii of H and O, 2.6 Å. Therefore, the energy of the hydrogen bond between two water molecules is stronger than the van der Waals force. Bernal and Fowler (1933), Verwey (1941), Pople (1951), Rowlinson (1951), and Bjerrum (1952) calculated the hydrogen bond energy of ice on the basis of point-charge models and obtained 5.7, 4.7, 5.95, 5.8, and 5.8 kcal/mol, respectively, which are very close to the experimental values. With the development of the electronic computer, many calculations of the hydrogen bond energy have been made using the SCF molecular orbital technique (Weissmann and Cohan, 1965; Morokuma and Pedersen, 1968; Del Bene and Pople, 1969; Kollman and Allen, 1969; Hankins et al., 1970).

The decrement in entropy of a system due to hydrogen bonds is large compared to that due to physical and electrostatic bonds, because the hydrogen bond has a higher directional quality in binding.

Structures of Ice and Water

The structure of ice is represented in Fig. 2.3, using the tetrahedral model shown by the dashed lines in Fig. 2.2. Each oxygen atom is at the center of a tetrahedron formed by its four closest oxygen atoms (numbered 2, 3, 4, and 5 in Fig. 2.3). The dimensions of water molecules in ice are not very different from those of isolated molecules. The distance of O-H···O is about 2.76 Å and that of O-H is about 1.01 Å. Such a structure is very bulky; the density of ice is 0.917 g cm^{-3}, which is lower than that of water. Every water molecule is hydrogen-bonded to its four closest neighbors.

The heat of sublimation of ice is 12.2 kcal/mol. One-fourth of the heat may be due to the van der Waals force and the remaining 9 kcal/mol may be used to break two hydrogen bonds per molecule. On the other hand, the heat of fusion of ice is only 1.4 kcal/mol. This fact indicates that only 15% of the hydrogen bonds in ice are broken on melting. It is believed that a bulky, ice-like structure remains in considerable quantity in water, while a closely packed structure forms in the volumes where the hydrogen bonds are broken (Bernal and Fowler, 1933;

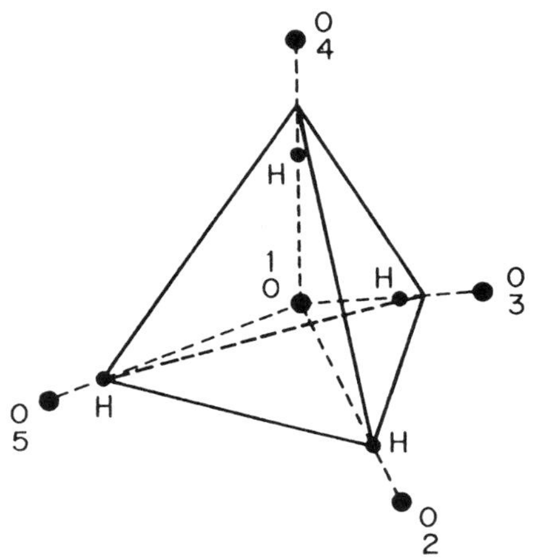

Fig. 2.3. Model for the structure of ice.

Morgan and Warren, 1938; Narten et al., 1967). More detail on the structure andproperties of water are contained in *The Structure and Properties of Water* by Eisenberg and Kauzmann (1969).

2.1.2 Hydration of Ions

It is thought that the intensity of the electric field in the vicinity of an ion is about 10^6 V/cm, and that water molecules with a dipole moment are oriented radially around the ion. According to the flickering cluster model of Frank and Wen (1957), water around an ion is divided into three structurally different regions (Fig. 2.4a). In region A, water molecules are attracted strongly by the ion and are oriented. Both the mobility and dielectric constant of water in this region are much lower than for bulk water. The extent of region A depends on the radius and valence of the ion. Water in region C is bulk water. Region B is a region of structure breaking; most of the water molecules are monomers. Consequently, the viscosity of water in region B may be smaller than that of bulk water, while the entropy may be larger. Bockris and Saluja (1972) proposed a new model on the basis of the Frank-Wen model (Fig. 2.4b). Water molecules in direct contact with an ion in the primary shell are divided into solvated and coordinated water (SCW) and nonsolvated coordinated water (NSCW).

Jones and Dole (1929) proposed an equation showing the dependence of electrolyte solution concentration on viscosity.

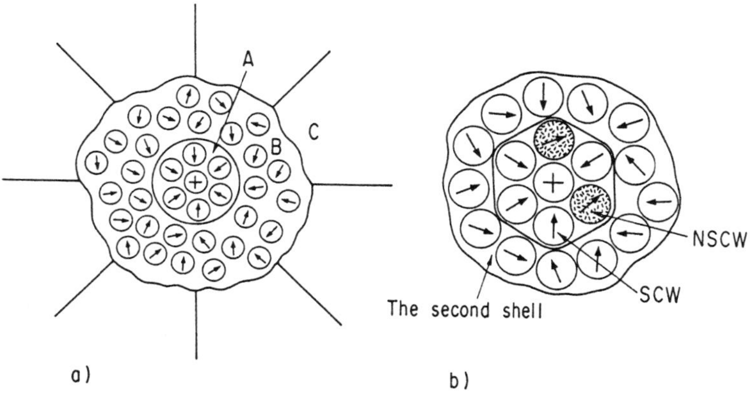

Fig. 2.4. Water structure zones around an ion. (From Zuzuki, 1980)

$$\eta/\eta_0 = 1 + A\, C^{1/2} + B + C \tag{2.1}$$

where η and η_0 are the viscosity coefficients of solution and liquid water, respectively, C is the molarity of the ion, A is the constant determined by the magnitude of ion-ion interaction, and B is the constant defined by both size of the ion and ion-water molecule interaction. The values of B and the activation energies, ΔE, of viscous flows for various ions have been given by Nightingale (1966) (refer to Table 2.1). When the values of B and activation energy for an ion are both negative, the ion is a "structure breaking ion." The existence of a structure breaking ion lowers the viscosity of its electrolyte solution. NH_4^+, K^+, Cs^+, Cl^-, or NO_3^- are structure breaking ions. Since the radius of each of these ions is large and they are monovalent, the ion has only region B, without region A. In Table 2.1, the ratio of the Stokes radius, r_s, of each ion in water solution to its ion radius, r_e, is also shown (Arakawa, 1989).

2.1.3 Physical and Chemical Properties of Clays

The term "clay" is defined as soil particles of less than 2 μm effective diameter, measured from their settling velocity in water. Clay, consisting of clay minerals and amorphous materials, is the most active constituent of soils. Although there are many clay minerals, we have selected kaolinite, montmorillonite, and allophane since their structures are characteristic of many others. We will their physical and chemical properties briefly. For further details, the reader should refer to books such as *Soil Components* by Gieseking (1975), *Minerals in Soil Environments* by Dixon and Weed (1977), or *The Colloid Chemistry of the Silicate Minerals* (Marshall, 1949).

Table 2.1. Stokes radii, B coefficients and activation energies for ions. (From Nightingale, 1966 and Arakawa, 1989)

Ion	r_e (pm)	r_s (pm)	$r_s \backslash r_e$	B(1/mol)	ΔE(kJ/mol)
Li^+	73	240	3.3	+0.150	+0.879
Na^+	116	180	1.6	+0.086	+0.084
K^+	152	130	0.86	-0.007	-0.920
Cs^+	181	120	0.66	-0.045	-1.80
NH_4^+	148	130	0.88	-0.007	-0.418
Mg^{++}	86	350	4.1	+0.385	+1.76
Ca^{++}	114	310	2.7	---	---
Cl^-	167	121	0.7	-0.007	-0.921
NO_3^-	---	---	---	-0.046	-1.00

Structure

Tetrahedral and Octahedral Layers Silica and alumina units are the two main structural units in clay minerals. The silica unit consists of a silicon atom, Si, surrounded by four oxygen atoms at the corners of a tetrahedron whose center is the Si atom (Figs. 2.5a and 2.6). Each of three oxygens at the base of the tetrahedron is shared by two silicons of adjacent units (Fig. 2.5a). This sharing forms a sheet, with hexagonal holes (Fig. 2.5b) of thickness 4.93 Å. Silicon has a positive valence of 4 and oxygen a negative valence of 2. Consequently, the top oxygen, bonded by only one silicon atom, has a negative charge of 1. When the top oxygen takes on a hydrogen with a positive valence of 1 to become hydroxyl, OH, the silica unit becomes neutral.

The alumina unit is composed of an aluminum atom, Al, and six oxygen or hydroxyls equidistant from the aluminum in octahedral coordination (Figs. 2.5c and 2.6). Each of the six oxygens is shared by two aluminum ions of adjacent units. This sharing forms a sheet of two layers of oxygen or hydroxyl (Fig. 2.5d). Aluminum exists at only two-thirds of the possible octahedral centers. This sheet is 5.05 Å thick in clay minerals.

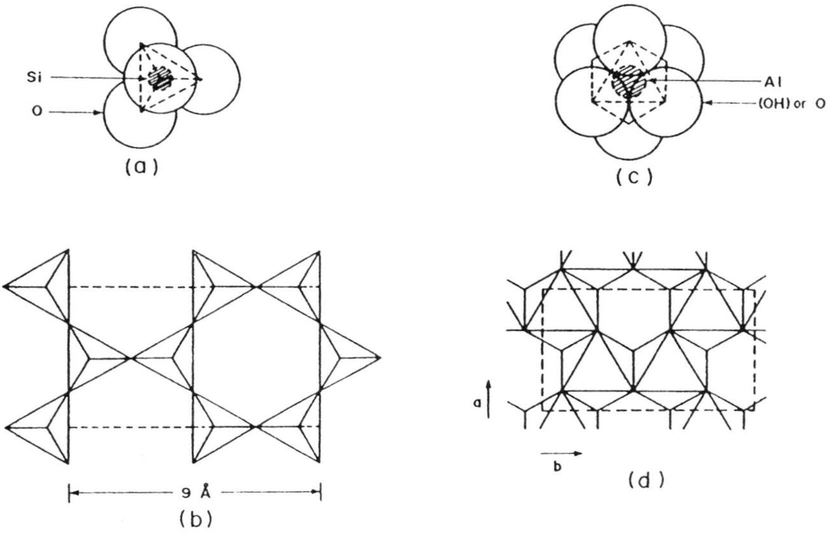

Fig. 2.5. Main structural units in clay minerals.

Kaolinite The elementary layer of kaolinite consists of a silica sheet and an alumina sheet (Fig. 2.7) sharing the top oxygens of the silica tetrahedra (Fig. 2.5a). The elementary layer is three oxygen atoms thick; one face consists of a hydroxyl layer from the alumina sheet, and the other an oxygen layer from the silica sheet (Fig. 2.7). A kaolinite particle is composed of layers held together by hydrogen bonding between the hydroxyl and the oxygen layer (Fig. 2.7). As the force due to the hydrogen bond is strong, hydration between layers does not occur naturally.

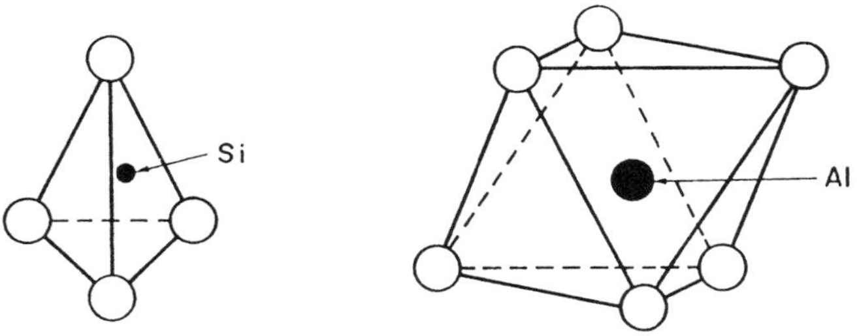

Fig. 2.6. Schematic representation of silica tetrahedron and alumina octahedron in clay minerals.

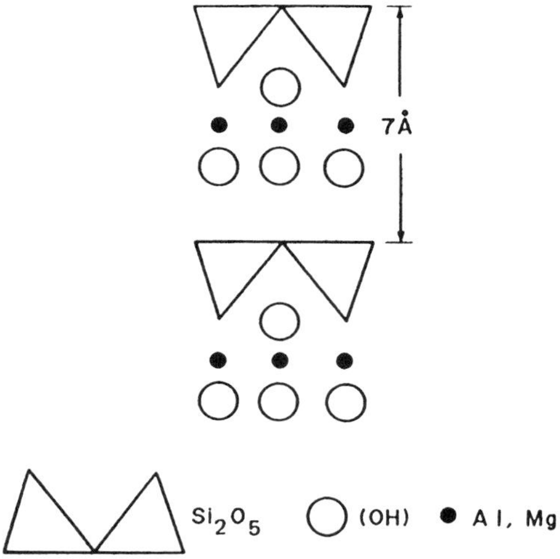

Fig. 2.7. Schematic representation of kaolinite structure.

Montmorillonite The elementary layer of montmorillonite consists of an alumina sheet sandwiched between two silica sheets with shared oxygens (Fig. 2.8). A montmorillonite particle is composed of elementary layers joined through a variable number of water layers (Fig. 2.8). The volume of a montmorillonite particle increases to several times its dry volume when it is placed in contact with water. This phenomenon is due to water adsorbed between elementary layers, pushing them apart and causing swelling. Relatively weak van der Waals forces join elementary layers to each other. The clay layers separated by several water layers can be considered to act as individual particles.

Allophane Ross and Kerr (1934) defined allophane as follows: "Allophane has no definite atomic structure or chemical composition and is a neutral solution of silica, alumina, water and minor bases and accessory acid radicals." However, in recent years, the idea that allophane does have a definite atomic structure has gradually found support among researchers. Most of the silica-alumina ratios, SiO_2/Al_2O_3, in various allophanes are between 1.0 and 2.0 (Yoshinaga, 1966; Kitagawa, 1973), indicating a preferred composition. Electron micrographs show that allophane is composed of unit particles, each of which is a hollow sphere with a diameter between 30 and 50 Å and wall thickness 7 to 10 Å (Fig. 2.9) (Kitagawa, 1971; Henmi and Wada, 1976). Water molecules diffuse in and out of the hollow volume of the unit particle (Wada and Wada, 1976).

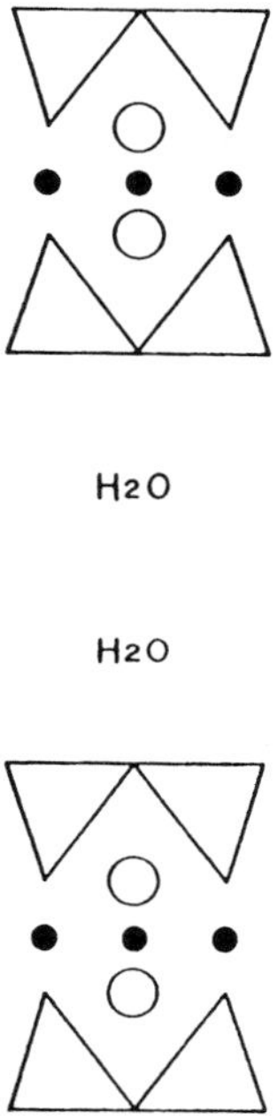

Fig. 2.8. Schematic representation of montmorillonite structure.

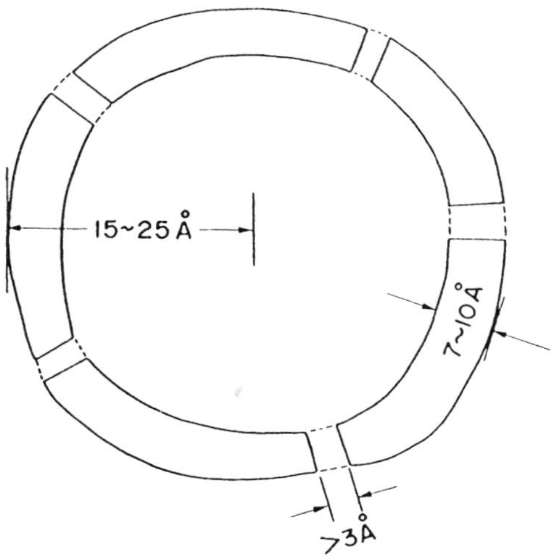

Fig. 2.9. Schematic representation of a unit particle of allophane.

Although the atomic structure of allophane has not yet been made sufficiently clear, it may be considered fundamentally similar to that of kaolinite minerals, but with a disordered layer structure. For further details, we refer the reader to reviews by Wada and Harward (1974) and Wada (1985).

Specific Surface Area

The shape and size of a clay particle determine specific surface area and influence the strength of the interaction between water and the particle. As the specific surface area of a soil particle increases, the amount of water and solutes that interact with the soil particles increases; in other words, the soil becomes more active. The shapes and sizes of kaolinite, montmorillonite, and allophane are given in Table 2.2.

The specific surface area of a clay (Table 2.2) is generally determined by measuring the amount of a liquid or gas required to cover its surface. Water vapor, nitrogen, and organic liquids have been used. The measured specific surface area of a clay varies with the materials used to cover the surface. For example, the value for montmorillonite obtained using N_2 is much lower than that obtained using glycerol. Nitrogen molecules, because they are not polar, are unable to enter the space between the elementary layers. Large molecules used to measure specific area of allophane would not include the surface area inside the hollow unit particle. A complete description of a clay surface may require the analysis of measurements using various materials covering the surface.

Electric Charge

Two groups of mechanisms produce electric charges on clay mineral surfaces.

Isomorphous Substitution When silicon atoms in a tetrahedral layer are substituted by aluminum atoms, or when aluminum atoms in an octahedral layer are replaced by magnesium or divalent iron atoms, the tetrahedral layer or the octahedral layer becomes negatively charged. This is called isomorphous substitution. Silicon and aluminum atoms have positive valences of 4 and 3, respectively, while magnesium and divalent iron have positive valences of 2. Each substitution leads to one negative elementary charge (4.8×10^{-10} esu) in the layer.

Charge from Broken Bonds If a clay mineral were considered as an ionic crystal, the negative valence of 2 from O^{2-} would be saturated by bonding to two Si^{4+} or one Si^{4+} and two Al^{3+}, and a negative valence of 1 from OH^- would be saturated by coordinating to two Al^{3+}. But at the edges of the layers there must

Table 2.2. Physical properties and electrical characteristics of typical clay minerals.

Clay minerals	Shape	Size	Specific surface area (m^2/g)	CEC (meq/100 g)	AEC (meq/100 g)	Electric density (meq/cm^2)
Kaolinite	Hexagonal laminar	Length 0.07-3.5 μm Width 0.5-2.1 μm Thickness 0.03-0.5 μm	10-20	3-15	--	Order of 10^{-6} for the edge surface
Montmorillonite	Disordered laminar	Thickness 0.02-0.002 μm	800	80-150	5-10	$1\text{-}2 \times 10^{-7}$ for the flat surface
Allophane	Disklike[a]	Several hundred Ångstrom	>600	20-30	10-20	1×10^{-8} (negative charge) 2×10^{-8} (positive charge) at pH 5.4[b]

[a]From Egashira (1977)

[b]Values calculated from Harada and Wada (1973)

be broken bonds between oxygen and silicon atoms, between oxygen and aluminum atoms, and between hydroxyl and aluminum atoms. Oxygens and hydroxyls with unsatisfied valences exist on broken edges:

$$Si] - O^- \qquad {}^{Si]}_{Al]}{>}O^{1/2-} \qquad Al] - O^{1/2-} \quad or \quad Al] -(OH)^{1/2-}$$

The charges on these oxygens and hydroxyls depend on the pH of the outer solution. The reactions are represented by

$$Si] - O^- + H^+ \rightarrow Si] - OH \qquad Si] - OH + OH^- \rightarrow Si] - O^- + H_2O$$

$${}^{Si]}_{Al]}{>}O^{1/2-} + H^+ \rightarrow {}^{Si]}_{Al]}{>}(OH)^{1/2+} \qquad {}^{Si]}_{Al]}{>}(OH)^{1/2+} + OH^- \rightarrow {}^{Si]}_{Al]}{>}O^{1/2-} + H_2O$$

$$Al] - O^{1/2-} + H^+ \rightarrow Al] - (OH)^{1/2+} \qquad Al] - (OH)^{1/2+} + OH^- \rightarrow Al] - O^{1/2-} + H_2O$$

$$Al] - (OH)^{1/2-} + H^+ \rightarrow Al] - (OH_2)^{1/2+} \qquad Al] - (OH_2)^{1/2+} + OH \rightarrow Al] - (OH)^{1/2-} + H_2O$$

As the pH of the outer solution decreases (i.e. as the hydrogen ion concentration of the outer solution increases), the reactions on the left are dominant. Consequently, the positive charges due to broken bonds increase as the pH decreases. As the pH of the outer solution increases, the reactions on the right are dominant and the negative charges due to broken bonds increase. At a certain pH, the amount of positive charge becomes equal to the negative charge. We call this value of pH the isoelectric point. Figure 2.10 shows the dependence of the charge of an alumina-silica gel on the pH of the outer solution. The isoelectric points of alumina gel and silica gel are pH 5.0 and 7.0, respectively.

Both kaolinite and montmorillonite possess charges due to isomorphous substitution and to broken bonds. As the ratio of isomorphous substitution to total charge is generally low in kaolinite, the charge of kaolinite depends strongly on the pH of the outer solution. In montmorillonite, the amount of charge due to isomorphous substitution dominates that due to broken bonds. In general, the amount of net charge of a clay mineral is obtained by subtracting the anion exchange capacity (AEC) from the cation exchange capacity (CEC). Typical

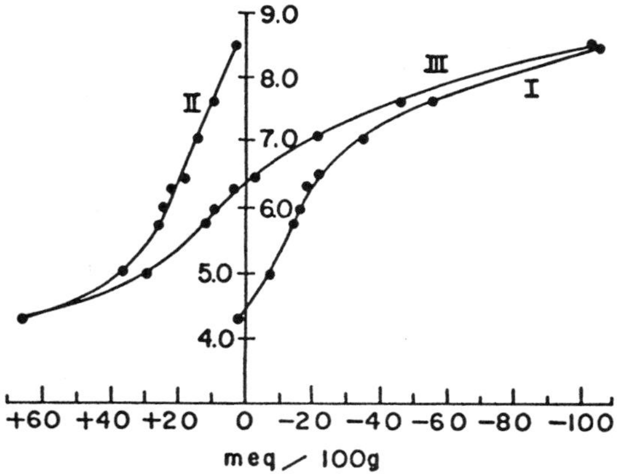

Fig. 2.10. Dependence of electric charge on coprecipitated silica-alumina gel on pH of the equilibrium solution. I, Amount of negative charge; II, amount of positive charge; III, II + I. (From Iimura, 1966)

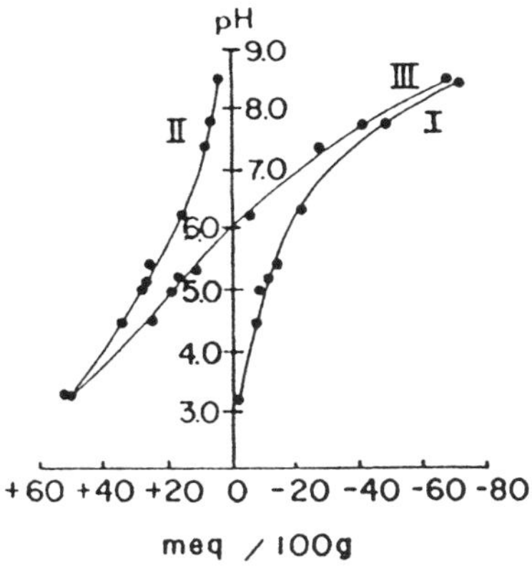

Fig. 2.11. Dependence of electric charge of allophane on pH of the equilibrium solution. I, Amount of negative charge; II, amount of positive charge; III, II + I. (From Iimura, 1966)

values of CEC and AEC for kaolinite, montmorillonite, and allophane are given in Table 2.2.

Allophane has no charge due to isomorphous substitution; all charge results from broken bonds, dependent on the pH of the outer solution. Figure 2.11 shows the dependence of charge of allophane on pH (Iimura, 1966). Allophane has a large positive charge. Generally, the isoelectric point of allophane is between pH 6.0 and 7.0.

The point of zero charge for the edge surfaces of montmorillonite has been regarded to be between 7.3 and 7.7 (Gayer et al., 1958; Parks, 1965), and the isoelectric point of edge surfaces of kaolinites to be around 7.0 (e.g. Rand and Melton, 1977).

2.2 INTERACTION BETWEEN WATER AND AN UNCHARGED SOIL PARTICLE

It is difficult with present knowledge to calculate theoretically the potential energy of interaction between an uncharged clay surface and a water molecule. Therefore, we obtain the potential energy from experimental data.

Jurinak (1963), applying the equations of Frenkel (1946), Halsey (1948), and Hill (1952) to absorption by Na kaolinites, obtained the equation

$$\log_{10} \frac{p}{p_0} = -\frac{0.88}{n^{1.80}} \tag{2.2}$$

where n is a statistical number of layers. Multiplying both sides of Eq. (2.2) by RT/M, we have

$$\frac{RT}{M} \log_{10} \frac{p}{p_0} = \frac{RT}{2.3\ M} \ln \frac{p}{p_0} = -\frac{RT}{M} \frac{0.88}{n^{1.80}} \tag{2.3}$$

The chemical potential of water in equilibrium with water vapor whose pressure is P is given by

$$\mu = \frac{RT}{M} \ln \frac{P}{P_0} \tag{2.3'}$$

Comparing Eq. (2.3) with Eq. (2.3′), we have

$$\mu = \psi = -\frac{RT}{M} \frac{2.3 \times 0.88}{n^{1.80}} \tag{2.4}$$

The flat surface of kaolinite has only a few charges. In addition, the area of the edge surface is one-tenth or less than that of the flat surface. Consequently, Eq. (2.4) gives an approximation of the energy of interaction between an uncharged clay surface and a water molecule.

2.3 DISTRIBUTION OF IONS IN THE NEIGHBORHOOD OF A SOIL PARTICLE

Ions in the solution around sand or silt particles with no electric charge are distributed uniformly, and the solutions can be considered as homogeneous systems. However, the distribution of ions around the charged surface of a clay is not uniform. Most physical and chemical properties of clay soils are influenced by the existence of the solution in which ions are distributed nonuniformly. The distribution of ions in such a solution has been calculated using the Gouy theory.

2.3.1 Balance of Forces on Ions, and the Ionic Distribution

The balance of forces acting on ions in a solution in contact with a charged surface can be explained using a simple model. Let us suppose a box with a square base of length a. The surface of the inner base is charged negatively, with a surface charge density of σ electrons/cm^2. In the box, there is an ideal gas of positive ions; the number of ions, N_p, satisfies the condition

$$N_p \, ve = a^2 \, \sigma$$

where v is the valence of the ions and e the elementary electric charge.

Assuming that a is much larger than the height of the box, b, the strength of the electric field due to the charged base is considered to be a function only of the distance from the base, z. If σ is zero, ions are distributed uniformly in the box by their thermal motion (Fig. 2.12b). This condition corresponds to the maximum entropy of the ion gas. When σ is not zero and the thermal motion of the ions ceases, the ions are attracted to the charged base (Fig. 2.12a). This condition corresponds to the minimum potential energy of the ion gas.

The ions are under two opposing forces; an electrostatic force and a diffusion force (an entropy force) due to thermal motion. The resulting equilibrium distribution corresponds to a condition where the sum of the forces acting on each ion is zero, in other words, a condition of minimum free energy. The curve of the equilibrium distribution has the characteristics shown in Fig. 2.12c. As the distance from the charged base increases, the concentration of ions

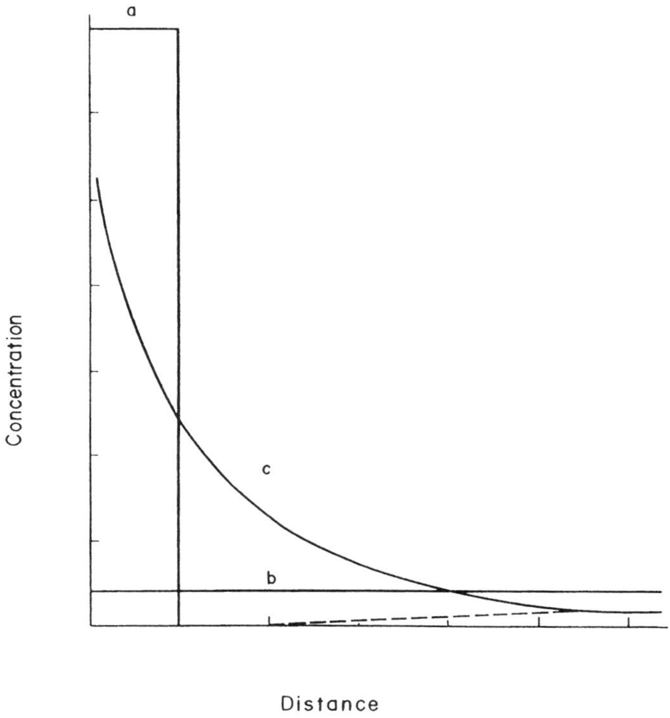

Fig. 2.12. Schematic representation of cation distribution in a diffuse ion layer.

at first decreases rapidly, then gradually, and finally becomes constant. The layer between the charged base and the plane on which the ion concentration reaches constant value is named the diffuse ion layer or diffuse double layer.

Where negative ions also exist, their distribution has a shape shown by the dashed line in Fig. 2.12, because the electric force acting on the negative ions is opposite to that acting on the positive ions. The number of negative ions, N_n, and positive ions, N_p, must satisfy the condition

$$\sigma = N_p - N_n \tag{2.5}$$

The outer, or equilibrium, solution, where the concentration of negative ions is equal to that of positive ions, can be dealt with as a homogeneous system.

The diffusion force is considered to be proportional to the gradient of the pressure of the ion gas. Then the balance of forces acting on an ion is expressed as

$$-\frac{1}{n(z)}\frac{dp}{dz} = \nu e\frac{d\psi(z)}{dz} \tag{2.6}$$

where n and ψ represent the number of ions per cm^3 and the electric potential at z, respectively. For an ideal gas

$$p = nkT \tag{2.7}$$

Substitution of Eq. (2.7) into Eq. (2.6) yields

$$-kT\frac{1}{n}\frac{dn}{dz} = \nu e\frac{d\psi}{dz} \tag{2.8}$$

Solving Eq. (2.8), we obtain

$$n = A\exp\left(\frac{-e\nu\psi}{kT}\right) \tag{2.9}$$

where A is a constant. Assuming that the concentration at $\psi = 0$ is n_0, Eq. (2.9) becomes

$$n = n_0\exp\left(\frac{-e\nu\psi}{kT}\right) \tag{2.10}$$

This is Boltzmann's equation of energy distribution.

The distribution of ions can also be obtained by thermodynamic analysis. The chemical potential of an ion at z, $\mu(z)$, is written

$$\mu(z) = \mu_0 + e\nu\psi + kT\ln\frac{n}{n_0} \tag{2.11}$$

where μ_0 is the chemical potential of an ion at $\psi = 0$ ($z = \infty$). Rewriting Eq. (2.11), we again have Boltzmann's equation:

$$n = n_0\exp\left(\frac{-e\nu\psi}{kT}\right) \tag{2.12}$$

2.3.2 Gouy Theory

In 1910, Gouy gave a theoretical description of the distribution of ions near a charged surface. The Gouy theory assumes that the electric potential is determined by the Poisson equation, and that the distribution of the ions is governed by the Boltzmann equation of energy distribution. Thus we have the two equations

$$\nabla\psi = -\frac{4\pi}{\epsilon}\rho \tag{2.13}$$

$$n_i = n_{i0}\exp\left(-\frac{\nu e\psi}{kT}\right) \tag{2.14}$$

where ψ, ρ, and n_i are the electrical potential, the density of volume charge, and the concentration of the ith ion (the number of ions per cm^3) at a point in the diffuse layer, whose coordinates are (X, Y, Z); ε is the dielectric constant of the solution in the diffuse layer; and n_{i0} is the concentration of the ith ion at $\psi = 0$. Since

$$\rho = \Sigma\nu_i e n_i \tag{2.15}$$

Combining Eqs. (2.13), (2.14), and (2.15), we have

$$\nabla\psi = -\frac{4\pi}{\epsilon}\Sigma\nu_i e n_{i0}\exp\left(-\frac{\nu_i e\psi}{kT}\right) \tag{2.16}$$

Equation (2.16) represents the Gouy theory constructed on the assumption that all ions are regarded as point charges, and the interaction between water and ions is neglected, as described later in detail. The system is considered as a charged surface and an ideal gas of ions; water plays only the role of offering a space with a dielectric constant.

For an infinite flat surface with a uniform surface charge density, immersed in a single symmetrical electrolyte, Eq. (2.16) becomes

$$\frac{d^2\psi}{dz^2} = -\frac{4\pi}{\epsilon}en_0\left[\exp\left(\frac{+e\psi}{kT}\right) - \exp\left(\frac{-e\psi}{kT}\right)\right] \tag{2.17}$$

under the boundary conditions $\psi = 0$ and $d\psi/dz = 0$ at $z = \infty$. Integrating Eq. (2.17), we obtain

$$\frac{d\psi}{dz} = -\sqrt{\frac{8\pi\, n_0 kT}{\epsilon}} \left[\exp\left(\frac{e\psi}{2kT}\right) - \exp\left(\frac{-e\psi}{2kT}\right)\right] \tag{2.18}$$

Infinite Distance Between Plates

When the distance between two plates is infinite, the boundary conditions for Eq. (2.17) are

$$\psi = 0 \quad \textit{and} \quad \frac{d\psi}{dz} = 0 \qquad \textit{at } z = \infty$$

and

$$\psi = \psi_0 \qquad \textit{at } z = 0$$

Integrating Eq. (2.18) for these boundary conditions, we obtain

$$\kappa z = \ln \frac{[\exp(e\psi/2kT) + 1]\,[\exp(e\psi_0/2kT) - 1]}{[\exp(e\psi/2kT) - 1]\,[\exp(e\psi_0/2kT) + 1]} \tag{2.19}$$

where κ, the Debye κ, is expressed by

$$\kappa = \sqrt{\frac{8\pi n_0 e^2 v^2}{\epsilon kT}}$$

where v is the valence of the counterion.

The characteristic length $1/\kappa$ is a measure of the thickness of the diffuse ion layer (Verwey and Overbeek, 1948). The dependence of $1/\kappa$ on concentration of the outer solution and valence of the counterion is

C (mol/liter)	Monovalent ions (cm)	Divalent ions (cm)	Trivalent ions (cm)
10^{-5}	10^{-5}	0.5×10^{-5}	$1/3 \times 10^{-5}$
10^{-3}	10^{-6}	0.5×10^{-6}	$1/3 \times 10^{-6}$
10^{-1}	10^{-7}	0.5×10^{-7}	$1/3 \times 10^{-7}$

where C is the ionic concentration in the outer solution.

The form of Eq. (2.19) becomes simpler where $e\psi_0/kT << 1$ is applicable. Then

$$\psi = \psi_0 \exp(-\kappa z) \tag{2.20}$$

For ψ = 25 mV, and using k = 1.38 x 10^{-16} erg/K, T = 300 K, and e = 1.6 x 10^{-19} C, $e\psi/kT$ becomes nearly 1 (1 C • V = 1 J = 1 x 10^7 ergs). Solving the Poisson Equation (2.13) for ρ and integrating, we obtain

$$\sigma = \int_0^\infty \rho \, dz = -\int_0^\infty -\frac{\epsilon}{4\pi}\frac{d^2\psi}{dz^2}\,dz = \frac{\epsilon}{4\pi}\frac{d\psi}{dz}\bigg|_0^\infty = -\frac{\epsilon}{4\pi}\left(\frac{d\psi}{dz}\right)_{z=0} \tag{2.21}$$

Substitution of Eq. (2.18) into Eq. (2.21) yields

$$\sigma = \sqrt{\frac{\epsilon n_0 kT}{2\pi}}\left[\exp\left(\frac{e\psi_0}{2kT}\right) - \exp\left(\frac{-e\psi_0}{2kT}\right)\right] \tag{2.22}$$

where σ is the surface charge density. If $e\psi_0 << kT$, the equation becomes

$$\sigma = \frac{\epsilon\kappa}{4\pi}\psi_0 \tag{2.23}$$

Where σ is known, we can calculate ψ_0 using Eq. (2.22) or (2.23). Figure 2.13 is an example of a distribution of the electric potential. In Fig. 2.14, the concentration of cations and anions are plotted for two values of n_0.

Panjukov (1986) provides an approximate analytic expression for the relationship between surface charge density and surface potential of a spherical or cylindrical colloidal particle in a general electrolyte. Schuhmann and D'Epenoux (1987) studied the effect of electrical nonhomogeneities of particles on mean surface potential.

Chan et al. (1984) presented equations that permit calculation of double layer potentials for the clay-solution interfaces from negative adsorption (co-ion exclusion) results when the area of the interface is known. Using these equations, they re-analyzed previously reported experimental results on co-ion (chloride) exclusion from well-characterized homoionic illite and montmorillonite in the presence of alkali and alkaline earth counterions (Edwards and Quirk, 1962, 1965a,b,c) to determine their surface potentials in a range of electrolyte solution. The results showed that 1) the less hydrated monovalent ions (i.e. NH_4^+, Cs^+, and Rb^+) bind more strongly to the clay surface and therefore produce low double-layer potentials, 2) the more hydrated ions, Li^+ and Na^+ bind less and

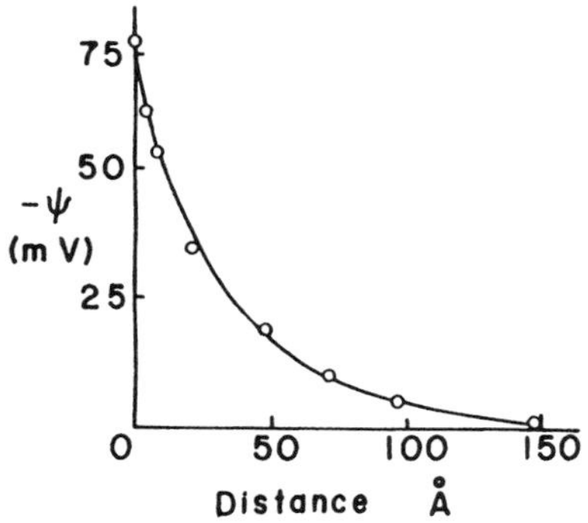

Fig. 2.13. Electric potential as a function of distance from a charged surface with $\sigma = 5 \times 10^3$ esu/cm^2 and $n_0 = 0.005$ N. (From Babcock, 1963)

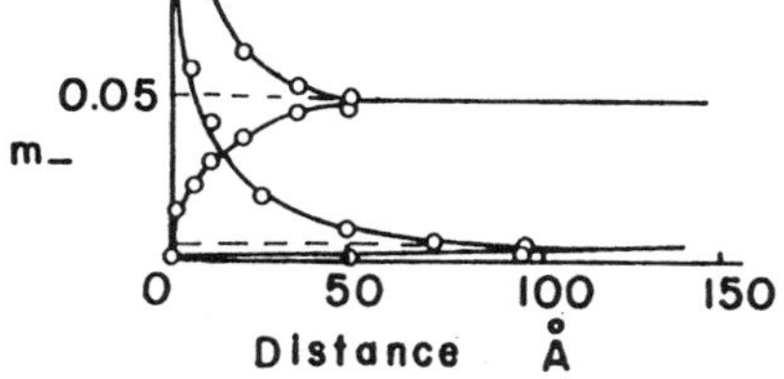

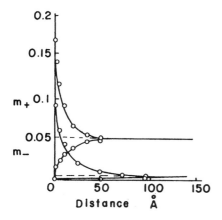

Fig. 2.14. Cation, m^+, and anion, m^-, concentrations as a function of distance from a charged surface with $\sigma = 5 \times 10^3$ esu/cm^2 and $n_0 = 0.05$ N, 0.01 N. (From Babcock, 1963)

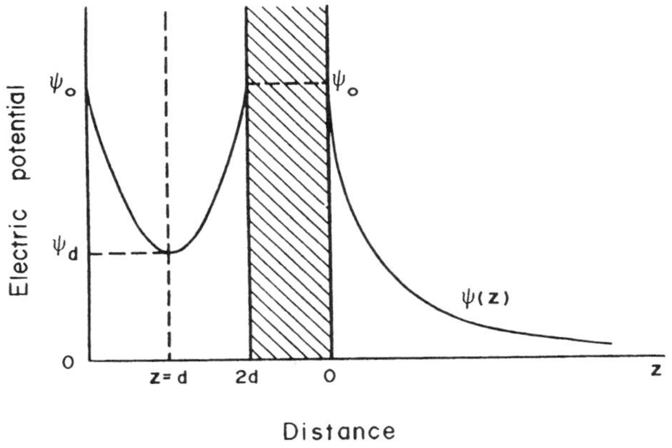

Fig. 2.15. Schematic representation of electric potential distribution for two interacting diffuse ion layers.

give higher potentials, 3) divalent alkaline earth ions also produce very low potentials on illite and montmorillonite, presumably due to stronger electrostatic attraction, and 4) in all samples studied, the surface-potential is roughly independent of electrolyte concentration over a range of about 0.3 to 0.003 molar.

Result 4) may show that clay surfaces appear to behave more like constant-potential than constant charge surfaces (see Section 3.1.1). Horikaw et al. (1988) obtained the same result from measuring zeta potentials of homoionic montmorillonite and illite. Miller and Low (1990) also concluded that smectites are colloids with a constant surface potential rather than a constant surface charge through measurements of zeta potentials of 34 different Na-smectites.

Finite Distance Between Plates

When two adjacent plates are close enough for the two diffuse layers to overlap, the distribution of the electric potential between plates is expressed by a symmetrical curve as shown in Fig. 2.15. The curve has a minimum at the midpoint between plates, but the electrical potential at that point is not zero.

Equation (2.17) should be solved under the boundary conditions $\psi = \psi^0$ at $z = 0$, $d\psi/dz = 0$ at $z = d$ (2d is the distance between plates). The solution is

$$\int_y^u \frac{dy}{\sqrt{2(\cosh y - \cosh u)}} = -\int_{\kappa z}^{\kappa d} d(\kappa z) = -\kappa(d - z) \tag{2.24}$$

where u and y denote $e\psi_d/kT$ and $e\psi/kT$, respectively; ψ_d is the electric potential at z = d; and ψ is the electric potential at z. Then the surface charge density σ is given by

$$\sigma = -\int_0^d \rho \, dz = -\frac{\epsilon}{4\pi}\left(\frac{d\psi}{dz}\right)_0 = \sqrt{\frac{n_0 \epsilon kT}{2\pi}}\sqrt{2(\cosh w - \cosh u)} \qquad (2.25)$$

where w is $e\psi_0/kT$. If the integration is carried out between z = 0 and z = d in Eq. (2.24), we obtain the desired expression giving us the potential midway between plates, u, as a function of the plate distance, 2d. That is,

$$\int_w^u \frac{dy}{\sqrt{2(\cosh w - \cosh u)}} = -\kappa d \qquad (2.26)$$

The electric potential and ion concentrations are plotted as a function of distance for $\sigma = 5 \times 10^3$ esu/cm^2, $n_0 = 0.0005$ M, d = 120 Å, and $\psi_d = -60$ mV in Figs. 2.16 and 2.17. Solutions for other cases are given in *Theory of the Stability of Lyophobic Colloids* by Verwey and Overbeek (1948).

Numerical solutions under more complicated conditions were also obtained by many researchers [e.g. Loeb et al. (1961) for spherical colloids; Philip (1970) for the three-dimensional case; Teubner and Frahm (1981) for colloids of

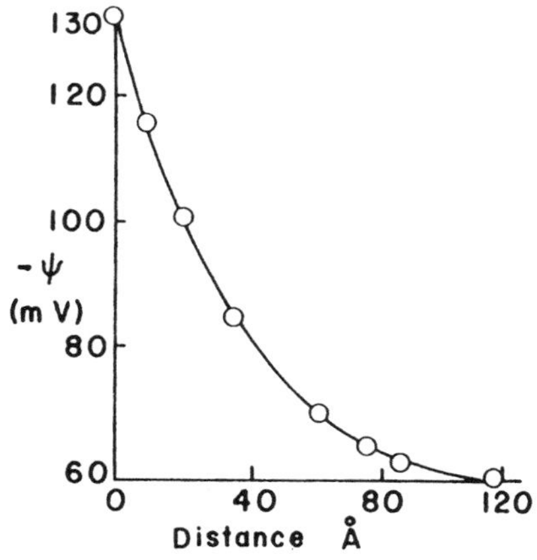

Fig. 2.16. Electric potential as a function of distance from a charged surface with $\sigma_0 = 5 \times 10^3$ esu/cm^2 for a plate distance of 240 Å. n = 0.0005 N. (From Babcock, 1963)

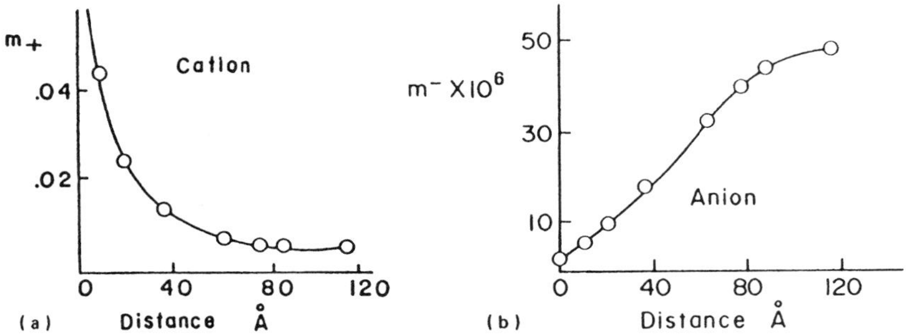

Fig. 2.17. Cation (a) and anion (b) concentrations as a function of distance from a charged surface with $\sigma_0 = 5 \times 10^3$ esu/cm^2 for a plate distance of 240 Å. n = 0.0005 N. (From Babcock, 1963)

arbitrary shape; Bentz (1982) for any combination of mono-, di-, and trivalent counterions; Barough and Kulkarni (1986) for two spheres with fixed surface potentials; Barak (1989) for the case of diffuse electric double layers containing mixed divalent and trivalent electrolytes]. In addition, Sridharan and Jayadeva (1980) proposed simplified graphical procedures using the Gouy theory to determine simply, accurately, and rapidly the surface potential of a clay platelet and the variation of potential with distance.

Unusual Distribution of Electrical Potential Near an Edge Surface of Montmorillonite

A surface having positive charges generally produces a positive double layer near the surface. However, when negatively and positively charged parts exist on adjacent surfaces, what is the distribution of electrical potential in the vicinity?

Under nonalkaline conditions, an elementary particle of montmorillonite has two flat surfaces with negative charges due to isomorphous substitution and edge faces charged positively (pH dependent charge). The total area of the flat surfaces is much larger than that of edge surfaces. Distribution of electrical potential near the edge surface must be known to understand modes of particle association when a suspension of montmorillonite flocculates (edge-to-face, edge-to-edge, face-to-face) (see Section 3.2.1).

Secor and Radke (1985) estimated theoretically the distribution of electrical potential near an edge face of montmorillonite. Their assumptions were: 1) A montmorillonite particle is a thin disk as shown in Fig. 2.18. The disk half thickness is δ (0.5 nm), and its radius is R. The coordinate frame is cylindrical with the origin located at the particle edge. 2) The flat surface charge density

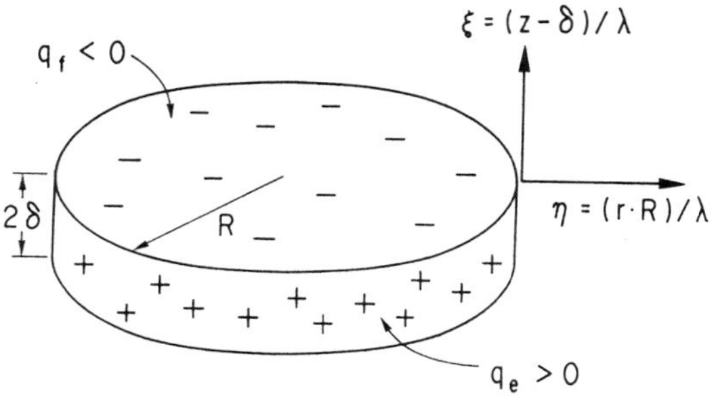

Fig. 2.18. Schematic of the montmorillonite particle. The aspect ratio is R/δ, with negative face and positive edge surface charge. (From Secor and Radke, 1985)

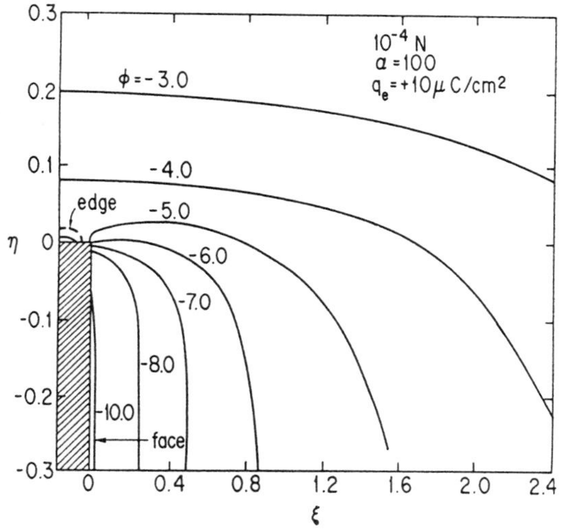

Fig. 2.19. Electrostatic potential around a montmorillonite particle. The face charge density q_f = -10 μC/cm^2, and the edge charge density q_e = +10 μC/cm^2. (From Secor and Radke, 1985)

is -10 $\mu C/cm^2$ on the basis of a cation exchange capacity of 80 meq/100 g and specific surface area of 800 m^2/g. The positive edge face charge density is set at 10 or 50 $\mu C/cm^2$ based on the study of Quirk (1960) with kaolinite. 3) All concentrations refer to a 1-1 indifferent electrolyte.

The Poisson-Boltzmann equation (Eq. 2.16) was solved numerically for the model using the Galerkin finite-element method with a Newton-Raphson iteration scheme (Strang and Fix, 1973). Fig. 2.19 shows equipotential lines of electrical potential for a platelet with an aspect ratio, $\alpha = R/\delta$, of 100 immersed in a 10^{-4} N aqueous salt solution. The reduced potential is expressed by $\Phi = ze\psi/kT$ where z is the valence of the ions, e the elementary electric charge, ψ the electrical potential, k the Boltzmann constant, and T is the temperature. A dramatic effect of the negative surface on potentials near the edge surface is recognized from the figure; that is, Φ is everywhere negative including directly up to the positive edge. If there were no influence of the flat surface, the edge would exhibit a monotonically decaying positive potential characterized by a reduced potential of +10.3.

Figure 2.20 indicates the role of electrolyte concentration on the potential along the disk midplane, $\Phi(\eta, -\delta)$. To permit direct comparison among different electrolyte concentrations the abscissa is expanded by a base-10 logarithmic scale. For salt content below 5 x 10^{-3} N, more and more of the midplane potential profile becomes positive. Nevertheless, there is still a significant influence of the flat surface charge. This is established by comparison to the dashed line in Fig. 2.20 which gives the one-dimensional edge potential.

2.3.3 Problems in Application of the Gouy Theory to Clay-Water Systems

Assumption of Point Charge

In the Gouy theory, ions are considered as point charges, distributed continuously in a diffuse layer. This assumption often causes an overestimation of the concentration of ions near the clay surface. Suppose that σ = 1 electron per 100 $Å^2$ = 10^{14} electrons/cm^2 = 4.8 x 10^4 esu/cm^2, ε = 80, v = 1, T = 300°C, and $n_0 = 10^{-4}$ mol/liter = 6 x 10^{16} ions/cm^3. Substituting these values into Eq. (2.22), we obtain $e\psi_0/kT$ = 11.2. Consequently, the cation concentration (the volume charge density) directly above the surface, n_s^+, is 7.3 mol/liter.

Next we will calculate the value of the cation concentration when all excess cations in the diffuse layer are absorbed in one layer on the surface. The amount of excess ions is equal to σ, so 10^{14} cations/cm^2 exist in the first absorbed water layer. The number of water molecules is about 9 x 10^{14} cm^{-2} if the area of one

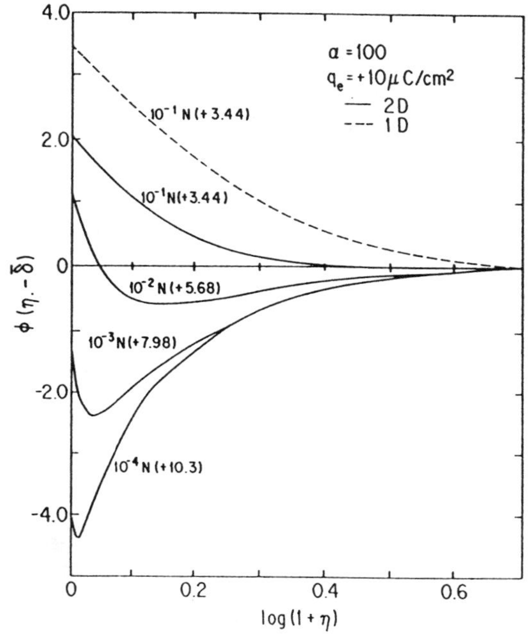

Fig. 2.20. The electrostatic potential along a montmorillonite particle midplane for varying indifferent electrolyte concentrations. The dashed line corresponds to the one-dimensional, edge potential profile, and numbers in parentheses refer to the one-dimensional edge surface potential. (From Secor and Radke, 1985)

water molecule is 10.8 Å^2. Therefore, the concentration of cations in the first layer, n_s^+, is

$$n_s = \frac{10^{14}}{9 \times 10^{14}} \times 55.5 = 6 \ mol/liter$$

Only about half of the excess ions might be absorbed directly onto the surface (Shainberg and Kemper, 1966). Thus it is concluded that the Gouy theory greatly overestimates the ion concentration near the surface.

Distribution of Electric Intensities in the Vicinity of a Clay Surface

The Gouy theory has been developed on the basis of macroscopic electromagnetism. According to electromagnetism, the intensity of an electric field in a space in contact with an infinite plane with surface charge density of σ is given by $2\pi\sigma/\varepsilon$. This equation is not valid for electric intensities in the range where the distance from the charged surface is smaller than the average distance between adjoining electrons. Assuming an electron distribution on the plane as shown in Fig. 2.21, the distribution of the electric intensities along the line perpendicular to the charged surface through point A is represented in Fig. 2.22 (Iwata, 1974b). We can read from Fig. 2.22 that although the electric intensity initially increases with distance from the charged surface, it maintains a constant value, $2\pi\sigma/\varepsilon$, in the range where the distance is larger than 1.0 a. This must be noted in calculating the distribution of electric intensities near a clay surface.

Diffuse Layer Terminated by an Air-Water Interface

It is difficult to calculate the electric potential distribution (the ion distribution) in a diffuse layer terminated by a flat air-water interface, because the solution is not in equilibrium with an outer solution. The surface charge density, σ, and the thickness of the water film, d, can be calculated from the physical quantities usually measured: specific surface area, amount of adsorbed water, cation exchange capacity, species of ions contained in adsorbed water, and contents of anions per cm^2 of solid-liquid interface, $\bar{\sigma}$.

If σ, d, and $\bar{\sigma}$ are known, we can adopt the following tedious procedure to calculate the electric potential distribution on the basis of the Gouy theory. The assumptions are:

1. Cations and anions in a diffuse layer terminated by the flat air-water interface are composed of one type of monovalent cation and anion (e.g. Na^+ and Cl^-).

2. The electric potential distribution in the diffuse layer terminated by the flat air-water interface is essentially identical with the symmetric part of the electric potential distribution in the solution between the mid plane and one plate in a system, as shown in Fig. 2.15. This is in equilibrium with an outer solution between parallel, flat plates separated by 2d if the values of σ and $\bar{\sigma}$ in both systems are the same, that is, if the electric potential distribution in the solution between parallel, flat plates is symmetric with respect to the plane midway between plates. This comparison breaks down if the pressure at the central plane in significantly larger than atmospheric pressure, which exists at the flat air-water interface (Bolt and Miller, 1958).

The procedure is as follows:

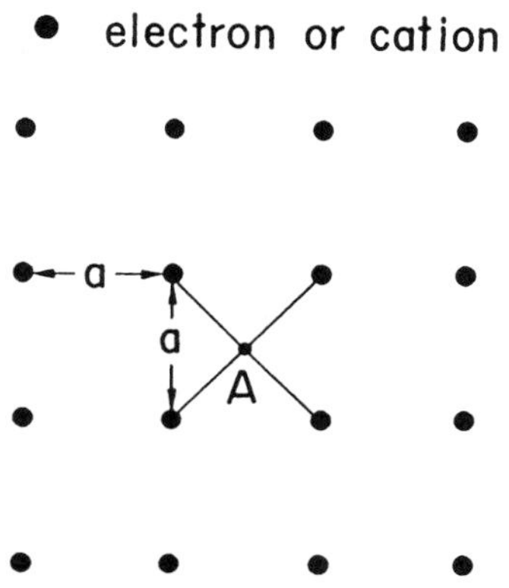

Fig. 2.21. Assumed distribution of electrons or cations on a charged plane.

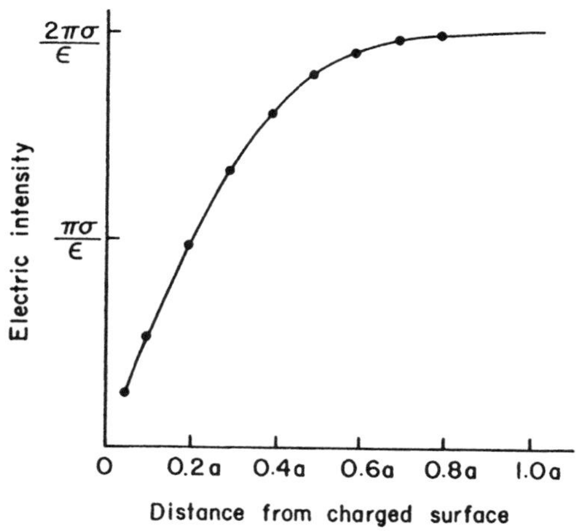

Fig. 2.22. Distribution of electric intensities along the line perpendicular to the charged plane through point A in Fig. 2.21. (From Iwata, 1974b)

1. Assume a given value for the concentration of the outer solution, n_0.
2. Substitute the value of the surface charge density, σ, and the assumed value of n_0 into Eq. (2.25) to obtain various sets of ψ_0 and ψ_d satisfying the equation.
3. Substitute the value n_0 and the obtained sets of ψ_0 and ψ_d into Eq. (2.26) to get the set of ψ_0 and ψ_d satisfying the equation (κ is a function of n_0).
4. Calculate the electric potential distribution corresponding to d, n_0, and the set of ψ_0 and ψ_d by using Eq. (2.24).
5. Calculate the content of anions per cm^2 of solid-liquid interface from the equation $\int_0^d n_0 \exp(e\psi/kT)\, dz$.
6. If the calculated content of the anions is equal to the value of $\bar{\sigma}$, the electric potential distribution obtained in step 4 is the desired one. Then we can calculate the desired ion distribution using the electric potential distribution. The calculation should be continued for various values of n_0 until the condition above is satisfied.

Iwata (1974a) proposed a method to obtain the ion distribution by establishing an adsorption model. In the following description, we will introduce this method by selecting a simple case through which it can readily be understood. The conditions are:

1. The location of the electric charges is 4.46 Å below the surface and the charge density (negative) is Γ.
2. The thickness of water film, d_t, is 2.76 x n Å, where n is the number of adsorbed water molecular layers.
3. There are only Na^+ ions in the water film.

The adsorption model shown in Fig. 2.23 is established on the following assumptions:

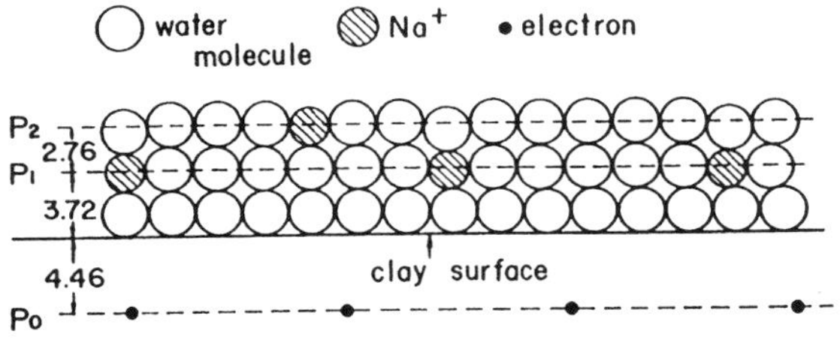

Fig. 2.23. Model of water adsorption on a clay surface. (From Iwata, 1974a)

1. Ions are distributed diffusely according to the Maxwell-Boltzmann law of energy distribution.
2. The water molecule is of spherical shape with a radius of 1.38 Å.
3. Na^+ ions are distributed in such a way that they do not overlap each other in a direction perpendicular to the surface. This means that Na^+ ions are distributed only on the planes which are parallel to the surface and are separated at a distance of 2.76 Å from each other.

For calculation of the distribution of Na^+ ions and electric potential, the charged plane on the clay surface is taken as P_0, the known density of charge as σ_0, the next plane as P_1, with a charge density of σ_1 and the rest in succession as P_2, P_3, ..., P_n, σ_2, σ_3, ..., σ_n (Figs. 2.23 and 2.24). According to electromagnetism, the intensity of the electric field in the space in contact with an infinite plane with surface charge density σ is given as $(2\pi/\varepsilon)\sigma$. Accordingly the electric intensities between planes are obtained as follows:

$$E_{01} = \frac{2\pi}{\varepsilon_{01}} \sigma_0 + \sum_{i=1}^{n} \frac{2\pi}{\varepsilon_{01}} \sigma_i$$

$$= \frac{4\pi}{\varepsilon_{01}} \sigma_0 \left(\sum_{i=1}^{n} \sigma_i = \sigma_0 \right)$$

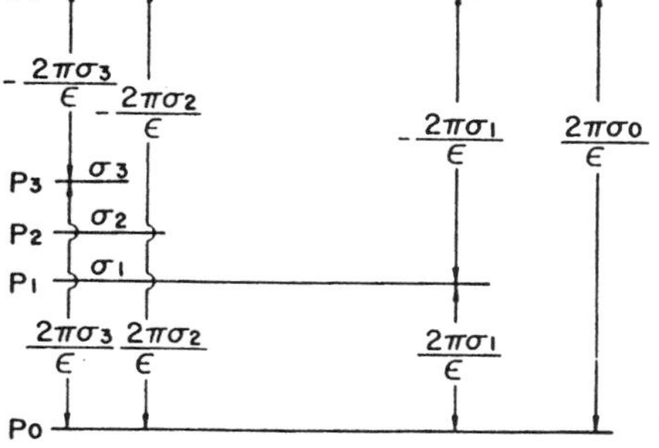

Fig. 2.24. Schematic representation of electric intensities between each plane from Fig. 2.23 on which cations or electrons exist. (From Iwata, 1974a)

$$E_{01} = \frac{2\pi}{\epsilon_{01}} \sigma_0 + \sum_{i=1}^{n} \frac{2\pi}{\epsilon_{01}} \sigma_1 = \frac{4\pi}{\epsilon_{01}} \sigma_0 \left(\sum_{i=1}^{n} \sigma_i - \sigma_0 \right) \tag{2.27}$$

where E_{01} is the electric intensity between P_0 and P_1, n the number of adsorbed water layers, and ε_{01} the dielectric constant of water existing between P_0 and P_1.

$$E_{12} = \frac{2\pi}{\epsilon_{12}} \sigma_0 - \frac{2\pi}{\epsilon_{12}} \sigma_1 + \sum_{i=2}^{n} \frac{2\pi}{\epsilon_{12}} \sigma_i = \frac{4\pi}{\epsilon_{12}} (\sigma_0 - \sigma_1) \tag{2.28}$$

$$E_{i \cdot i+1} = \frac{4\pi}{\epsilon_{i \cdot i+1}} \left(\sigma_0 - \sum_{i=1}^{i} \sigma_i \right) = \frac{4\pi}{\epsilon_{i \cdot i+1}} \left(\sum_{i=i+1}^{n} \sigma_i \right) \tag{2.29}$$

The electric potentials on each plane, P_1, P_2, ..., P_n, are calculated using the electric intensities shown in Eqs. (2.27), (2.28), and (2.29). That is,

$$\begin{aligned} \psi_n &= 0 \\ \psi_{n-1} &= \frac{-4\pi}{\epsilon_{n \cdot n-1}} \times \sigma_n \times 2.76 \\ \psi_2 &= \frac{-4\pi}{\epsilon_{2.3}} \times \left[\sigma_n(n - 2) + \sigma_{n-1}(n - 3) + \ldots + \sigma_3 \right] \times 2.76 \\ \psi_1 &= \frac{-4\pi}{\epsilon_{12}} \times \left[\sigma_n(n - 1) + \sigma_{n-1}(n - 2) + \ldots + \sigma_2 \right] \times 2.76 \end{aligned} \tag{2.30}$$

where ψ_i is the electric potential on P_i. From the Boltzmann law of energy distribution

$$\sigma_i = \sigma_n \exp\left(\frac{-\psi_i e}{kT} \right) \tag{2.31}$$

where e is the elementary charge. If the value of σ_n is assumed, the distribution of cations can be obtained using Eqs. (2.30) and (2.31). When these values of σ_i meet the condition $\Sigma_{i=1}^{n} \sigma_i = \sigma_0 = \Gamma$, the distribution is the desired one. Calculations should be continued until this condition is obtained.

The proposed theory is also applicable to cases where anions as well as cations are present (Iwata, 1974a). The distributions of cations at 20°C calculated by the theory are shown in Fig. 2.25 (Γ = 1.16 x 10^{-7} meq/cm^2, c =

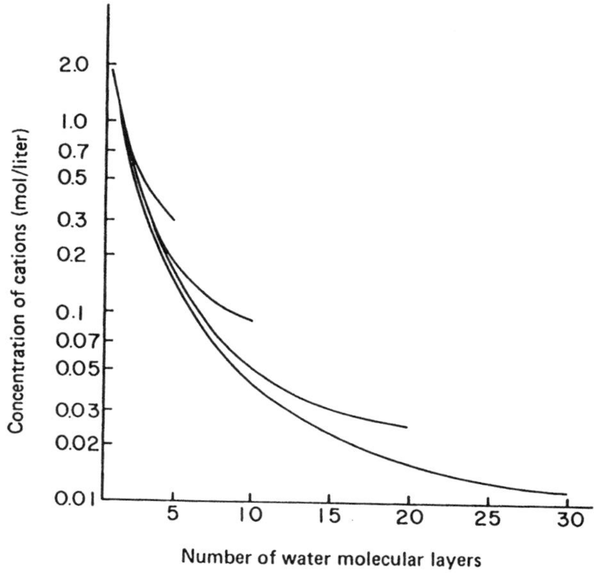

Fig. 2.25. Distribution of Na ions in a diffuse ion layer terminated by an air-water interface. $\Gamma = \sigma_0 = 1.16 \times 10^{-7}$ meq/cm^2 and c = 0. (From Iwata, 1974a)

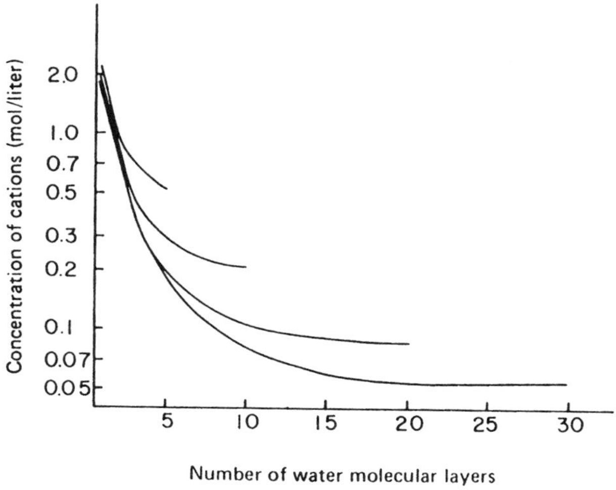

Fig. 2.26. Distribution of Na ions in a diffuse ion layer terminated by an air-water interface. $\Gamma = \sigma_0 = 1.16 \times 10^{-7}$ meq/cm^2 and c = 0.35×10^{-7} meq/cm^2. (From Iwata, 1974a)

0) and Fig. 2.26 (Γ = 1.16 x 10^{-7} meq/cm^2, c = 0.35 x 10^{-7} meq/cm^2), where c denotes the amount of monovalent anion per cm^2 of solid-liquid interface. Figure 2.27 shows the distribution of electric potential (c = 0).

These distribution curves are similar to those calculated by Babcock (1963) and Bolt and Miller (1958), but there is a difference in the quantitative relationship between them. Figure 2.28 shows cation concentrations at a flat air-water interface bounding a double layer as a function of distance from the clay surface calculated by Bolt and Miller (1958) and the proposed theory assuming that c = 0. Bolt and Miller did not indicate the value of the surface charge density which was used to calculate the cation concentration, but it is certain that the value was smaller than 4.0 x 10^{-7} meq/cm^2. As shown in Fig. 2.28, the cation concentrations calculated by Bolt and Miller at any film thickness are always higher than those calculated by Iwata. When a monolayer is adsorbed (film thickness is 2.76 Å), the cation concentrations are 13.4 mol/liter for Γ = 4.0 x 10^{-7} meq/cm^2 and 3.9 mol/liter for Γ = 1.17 x 10^{-7} meq/cm^2, respectively. The values calculated by Iwata agree with this.

Taylor (1959) also calculated the concentration of ions at a clay surface. He obtained a value of 2.68 mol/liter assuming $\sigma_0 = \Gamma$ = 1.0 x 10^{-7} meq/cm^2 and c = 0. This means that about 80% of the counterions exist at the clay surface. The result by Iwata's method gives the cation concentration in the plane closest to the clay surface of 1.76 mol/liter where $\sigma_0 = \Gamma$ = 1.0 x 10^{-7} meq/cm^2 and c = 0. This means that 45% of counterions exist on the closest plane. The reasons

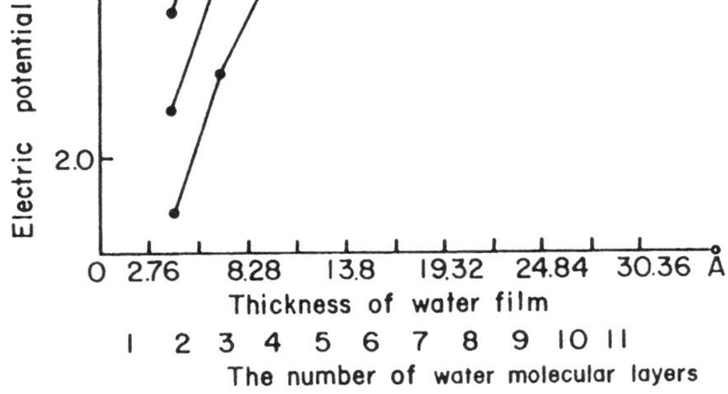

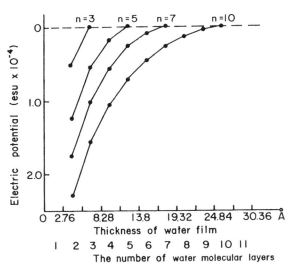

Fig. 2.27. Distribution of electric potential in a diffuse ion layer terminated by an air-water interface. $\Gamma = \sigma_0$ = 1.16 x 10^{-7} meq/cm^2 and c = 0. (From Iwata, 1974a)

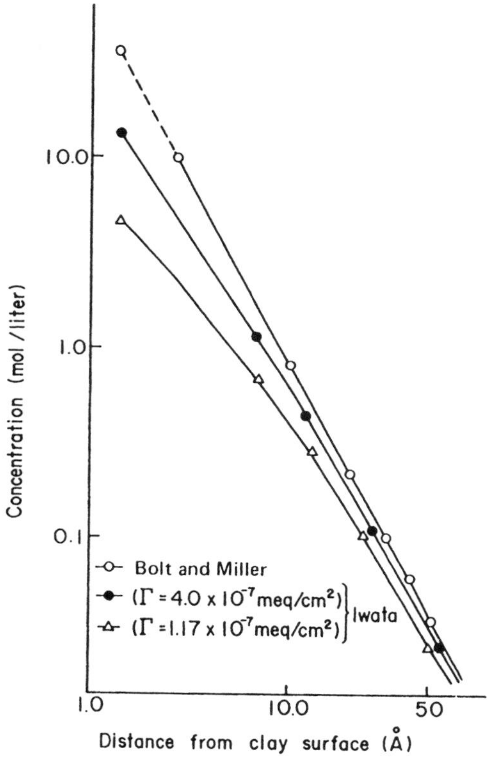

Fig. 2.28. Concentrations at a flat-air water interface bounding a diffuse ion layer as a function of film thickness (distance from the clay surface to an air-water interface). (From Iwata, 1974a)

why the values calculated by Taylor and by Bolt and Miller are high may be due to the potential and point-charge errors considered in the preceding section.

There is an excellent report by Shainberg and Kemper (1966) on the number of cations directly in contact with a clay surface. Ravina and Gur (1978) corrected the over-estimation of ionic concentration near charged surfaces by introducing a Poisson-Boltzmann equation including hydration forces and the dielectric saturation effects.

Fujii and Nakano (1984) improved the theory to make the adsorption model proposed by Iwata (1974a) suitable for water adsorption to internal surfaces of montmorillonite. They calculated the distribution of cation and electric potential between parallel clay particles. Using these values, the decrements of chemical potential of water adsorbed on the internal surfaces due to the electric field and the solute could be obtained. In addition, the decrements due to the intermolecular force were calculated at the center and at the edge of the clay particle. They showed that the intermolecular force played the most important role of the three factors affecting chemical potential. The water characteristic curve of the montmorillonite was derived on the basis of the relationship obtained between the number of water molecule layers and the chemical potential.

Other Factors Neglected in the Gouy Theory

In addition to the problems described above, the Gouy theory neglects contributions such as (1) the dependence of the dielectric constant on the electric field strength, (2) the short-range forces between ions, (3) the polarizability of ions and their hydration sheath, (4) the atmosphere around the ions imposed by their charge, (5) hydration forces, and (6) the compressibility of the solvent (Frahm and Diekmann, 1979). Modified equations for Eq. (2.16) (Poisson-Boltzmann equation), derived taking some of these contributions into account, were proposed and solved numerically by many researchers [e.g. Levine and Bell (1960) for (2) and (3); Levine and Bell (1964) for (4) and (6); Gur et al. (1978) for (5); Frahm and Diekmann (1979) for the influence of finite ion sizes].

2.3.4 Force Acting on Water Molecules in a Diffuse Ion Layer

As described in Chapter 1, the chemical potential of water in a diffuse ion layer is given by

$$\mu^{(\alpha)} = \mu_0^{(\alpha)} + \mu_p^{(\alpha)} + \mu_f^{(\alpha)} + \mu_e^{(\alpha)} + [\mu] \tag{2.32}$$

where $[\mu]$ is the chemical potential of bulk water at an external pressure P. The chemical potential of the outer solution, μ_{os}, is

$$\mu_{os} = \mu_0^{(\alpha)} + \mu_p^{(\alpha)} + \mu_f^{(\alpha)} + \mu_e^{(\alpha)} + [\mu] \tag{2.33}$$

At equilibrium, μ_{os} is constant and $\mu_o^{(\alpha)}$, $\mu_p^{(\alpha)}$, $\mu_f^{(\alpha)}$, and $\mu_e^{(\alpha)}$ may be regarded as functions of distance z from the clay surface. Differentiating Eq. (2.33) with respect to z yields

$$0 = \frac{\partial}{\partial z}\left(\mu_0^{(\alpha)}\right) + \frac{\partial}{\partial z}\left(\mu_p^{(\alpha)}\right) + \frac{\partial}{\partial z}\left(\mu_f^{(\alpha)}\right) + \frac{\partial}{\partial z}\left(\mu_e^{(\alpha)}\right) \tag{2.34}$$

Substituting Eqs. (1.29), (1.31), (1.33), and (1.45) into Eq. (2.34), and assuming that $\bar{v}_w$ is constant, we get

$$0 = \frac{\partial \pi}{\partial z} + \frac{\partial P_{in}}{\partial z} + \frac{\partial}{\partial z}\left[\left(\frac{1}{\epsilon} - 1\right)D^2\right] + \frac{\partial}{\partial z}\left(\frac{A}{z^n}\right) \tag{2.35}$$

This equation indicates that the gradient of osmotic pressure, the gradient of internal pressure, the intermolecular force, and the electric force acting on water molecules are balanced.

Figure 2.29 shows that the decrement of chemical potential of the outermost layer due to the solute is negligible compared to the force field and the electric field (Iwata, 1974b). The decrement of chemical potential due to the force field is larger than that of the electric field for thin films, but the difference decreases as the film thickness increases.

2.3.5 Significance of the Interaction Between Soil Particles and Soil Solution on Soil Properties

The Gouy theory shows that the ionic concentration in the soil solution depends on distance from the clay surface. The extent of the diffuse layer depends on water content and the amount and valence of ions. For a soil with a specific surface area of 100 m^2/g and a water content of 50% by weight, the average film thickness is about 50 Å. If the thickness of the diffuse layer is 30 Å (a reasonable value), 60% of the total liquid phase is of a different ionic composition from that of the equilibrium solution. This fact is very important for plant physiology and pedology.

Figure 2.30 shows that the values of pH near the surface are much lower than in the equilibrium solution. This fact is important in considering reactions that take place at the clay surface. The relevance of a measured pH must also be considered critically. Does the measured pH value represent that of the equilibrium solution, that at a certain point in the diffuse layer, or an average value of the soil solution? At the present time, we may not be able to answer this question exactly. It is an important problem remaining to be solved. The reader is referred to the fine description of this material by Bolt and Bruggenwert (1976).

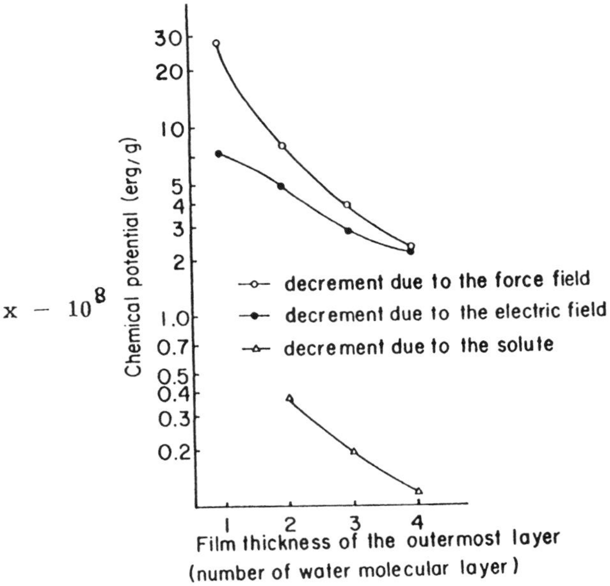

Fig. 2.29. Decrement of chemical potential of the outermost water layer due to the solute, the force field, and the electric field. (From Iwata, 1974b)

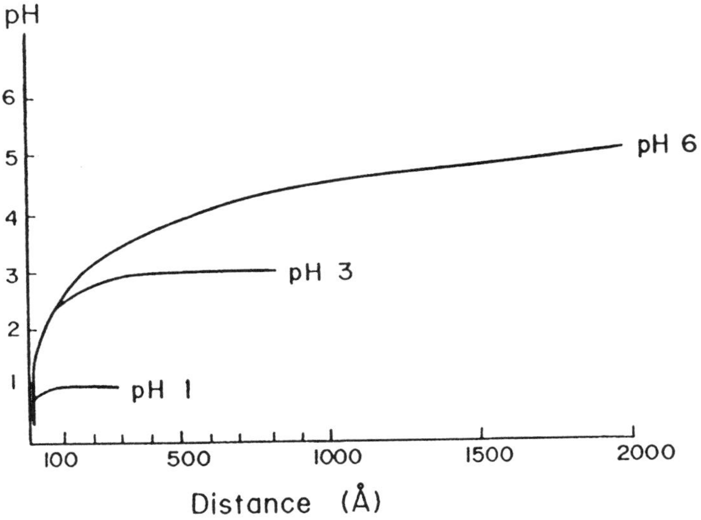

Fig. 2.30. Dependence of pH on distance from a soil surface for three equilibrium pH values. (From Yoshida and Iwata, 1970)

2.4 SOIL FREEZING

Soil freezing is an important phenomenon for agriculture and engineering uses of soils in cold regions. The mechanism of soil freezing will be discussed in this section.

2.4.1 Water-Ice Equilibria

Chemical Potentials of Water and Ice

Suppose a system in which pure water is in equilibrium with ice at pressure P and temperature T (Fig. 2.31). The chemical potential of water, μ_w, is equal to that of ice, μ_i:

$$\mu_w(T,P) = \mu_i(T,P) \tag{2.36}$$

If P or T is changed, the equilibrium is disturbed. When the change satisfies the conditions

$$(d\mu_w)_P = \left(\frac{\partial \mu_w}{\partial T}\right)_P dT > (d\mu_i)_P = \left(\frac{\partial \mu_i}{\partial T}\right)_P dT$$

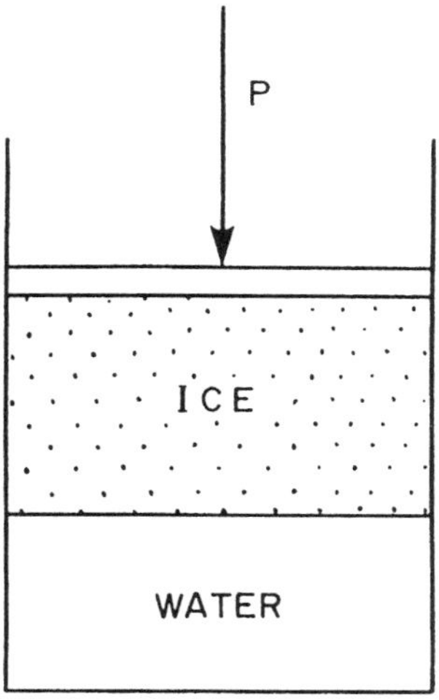

Fig. 2.31. Diagram illustrating the equilibrium between water and ice.

or

$$(d\mu_w)_T = \left(\frac{\partial \mu_w}{\partial P}\right)_T dP > (d\mu_i) = \left(\frac{\partial \mu_i}{\partial P}\right)_T dP$$

water is converted into ice because the chemical potential of water becomes larger than that of ice. When the inequalities are reversed ice changes to water.

The chemical potential of ice is always lower than that of super-cooled water below 0°C at 1 atm (Fig. 2.32). Consequently, when the external pressure is at 1 atm, ice is more stable than water below 0°C. The effect of pressure on chemical potentials of water and ice is described later.

Super-cooling

When a small quantity of water is cooled below its freezing temperature, it does not freeze instantly but remains in a super-cooled condition for some time. The degree of super-cooling increases with purity of the water. The phase transition to ice is completed by going through the processes of formation of small crystal embryos in the liquid, growth of the crystal embryos, and finally, freezing of the liquid. Nucleation in perfectly pure water is called homogeneous nucleation, and nucleation in water containing extraneous particles is heterogeneous nucleation.

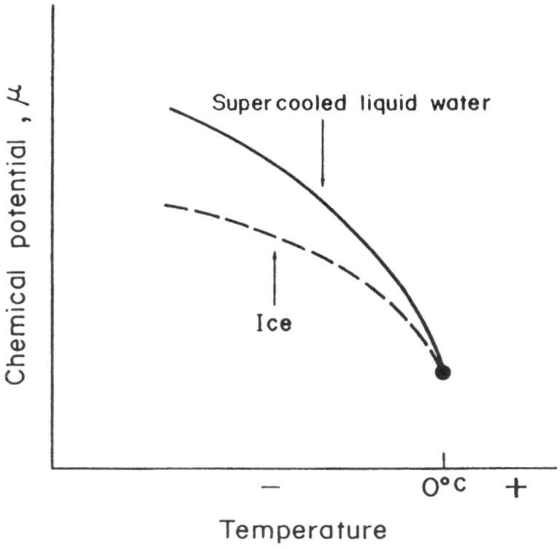

Fig. 2.32. Chemical potentials for super-cooled water and ice.

The small dimensions of the embryos makes the surface free energy at the liquid-crystal interface an important influence on stability of the crystal embryo. A small crystal embryo is unstable, which induces the phenomenon of super-cooling. A discussion of super-cooling and nucleation can be found in Fletcher (1970).

Freezing-Point Depression

Pressure Effect When T = const, the changes of chemical potentials of water and ice, $\Delta\mu_w$ and $\Delta\mu_i$, due to the increase of pressure ΔP, are given by the equations

$$\Delta\mu_w = \left(\frac{\partial\mu_w}{\partial P}\right)_T \Delta P = v_w \,\Delta P \tag{2.37}$$

$$\Delta\mu_i = \left(\frac{\partial\mu_i}{\partial P}\right)_T \Delta P = v_i \,\Delta P \tag{2.38}$$

where v_w and v_i are the specific volumes of water and ice, with values at 0°C of 1.0 and 1.091, respectively. Consequently, $\Delta\mu_w$ is smaller than $\Delta\mu_i$ and ice is changed into water. This means that an increase in pressure depresses the freezing point. The dependence of freezing-point depression on pressure is defined by the relative magnitude of v_w and v_i.

We can obtain the dependence of freezing-point depression on the increment of pressure from the Clausius-Clapeyron equation:

$$\left(\frac{\partial T}{\partial P}\right)_{\mu_i=\mu_w} = \frac{T(v_i - v_w)}{\Delta H} \doteq \left(\frac{\Delta T}{\Delta P}\right)_{\mu_i=\mu_w} \tag{2.39}$$

where ΔT is the freezing-point depression corresponding to the increase of pressure ΔP, and ΔH is the difference in enthalpy between ice and water. A pressure increase of 1 atm results in a freezing-point depression of 0.007°C.

Overburden pressure has the same effect on the freezing-point depression of water in soil. Equation (2.39) should be amended:

$$\left(\frac{\partial T}{\partial P}\right)_{\mu_i=\mu_w} = \frac{T(v_i - \bar{v}_w)}{\Delta \bar{H}} \doteq \left(\frac{\Delta T}{\Delta P}\right)_{\mu_i=\mu_w} \tag{2.40}$$

where $\Delta\bar{H}$ is the difference in partial enthalpy between ice and water in the outermost phase in contact with ice, and $\bar{v}_w$ is the partial volume of water in that phase.

Solutes The addition of solute decreases the chemical potential of water according to Eq. (1.29). Consequently, the freezing-point depression is

$$\left(\frac{\partial T}{\partial n}\right)_{\mu_i=\mu_w,\ P} = \frac{-RT}{1000(S_w - S_i)} = \frac{-RT^2}{1000\Delta H} \doteq \left(\frac{\Delta T}{\Delta n}\right)_{\mu_i=\mu_w,\ P} \qquad (2.41)$$

where ΔT is the freezing-point depression due to the addition of Δn moles of a solute per liter of water, S_w the specific entropy of water, S_i the specific entropy of ice, and ΔH the difference in specific enthalpy between water and ice (heat of solidification). Equation (2.41) indicates that a 1 M solution produces a freezing-point depression of 1.86°C.

Estimating the Freezing-Point Depression All factors that decrease the chemical potential of water depress the freezing-point. In addition to solutes and pressure, potential fields and surface tension effects lower the freezing point. Therefore, the freezing-point of water in soil is always lower than 0°C, and becomes even lower as the water content of soil becomes smaller.

When a soil with water content θ begins to freeze at temperature T,

$$\mu_i(T) = \mu_w(\theta,T) \qquad (2.42)$$

where $\mu_i(T)$ and $\mu_w(\theta,T)$ denote the chemical potentials of ice and water on the basis of super-cooled bulk water at T. The values of $\mu_i(T)$ can be obtained, using Eq. (2.3′), from tables of equilibrium vapor pressure versus temperature for ice and super-cooled water. If $\mu_w(T,\theta)$, including the super-cooled condition, was also given as a function of temperature and water content, we could determine the freezing temperature T_f of the water in the soil as that temperature which satisfies the condition $\mu_w(T,\theta) = \mu_i(T)$. If the dependence of chemical potential of water in a soil on temperature was known theoretically, it would be possible to estimate $\mu_w(T,\theta)$ at any temperature from the water-retentivity curve measured at a certain temperature. This cannot yet be done. However, in the authors' opinion, if θ is kept constant, the chemical potential of water measured at T_1 may be used as an approximation of that at T_2. That is,

$$\mu_w(T_1,\theta) = \mu_w(T_2,\theta) \qquad (2.43)$$

where $\mu(T_1,\theta)$ and $\mu(T_2,\theta)$ are the chemical potentials on the basis of those of bulk water at T_1 and T_2, respectively. Low et al. (1968) have introduced an equation relating the relative partial molar free energy (chemical potential) of water in a soil to its freezing-point depression and relative partial molar heat content on the basis of the assumption $\mu_w(T_1,\theta) = \mu_w(T_2,\theta)$.

2.4.2 Unfrozen Water in Frozen Soils

Soil water beings to freeze when $\mu_i(T)$ becomes equal to $\mu_w(T,\theta)$. If the chemical potential of the soil water were not changed by freezing, all the soil water would be frozen. The existence of unfrozen water shows that the chemical potential of water in soil decreases with freezing as shown in Fig. 2.33. Recently, Brown and Payne (1990) determined the relationship between temperature and unfrozen water content in seven clays between 0°C and -15°C to -24°C using a modified differential scanning calorimeter procedure. Their results for sodium Wyoming bentonite showed that the unfrozen water contents at -5°C and -15°C are 43% and 38%, respectively. The amount of unfrozen water decreases with decreasing temperature. Figure 2.34 shows interfacial water thicknesses, calculated by dividing the unfrozen water content by the surface area, as a function of temperature.

The decrease of chemical potential of water in the layer directly in contact with ice could result from several factors:

1. Ice formed in soils does not generally contain solutes; consequently, the concentration of solutes in the unfrozen water layer increases. The freezing-point depression for unfrozen water with a thickness of three water molecular layers calculated on the basis of the decrement of chemical potential due to solute shown in Fig. 2.29 is around -2°C.
2. If ice is formed within the range where intermolecular forces due to the particle (clay, silt or sand) are not negligible, the internal pressure in the unfrozen layer decreases because the density of ice is smaller than that of water (Iwata, 1974c; 1985). Iwata (1985) calculated the number of water molecular layers of unfrozen water adsorbed by a flat kaolinite surface with negligible surface charge as a function of temperature using values of decreases in internal pressure obtained from Eq. (2.4) and the vapor pressure tables for ice and super-cooled water. The calculated values and experimental values by Nersesova and Taytovich (1966) are in Table 2.3.
3. The intensity of the electric field in the layer increases as the unfrozen water film becomes thinner. Using the result calculated for montmorillonite shown in Fig. 2.29 and the vapor pressure tables of ice and super-cooled water, the

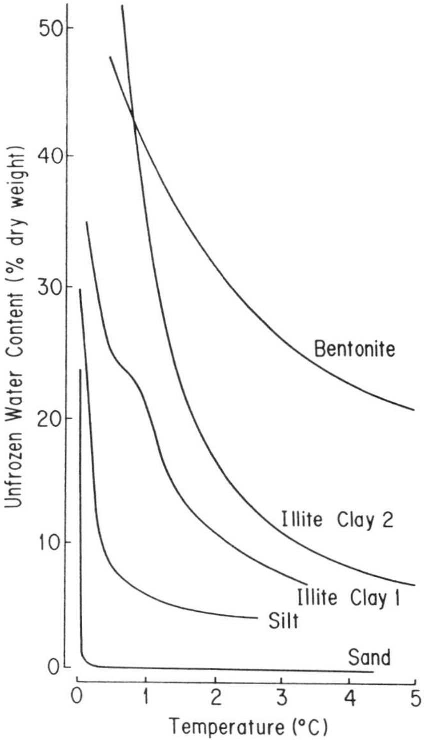

Fig. 2.33. Amount of water remaining unfrozen at temperatures below 0°C, for various soils. (From Williams, 1988)

value of the freezing-point depression for three water molecular layers of unfrozen water is estimated to be considerably lower than -15°C.

4. If the surface of water in direct contact with ice is concave, the pressure of water is lower than that of ice. This pressure difference produces a freezing-point depression. Koopmans and Miller (1966) indicated the surface tension between ice and water to be 1/2.2 of that between air and water. The estimated freezing-point depression for water with a radius of curvature of the ice-water interface of -10^{-4} cm is considerably larger than -0.1°C.
5. Assuming that ice in soil has a higher pressure than the water in direct contact with it, and applying the Clausius-Clapeyron relation to the system, we can obtain a differential equation showing a relationship between the difference in pressure of the ice and water and the freezing-point depression (Edlefsen and Anderson, 1943). The integration of an appropriate form of the differential equation was used to introduce transport equations of water

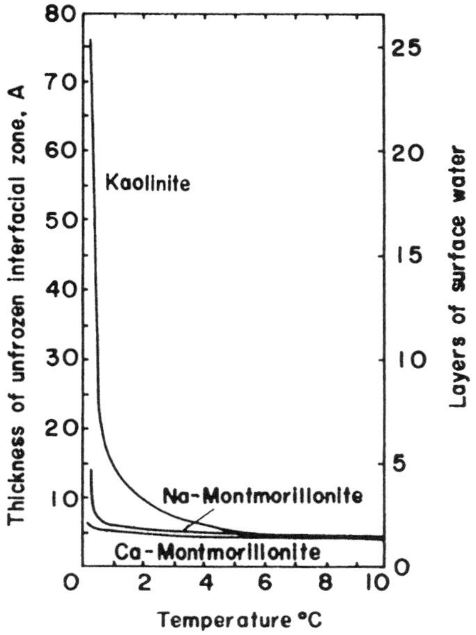

Fig. 2.34. Unfrozen water content plotted against temperature for three representative clays. (From Anderson and Morgenstern, 1973)

Table 2.3. The relationship between number of unfrozen water molecular layers and temperature.

Temperature °C	-0.74°	-1.53°	-5.38°
Calculated value	6	4	2
Measured value	5	3	1

and heat in a rigid ice model of frost heave (e.g. O'Neill and Miller, 1985). The higher pressure of the ice makes the co-existence of water and ice possible. However, it should be borne in mind that the freezing-point depression due to a pressure difference is very small, 0.007°C for a pressure increase of 1 atm.

A number of models of the unfrozen water film have been put forward to analyze various aspects of this phenomenon (Enüstün et al., 1978; Fletcher, 1973; Koopmans and Miller, 1966; Römkens and Miller, 1973; Vignes and Dijkema, 1974; Vignes-Alder, 1977; Drost-Hansen, 1977; Gilpin, 1979; 1980a; O'Neill and Miller, 1985; Kuroda, 1985; Black, 1991). However, a complete model, by which all phenomena can be understood, has not yet been established.

Dependence of Unfrozen Water on Original Water Content

For different soils at the same initial water content, the quantity of unfrozen water as a proportion of the total water content (unfrozen water percent) at a given temperature is dependent mainly on surface area and surface charge density. The former determines the amount of adsorbed water, and the latter defines the decrement in chemical potential of the adsorbed water. The difference between the clay and silt at 20% water in Fig. 2.35 is due to differences in these two factors (Yong, 1965). Anderson and Hoekstra (1965) have emphasized the importance of the total specific surface area in determining unfrozen water content at temperatures below -5°C.

Opinions differ on whether the unfrozen water content of a soil at a certain temperature is dependent on the initial water content. The experimental data by Yong (1965) in Fig. 2.35 show that the unfrozen water content of a soil at a given temperature increases with initial water content. Yong has not explained this phenomenon. Anderson and Morgenstern (1973) said: "Thermodynamic arguments inexorably lead to the conclusion that, at the freezing temperature of a soil at a given water content, this water content may be taken as or regarded as equal to the unfrozen water content of that soil at the same temperature, even though it may contain additional water in the form of ice." That is, they have not recognized the dependence of unfrozen water content on the original water content as shown in Fig. 2.35.

Figure 2.36 shows soil with three initial water contents, W_i, where Z_i is the thickness of ice plus unfrozen water, $Z_i(u)$ the unfrozen water, and Z_f the distance at which the potential of the van der Waals force field becomes negligible [$\psi(Z_f) = 0$]. Iwata (1974c) has obtained theoretically the result that where $Z_i > Z_f$ (Fig. 2.36a and b), the unfrozen water content of a soil at a certain temperature is not dependent on the initial water content, but where $Z_i < Z_f$ (Fig. 2.36c), the quantity of unfrozen water decreases with increase of initial water content at a given temperature (Fig. 2.35 and Table 2.4). Following the theory, the Anderson and Morgenstern statement is considered to be correct as long as the condition $Z_i > Z_f$, as shown in Fig. 2.36 a and b, is satisfied. On the other hand, this result is contrary to the experimental results obtained by Yong (1965). In the calculation of the values shown in Table 2.4 only the internal pressure due to the intermolecular forces, among the factors that make the existence of un-

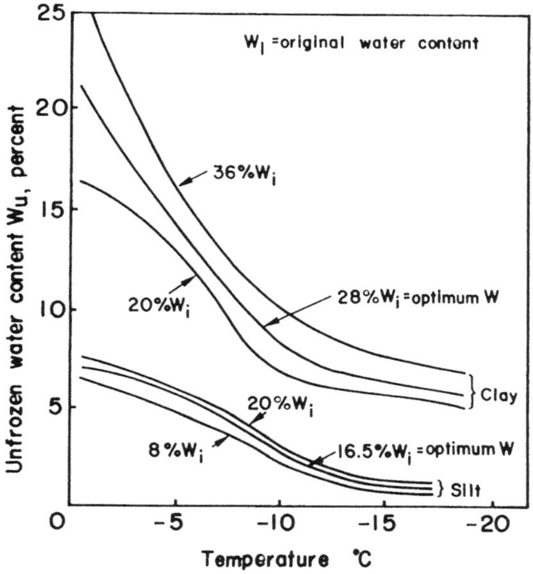

Fig. 2.35. Temperature and unfrozen water content relationships for clay and silt at different initial water contents. (From Yong, 1965)

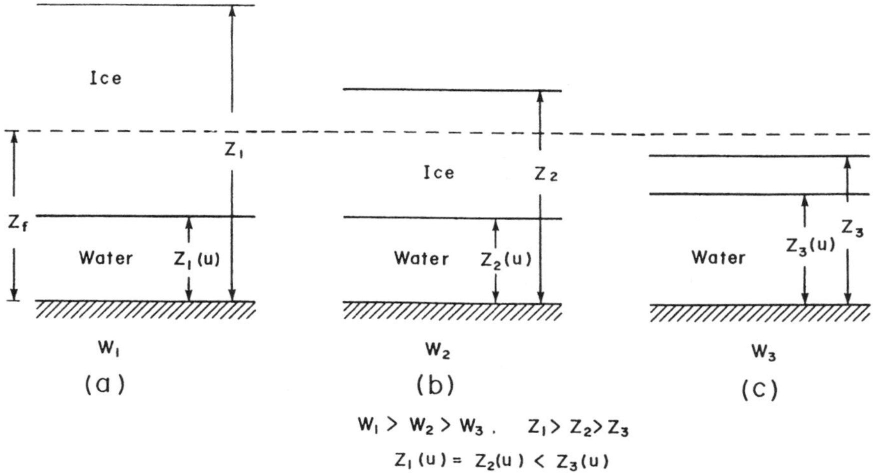

Fig. 2.36. Schematic representation illustrating the relationship between the thickness of unfrozen water and the initial water content at a certain temperature.

frozen water possible, was taken into consideration. However, the other two factors—the concentration of solutes and the intensity of the electric field in the adsorbed water layer—cannot be considered to have had as much influence on the unfrozen water content of the soil as shown in Fig. 2.35 under Yong's (1965) experimental conditions. It may be necessary to consider additional factors such as an external pressure acting on both phases of water, which is induced by the increase of water volume in the phase change from water to ice. Further investigation is required to clarify the dependence of unfrozen water content on the original water content.

2.4.3 Frost Heaving

The expansion of soils on freezing is called "frost heaving." Continued heaving is caused by growth of ice lenses. The pressure exerted against a load by this expansion is called "heaving pressure." When frost heaving occurs, the water content (including ice) in the frozen zone is larger than the saturated water content of the soil, and the water content of the unfrozen zone below the freezing front is very low (Fig. 2.37). Water moves from the unfrozen soil to the freezing front, and from the freezing front to the ice lens. The unfrozen water films between ice and soil particles have an equilibrium thickness that decreases with decreasing temperature. Water migration in the frozen zone is through these films.

Generally speaking, theories of frost heaving are divided into two groups, capillary force theories and adsorption force theories. The former are constructed on the basis of the Kelvin equation relating pressures in two phases that meet at a curved water-ice interface formed in a pore at the freezing front. They are analogous to the surface tension effects used to evaluate the pressure in water of an unfrozen soil (Gold, 1957; Miller et al., 1960; Everett, 1961; Yong, 1966; Penner, 1967; Williams, 1968; Sutherland and Gaskin, 1973). A number of the research findings on freezing soils and frost heaving based on this theory have been reviewed by Anderson and Morgenstern (1973) and Miller (1980).

The adsorption force theories are based on the importance of film water surrounding soil particles along the freezing front in causing frost heaving (Taber, 1930; Beskow, 1935; Derjaguin and Churaev, 1978; Takagi, 1980; O'Neill and Miller, 1985; Kuroda, 1985). Takagi (1980) revived and developed the adsorption force theory into a logical system by systematizing and improving on the concepts developed so far (Takagi, 1963, 1965, 1970, 1975, 1977, 1978, 1979; Vignes and Dijkema, 1974; Vignes-Alder, 1977). Miller (1980) discussed the unfrozen pore water-ice equilibrium in relation to elevated ice pressure. Kay and Perfect (1988) published a comprehensive review on heat and mass transfer in freezing soils.

Table 2.4. Dependence of unfrozen water content on the initial water content. (From Iwata, 1974c)

Original water condition (number of water molecular layers)	Θ_i[a]	$\Theta_o(Z_w)$[b]
$Z_w = \infty$ (saturated)	$\Theta_\infty = 0°C$	
$Z_w = 2.76 \times 10$ Å (10 water molecular layers)	$\Theta_{10} = -2.6°C$	-0.2°C
$Z_w = 2.76 \times 6$ Å (6 water molecular layers)	$\Theta_6 = -8.0°C$	-0.65°C
$Z_w = 2.76 \times 3$ Å (3 water molecular layers)	$\Theta_3 = -28.0°C$	-2.3°C

[a]Θ_i denotes the temperature at which the outermost water molecular layer begins to freeze when the water film thickness under the original water condition is Z_w.

[b]$\Theta_o(Z_w)$ indicates the temperature when the thickness of the unfrozen water of the soil whose original water content is $Z_w = \infty$ is Z_w, Z_w represents the thickness of unfrozen water.

Water Movement in Frost Heaving

The amount of frost heaving in a soil depends on the water flux densities from the unfrozen soil to the freezing front, and from the freezing front to the plane where ice lenses grow. If either flux density is very small, frost heaving does not occur. The first flux density is determined by the relationship between hydraulic conductivity and water content of the soil. For example, since hydraulic conductivity of a sandy soil becomes very small as the water content decreases, it is difficult for frost heaving to occur in sandy soil.

The flux density from the freezing point to the ice lens is governed by the hydraulic conductivity of the unfrozen water films in the frozen soil. The hydraulic conductivity is approximately proportional to the cross-sectional area of flowing water, and the cross-sectional area may be considered to be roughly proportional to the product of the thickness of the unfrozen water film and the specific external surface area of the soil. This is valid when the continuity of the unfrozen water film is not broken by growth of ice lenses.

Let us suppose a sandy soil whose specific external surface area is 1/100 that of a clay soil, and in which the thickness of the unfrozen water film is one-

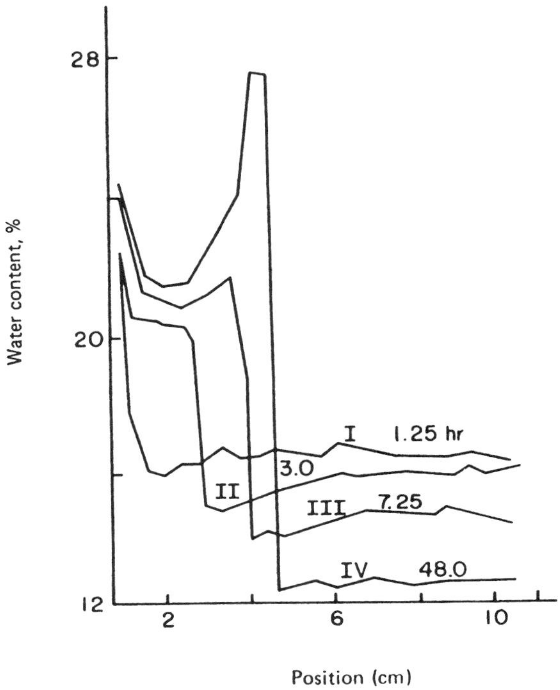

Fig. 2.37. Gravimetric water content (liquid plus ice) profiles of soil columns frozen for different periods of time, in hours. (From Dirksen and Miller, 1966)

fifth that in the clay soil at a given temperature. Then the hydraulic conductivity of the sandy soil is 1/500 that of the clay soil. Therefore, even if the water flux density from the unfrozen zone to the freezing front is sufficiently large in a sandy soil, significant frost heaving does not occur. Taylor and Luthin (1978) found that the water diffusivity in a frozen soil is 1/100 to 1/1000 that of unfrozen soil at the same total water content. Fukuda (1982) obtained a similar result for hydraulic conductivity. Van Loon et al. (1988) measured thermal conductivity in unsaturated frozen sands, from which they estimated hydraulic conductivity. The maximum values were 10^{-10} and 3×10^{-11} m/s.

In order to obtain unfrozen water content and hydraulic conductivity data for an air-free frozen silt, Black and Miller (1990) developed a new form of an ice sandwich dilatometer/permeameter that was designed to allow control of effective stress in the granular matrix through appropriate adjustments of pressure in the liquid surrounding a specimen, confined as in a triaxial test apparatus. The results showed 1) the pronounced drop in conductivity accompanied by the initial appearance of ice in the soil, and 2) the persistent decrease in apparent conductivity with time during measurements at a fixed value of the difference in pressures of ice and water. In addition, Barer et al. (1980) showed that the viscosity of unfrozen water interlayers between ice and molecular-smooth quartz surfaces was higher than that of bulk super-cooled water at the same temperature, but the difference was not as great as had earlier been supposed (Jellinek, 1967).

Thus a frost-susceptible soil must satisfy all the following conditions: the water table is high, hydraulic conductivity is not extremely low even at low water content, specific external surface area is large, and soil structure is not rigid.

Driving Forces for Water Migration in the Frozen Soil

Several models have been suggested to explain driving forces of water migration in frozen soils. The temperature gradient is the origin of the driving forces, but the mechanisms causing water migration in different models are different. The driving forces for water flow have been confirmed to be proportional to the temperature gradient (Loch and Kay, 1978; Konrad and Morgenstern, 1980, 1981; Ishizaki and Nishio, 1988). Konrad and Morgenstern (1980, 1981) empirically found that the water intake flux at the formation of the final ice lens is proportional to the average temperature gradient in the frozen fringe when a soil sample freezes under different cold-side step temperatures with the same warm-side temperature. Figure 2.38 shows a relationship between water intake rate and temperature gradient in the frozen zone of a saturated soil sample (Ishizaki and Nishio, 1988). The physical properties of the sample are shown in Table 2.5.

Soil Freezing

Groenevelt and Kay (1974) showed that according to the Clapeyron equation, the temperature gradient in the frozen zone was accompanied by a gradient of the local liquid pressure in the unfrozen film water. The difference of chemical potentials of ice at A′ and B′ in Fig. 2.39, $d\mu_i$, is given by

$$d\mu_i = -S_i \, dT \tag{2.44}$$

The difference in chemical potentials of the unfrozen film water at A and B, $d\mu_w$, is

$$d\mu_w = -\bar{S}_w \, dT + \bar{v}_w \, d\bar{P} \tag{2.45}$$

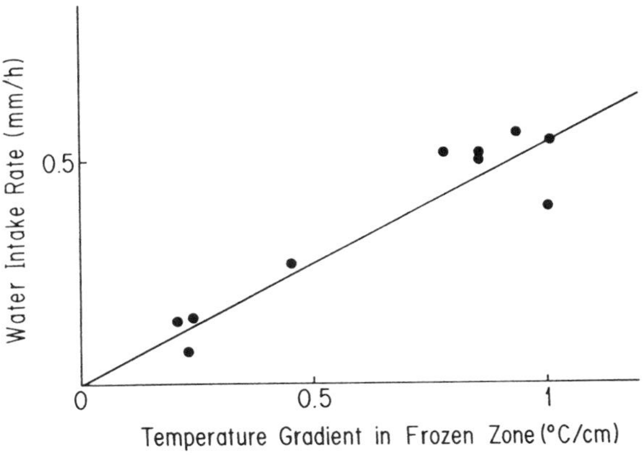

Fig. 2.38. Water intake rate vs. temperature gradient in the frozen zone. (From Ishizaki and Nishio, 1988)

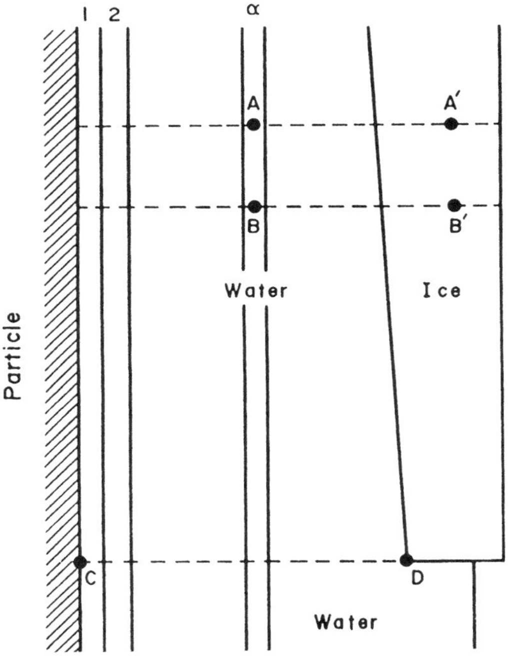

Fig. 2.39. Schematic representation illustrating the driving forces in freezing soil water.

Table 2.5. Properties of soil sample for results in Fig. 2.38. (From Ishizaki and Nishio, 1988)

silt (5 ~ 74 μm)	38%
clay (<5 μm)	62%
porosity	0.487
bulk density	1.99 g cm^{-3}
dry density	1.50 g cm^{-3}
hydraulic conductivity	1 ~ 5 x 10^{-1} cm sec^{-1}
specific surface area	20.8 m^2 g^{-1}

where $\bar{P}$ is the local pressure in the unfrozen film water. Considering $d\mu_w = d\mu_i$, we obtain

$$(\bar{S}_w - S_i)\,\nabla T = \frac{Q_w}{T}\,\nabla T = \nabla\bar{P} \tag{2.46}$$

where Q_w is the heat of solidification of water. Equation (2.46) shows that a temperature gradient of 1°C/cm produces a pressure gradient of 12 bar/cm. In addition, because the value of $\bar{S}_w$ in Eq. (2.46) changes with distance from the particle surface, the driving forces decrease with distance.

The driving force for water migration in frozen soils in this theory is the pressure difference produced by the temperature difference. However, the theory does not define the physical mechanism causing the pressure difference. This is a weak point recognized generally in applying the Clausius-Clapeyron relation to a system. In addition, it should be noted that such an application of the Clausius-Clapeyron relation is thermodynamically correct only when pressure and temperature are the only variables defining the state of the system, or when state variables other than temperature and pressure, such as the solute concentration, the electric strength, and the intermolecular force field, are kept constant.

Iwata (1980) assumed that 1) the driving forces are produced by the gradient of unfrozen water film thickness accompanied by a temperature gradient, and 2) the forces are given by the sum of the forces due to the change of internal pressure, the change of electric displacement, and the change of osmotic pressure (refer to Eq. 1.77), because the existence of a film thickness gradient necessarily changes the above three factors simultaneously. He calculated the driving forces for unfrozen water adsorbed by a montmorillonite particle (Table 2.6). The

Table 2.6. Components of the driving force (erg/g/cm) in unfrozen water in montmorillonite. (From Iwata, 1980)

	Internal Pressure X_{wp}	Electric Displacement X_{wD}	Osmotic Pressure X_{wn}	Total	$\nabla T(S_w - S_i)$
$\alpha = 8$	$4.99 \times 10^7\uparrow$	$-1.50 \times 10^7\downarrow$	$+0.08 \times 10^7\uparrow$	$3.57 \times 10^7\uparrow$	$3.4 \times 10^7\uparrow$
$\alpha = 5$	$4.01 \times 10^7\uparrow$	$-0.5 \times 10^7\downarrow$	$+0.06 \times 10^7\uparrow$	$3.57 \times 10^7\uparrow$	$3.4 \times 10^7\uparrow$
$\alpha = 3$	$5.5 \times 10^7\uparrow$	$-1.8 \times 10^7\downarrow$	$-0.13 \times 10^7\downarrow$	$3.57 \times 10^7\uparrow$	$3.4 \times 10^7\uparrow$
Mean	$4.38 \times 10^7\uparrow$	$-1.27 \times 10^7\downarrow$	$+0.003 \times 10^7\uparrow$	$3.57 \times 10^7\uparrow$	$3.4 \times 10^7\uparrow$

Temperature gradient in 1°C/cm. Arrows indicate direction of flow. α = number of molecular layers of unfrozen water.

change of internal pressure accompanying the change of unfrozen water film thickness is dominant among the three factors.

Gilpin (1980a) introduced a driving force with two terms, related to a temperature gradient and a pressure gradient in the unfrozen water film in the direction parallel to a plane solid surface. The pressure gradient is induced by a change in thickness of the unfrozen water film and the pressure is due to an assumed force of attraction of the solid for unfrozen water. The pressure, which increases as the distance from the solid surface decreases, is equivalent to the internal pressure discussed in Sec. 1.2.2. Loch (1978) derived an equation for a pressure difference, and inferred the existence of a driving force, a gradient in the pressure of the water phase across the system.

The adsorption force theory assumes that the thickness of the unfrozen film water existing between a heaving ice surface and a neighboring particle is determined by an equilibrium of the adsorption forces exerted by the particle and the ice. The film water, decreasing in thickness by loss to freezing ice, generates a suction force that draws water from the surrounding soil (Takagi, 1980). The mechanism by which the thickness of unfrozen water is kept constant has not been explained.

Miller et al. (1975) presented a new model related to the "liquid island" model of Philip and de Vries (1957). The ice phase is not immobile under conditions in which steady liquid transport is expected. Since ice is everywhere bounded by the liquid water film, movement can be accommodated by continuous transformation of ice to film water at the high-temperature interfaces while water is continuously transformed to ice at the low-temperature interface.

This motion involves appropriate movements of water in the film phase and internal exchanges of latent heat.

Derjaguin and Churaev (1978) presented a similar model based on nonequilibrium thermodynamics of steady-state processes and showed that the disjoining pressure (see Sections 3.1.4) was able to increase the phase transition rate, whether of crystallization or of melting. The driving forces for water are proportional to the temperature gradient.

O'Neill and Miller (1985) developed the model by Miller et al. (1975) for an incompressible soil, free of air, colloids, and solute, based on the rigid ice assumption. They described the mechanism of water migration based on the proposition that liquid water is attracted by grain surfaces more strongly than is rigid crystalline ice. If that is true, then at temperatures somewhat below 0°C the grain ought to be surrounded by a film of unfrozen liquid in equilibrium with the ice. If a temperature gradient is imposed, thermal equilibrium of water and ice at the interface is inconsistent with mechanical equilibrium in the hydrostatic field induced by surface adsorption forces. Whereas the thermal gradient induces asymmetry of film thickness, the action of adsorption forces is to center the grain within its liquid shell. Thus, the temperature field constantly acts to diminish the film thickness on the cold side, while surface forces seek to retain that film thickness by removing ice from the warm side and transporting the resulting unfrozen water to the cold side, where it refreezes.

Regelation is considered to be a decisive key for water migration in this model. They emphasized the uniqueness of their model in that pore ice in the frozen fringe is regarded to play the most important role in frost heave. This is opposite to most frost heave models, where segregated ice is assumed to form within the frozen soil. This assumption may be disputable (Williams, 1988). The second difference exists in considering ice within the fringe to be mobile by regelation. This is physically somewhat ambiguous.

It has been shown empirically that the gradient of temperature under uniform pressure fields and the gradient of unfrozen water pressure under isothermal conditions are two major driving forces for the transport of water in saturated frozen soils (Nakano, 1992). Nakano (1991, 1992) found that a hypothesis based on the independence of temperature and pressure in the frozen fringe is consistent with empirical data.

Driving Forces at the Freezing Front

In the adsorption force theory, the driving force for water migration from the unfrozen soil to the freezing point is considered to be generated by freezing of film water on and/or near the freezing front. The capillary force theory contends that the existence of curved water-ice interfaces in pores at the freezing point induces water movement from the unfrozen soil to the freezing front. A negative pressure of water in contact with ice, defined by Eq. (2.47), draws water from the unfrozen soil.

$$P = \frac{2\sigma_{iw}}{r} \tag{2.47}$$

where P is the negative pressure, r the radius of curvature of the ice-water interface, and σ_{iw} the surface tension between ice and water. It is assumed that ice is at atmospheric pressure. Although the theory may be valid where the unfrozen soil is wet, it is difficult to suppose the theory to be applicable where the unfrozen soil is dry.

Jackson and Chalmers (1958) described super-cooling of soil water as the energy source for frost heaving. Iwata (1980) introduced a driving force on the assumption that water near a freezing front was super-cooled. The existence of super-cooled water brings about a sudden change of chemical potential of water in the vicinity of line CD in Fig. 2.39. The magnitude of the driving force at the freezing front, induced by the difference of chemical potentials, depends on the super-cooling temperature. At -1°C, the driving force is on the order of 10^{13} ergs/g per centimeter, or 10^7 bar/cm. The observation that the water content ahead of the freezing front becomes very low can be understood from this large driving force.

Heaving Pressure

According to the capillary force theory, heaving pressure is given by substituting "effective radius" of pore necks for r in Eq. (2.47). Consequently, if one can know the effective radius for a given circumstance, one can predict the maximum heaving pressure that can be developed. However, Penner (1967, 1976), Sutherland and Gaskin (1973), and Loch and Miller (1975) showed that the measured frost heaving pressures were much larger than predicted by capillary theory.

The adsorption force theory views frost heaving pressure as follows. The suction force generated by freezing of some film water results in a stress in the film water. The film water sustains the weight of the overburden. Although the film water is a liquid, it should be treated as a solid. The average of the vertical components of stress taken over a sufficient number of film water layers and pores is the frost heaving pressure, which is related to the average segregation-freezing temperature (Takagi, 1980). In addition, the segregation-freezing temperature indicates the temperature of the freezing film water, and it is lower than the *in situ*-freezing temperature (i.e. the temperature of the freezing pore water).

A theory of frost heaving for clay soil has been proposed by Iwata (1980), in which heaving pressure is defined as the force that balances the driving forces. This means a no flow condition, where the maximum potential heaving pressure of a soil is determined by the temperature distribution through the soil profile. The difference in heaving pressure between sandy and clay soils is the difference in time required to reach maximum potential heaving pressure. The data obtained by Hoekstra (1969) show that the maximum heaving pressure for a clay increases as the temperature gradient increases.

Takashi et al. (1980) found in their laboratory experiments that the upper limit of heaving pressure in a clay is proportional to the temperature gradient. Førland et al. (1988) derived the following equation for frozen soils using irreversible thermodynamics in which frost heave is treated as a case of thermal osmosis.

$$J_w = -\overline{\ell}_{zz}\left(\Delta S_m \Delta T + V_{ice}\Delta P\right) \tag{2.48}$$

where J_w is the water flux and equal to the rate of growth for the ice lens, $\overline{\ell}_{zz}$ is the average hydraulic permeability for the total transport path for water, ΔS_m is the entropy of melting, ΔT and ΔP are differences in temperature and pressure at the lowest ice lens and the frozen front, respectively, and V_{ice} is the molar volume of ice. Maximum frost heaving pressure corresponds to ΔP at $J_w = 0$. Calculated values for maximum heave pressure agreed well with experimental values.

Mathematical models of frost heave have been proposed by many researchers (e.g. Harlen, 1973; Kinoshita, 1975; Taylor and Luthin, 1978; Jame and Norum, 1980; Hopke, 1980; Gilpin, 1980a; Nakano and Horiguchi, 1984; Blanchard and Fremond, 1985; O'Neill and Miller, 1985; Miyata, 1988; Fremond and Mikkola, 1991).

2.5 PHYSICOCHEMICAL PROPERTIES OF ADSORBED WATER

Physicochemical properties of water in the vicinity of clay surfaces are of great interest in various fields of science. Knowledge of the viscosity of water adsorbed by clay surfaces is important for understanding interparticle interactions and rheological processes such as flow of water through porous media. The ion product of water in the vicinity of a clay surface plays an important role in unusual chemical reactions occurring near the clay surface.

Low (1961) published a historical review titled "Physical Chemistry of Clay-Water Interactions," where he proposed a working hypothesis about the structure of water adsorbed by clays. His hypothesis was that the quasi-crystalline water structure at the clay surface may extend with considerable regularity for distances of 75 to 100 Å, by a mechanism similar to epitaxy. The degree of order decreases gradually with distance from the surface, but an attenuated structure may persist as far as 200 to 300 Å. The magnitudes of the specific volume and viscosity decrease continuously with distance from the surface. This idea was very important in stimulating scientific interest of young researchers to study the physical chemistry of adsorbed water.

2.5.1 Preliminary Knowledge Related to Physicochemical Properties of Adsorbed Water

Before introducing experimental results for physicochemical properties, it is useful to discuss general knowledge about these properties.

Physicochemical Properties of Adsorbed Water and Film Thickness

First, it will be necessary to determine whether physicochemical properties of adsorbed water layers change as the thickness of water adsorbed on an external clay surface increases. Without this discussion, the physical meaning of the experimental results may not be appreciated. For example, estimation of the energy of clay-water interaction as a function of initial thickness of adsorbed water on the basis of measured heats of immersion for a clay at different water contents may be in error if the physical state of adsorbed water layers near the surface is different before and after immersion. The value of a physicochemical property such as viscosity obtained under a condition where only two or three water molecular layers are adsorbed cannot be used for estimating that property with a very large film thickness if the property changes with film thickness.

Clearly, the chemical potential of water adsorbed at a clay surface increases as the film thickness grows, i.e. the equilibrium water vapor pressure increases with the film thickness. Factors defining the increase of the chemical potential are understood on the basis of Eq. (1.23). Considering that the change of vertical distance (h) is negligible and the principal radius of curvature at the air-water interface is infinite for this case, and assuming an isothermal condition, Eq. (1.23) becomes

$$d\mu_w = \bar{v}_w dP + \frac{\bar{v}_w}{4\pi}\left(\frac{1}{\epsilon} - 1\right)dD + \left(\frac{\partial\psi}{\partial z}\right)dz + \sum_{j=1}^{k-1}\left(\frac{\partial\mu_w}{\partial C_j}\right)dCj \qquad (2.49)$$

where μ_w is the chemical potential of water under consideration, $\bar{v}_w$ the partial volume of water, P the sum of the external pressure and the internal pressure, $\bar{\varepsilon}$ the partial dielectric constant, D the electric displacement, z the distance from the clay surface, Cj the concentration of the jth component and k the number of components.

Suppose that the difference in chemical potential of an adsorbed water layer that accompanies an increase of the film thickness is $\Delta\mu$, we then obtain

$$\Delta\mu = \int \bar{v}_w dP + \int \frac{\bar{v}_w}{4\pi}\left(\frac{1}{\bar{\epsilon}} - 1\right)dD + \int \left(\frac{\partial\psi}{\partial z}\right)dz + \int \sum_{j=1}^{k-1}\left(\frac{\partial\mu_w}{\partial Cj}\right)dCj \quad (2.50)$$

The first term on the right-hand side is the increment of chemical potential due to the increase of internal pressure produced by the existence of the van der Waals force field. Since a van der Waals force field is certainly formed around a clay particle by the interaction between the particle and a water molecule, internal pressure is increased with the thickness of water film. The exact functional relation between force and distance from the clay surface seems not yet to have been obtained. If Eq. (2.4) is used as an approximation of the potential energy of interaction between an uncharged clay surface and a water molecule, internal pressure acting on the first, second, and third water molecular layers is 394, 190, and 134 atms, respectively, in a case where the thickness of adsorbed water is larger than the distance from the surface to the line beyond which the force field is neglected. As seen in Eq. (2.39), an increase of external pressure on water results in a freezing point depression of the water, possibly by preventing formation of hydrogen bonds between water molecules. The internal pressure on the outermost water layer is zero, and the pressure is equal to the external pressure.

The second term indicates the change of chemical potential from the change of electric displacement accompanied by redistribution of ions produced by the increase of film thickness. Simultaneously, the distribution of electric intensities around a clay particle with constant charge also changes. At that time, the electric intensities near the surface increase (Fig. 2.7), i.e. the gradient of electric potential at a given distance decreases with thickness of water film. The third term is related to work done by the van der Waals force, and appears only when the partial volume of adsorbed water is changed with increments of internal pressure and electric intensity. The last term represents the increment in chemical potential caused by decreased ion concentrations with increased film thickness.

The effects of changes of these factors on physicochemical properties have not yet been estimated quantitatively. However, the changes must have some effects on the properties. If a structural change in adsorbed water is produced by formation, breakdown, or distortion of hydrogen bonds between water molecules in adsorbed water, the properties could be changed. This change is similar to phase changes of water, or to exothermic or endothermic reactions with a decrease or increase of entropy of adsorbed water. Since the change of entropy is compensated by internal energy, the structural change is not reflected in the change of chemical potential.

Distribution of Electric Intensity Near a Clay Surface

As understood from Figs. 2.22 and 3.6, the electric intensity is nonuniform vertically and horizontally at points where the distance from the charged surface is smaller than the average distance between adjacent charges on the surface. This distance for montmorillonite is generally 10 to 15 Å—three to five water molecular layers. Figure 2.40 shows the distribution of electric intensity directions in the vicinity of a montmorillonite surface for divalent cations, a surface charge density of one elementary electric charge/100 $Å^2$, with negative charges and cations distributed two-dimensionally and with 71 and 29% of cations existing in the second and third layers. The arrows in Fig. 2.40 show the directions of the electric intensities at the respective points. The direction of electric intensity changes greatly in space. Since the real distributions of cations and surface negative charges are three-dimensional, the directions of electric intensities at the respective points become more disordered in space. The electric intensity in this space is on the order of 100,000 volts/cm. Such a random distribution of electric intensity direction may influence formation, breakdown or distortion of hydrogen bonds in the adsorbed water.

Interaction Between an Ion and Water Molecules

The intensity of the electric field in proximity to an ion is around 10^6 volt/cm, and water molecules with a dipole moment are oriented radially around the ion (Sec. 2.1.2). In addition, it should be noted that most water molecules around a structure-breaking ion such as potassium are monomers, and their entropies may be larger than that of bulk water.

Interaction Between a Clay Surface and Water Molecules

A water molecule in direct contact with a clay surface can form a hydrogen bond. The questions are whether a hydrogen bond is formed, whether the bond is weak or strong, and whether the formation of a hydrogen bond depends on the kind of clay.

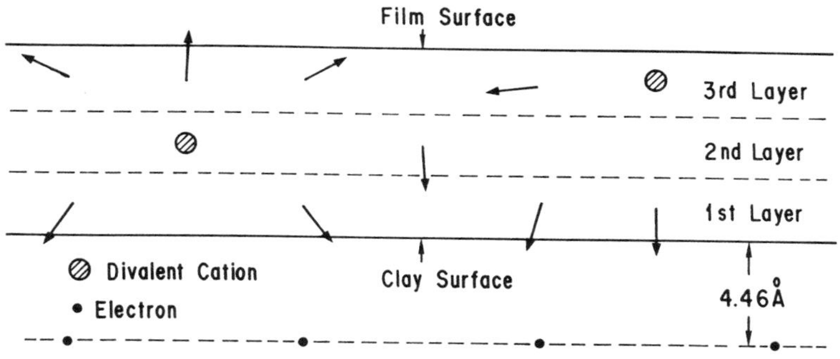

Fig. 2.40. A schematic representation of direction of electric intensity in water adsorbed on a clay surface.

The idea that a solid surface influences structure of a liquid in the vicinity of the surface through interaction between the surface and liquid molecules has been widely accepted by many researchers, e.g. Derjaguin (1954), Low (1961), and Drost-Hansen (1969). Water in proximity to a solid surface is assumed to be structurally modified, and is called vicinal water. The vicinal water is regarded to have physicochemical properties different from those of bulk water. Drost-Hansen (1969) has extensively reviewed the characteristics of vicinal water. The main interest here is whether the structure of water in the vicinity of a clay surface is different from that of bulk water.

Drost-Hansen proposed a three-layer model for the structure of vicinal water, with a layer of ordered structure near the surface, a disordered zone, and the bulk phase in the outermost layer. He presented schematic figures representing possible states of order of vicinal water on a nonpolar surface, a polar surface, and an ionic surface. Strong ion-dipole interactions may occur near an ionic surface (Fig. 2.41), giving rise to tightly bonded water adjacent to the interface. The nature and degree of structuring will, undoubtedly, depend on the nature of the solid and also on the diffuse double layer. Neglecting electric double layer effects, one might expect that the ordered structures of water near an ionic solid are the ones least compatible with the structure of bulk water, hence, decay most rapidly. If this dissimilarity is the case, the disordered zone might be most pronounced. However, the question now arises whether ordered structures can be induced by the innermost dipole-oriented layers. The answer to this question seems to be "yes."

Drost-Hansen (1969) stated that the thickness of various layers is more uncertain than the possible structures themselves. "The disordered region, in particular at some distance from a polar or ionic surface may, in turn, extend over only a few molecular diameters of water molecules, or perhaps, as many as 10 diameters. The thickness of the layers of oriented dipoles (effects by dipole-

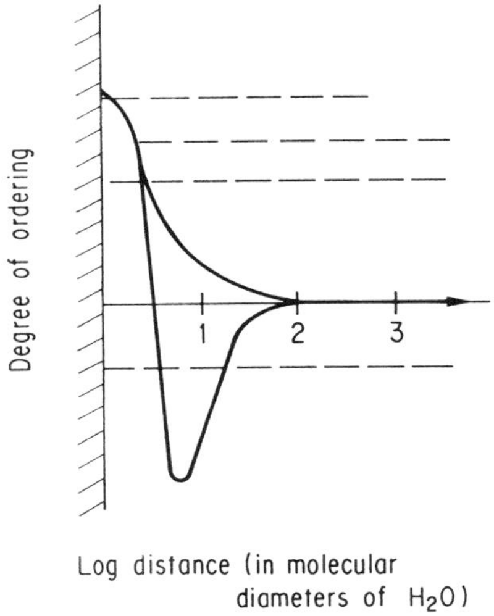

Fig. 2.41. A possible state of order for water near an ionic surface. (From Drost-Hansen, 1968)

dipole interactions between consecutive layers), should such dipole orientation actually occur, will most likely decay over a distance of, say 10 molecule diameters or less—probably less. However, the disordered zone in this case may prove considerably larger."

These models all relate to water adsorbed by a solid surface, but not to water trapped between two surfaces. Water sandwiched by two clay surfaces with electric charges is subject to a van der Waals force due to the other surface, a pressure produced by forces acting upon the two surfaces, and a structural force caused by overlapping of the two layers if a layer characterized by an ordered structure is formed in the vicinity of each surface. These are described in Chapter 3. It should be noted that the intermolecular force induced when one surface comes close to another may have a considerable effect on the physicochemical properties of adsorbed solvent through changing the arrangement of the solute molecules (Israelachvili, 1985).

2.5.2 Physicochemical Properties of Water Adsorbed by Clays

While it is not possible here to review all the measured physicochemical properties of water adsorbed by clays, we will call attention to several properties considered both interesting and important.

Degree of Dissociation of Adsorbed Water

What effect does the electric field near a montmorillonite surface have on the degree of dissociation of adsorbed water in the vicinity of the surface? The answer is very important for understanding chemical reactions occurring near the surface, such as the mechanism of catalysis.

Pickett and Lemcoe (1959) and Ducros and Dupont (1962) found evidence from nuclear magnetic resonance studies that water adsorbed by montmorillonite has a higher degree of dissociation than bulk water. The average "life" of dissociation in any particular molecule was so short that a degree of ionization at least 1000 times higher than usual is probable. Mortland et al. (1963) found that base-saturated montmorillonite that has reacted with ammonia showed the same infrared spectral features as NH_4-montmorillonite. They concluded that this proved chemisorbed NH_3 to be present as NH_4^+ ions rather than in molecular form, and residual water in the interlayer surfaces was dissociated to a greater extent than free water.

Fripiat et al. (1965) measured electric conductivity and dielectric constant associated with the first and second adsorbed water layers for Ca- and Na-montmorillonites, a silica gel, and a window glass powder. Symbols in Fig. 2.42 represent measured values of electric conductivity for pellets with different porosities. The conductivity σ (mho cm^{-1}) at a surface coverage larger than 1.3 for the Na-montmorillonite is of the order of 10^{-5} or more, while σ for the silica gel hardly reaches 1×10^{-7} mho cm^{-1}. The electric conductivity of bulk water is 4×10^{-8} mho cm^{-1}. Confirming that the metal cation counter ions were not the charge carriers, they concluded: 1) Electric conductivity is probably mostly protonic, and the protons originate from the dissociation of water molecules enhanced by strong surface electric fields. 2) The degree of dissociation is of the order of 10^{-2}. 3) The lower conductivity values observed for silica gel than those of montmorillonite reflect weaker electric fields, which arise from ionic Si-O bonds. Touillaux et al. (1968), from a pulsed magnetic resonance technique, suggested a dissociation degree 10^7 times higher in the water adsorbed by Na- and Ca-montmorillonites than free water.

Interactions Between a Clay Surface and a Water Molecule

Difficulties in estimating interactions between a clay surface and a water molecule are due to the existence of exchangeable ions balancing the charges held by the clay. Sposito and Prost (1982), in a fine review, stated that though the spatial arrangement of water molecules at the first stage of water adsorption by smectite indeed derives mainly from solvation of exchangeable cations, the role of the silicate surface of smectite may not be neglected. Suquet et al. (1977) proposed a model from X-ray diffraction and IR spectroscopy studies, in

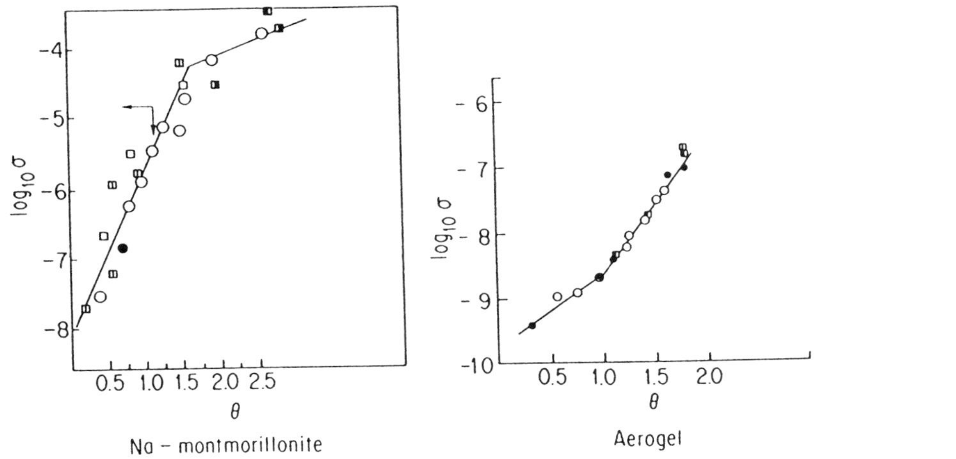

Fig. 2.42. Variation of the logarithm of conductivity σ against surface coverage θ. (From Fripiat et al., 1965)

which the adsorbed molecules form hydrogen bonds with the clay surface oxygen atoms at sites where Si^{4+} has been replaced by Al^{3+}. Mamy (1968) has suggested, from the study of dielectric relaxation time of adsorbed water on homoionic montmorillonites, that the one-layer hydrate consists of water molecules arranged in a strained, ice-like configuration, with bonds formed both with intermolecular interactions and with clusters of oxygen atoms in the silicate surface. As Sposito and Prost (1982) have pointed out, "If the charge deficit in smectite occurs in the tetrahedral sheet, there is a greater localization of the negative charge, and the formation of relatively strong hydrogen bonds between the adsorbed water molecules and surface oxygen atoms near sites of isomorphous substitution is favored. On the other hand, if the charge deficit occurs in the octahedral sheet, the negative charge on the surface oxygen atoms is delocalized and the adsorbed water molecules form only weak hydrogen bonds with these surface atoms."

Fripiat et al. (1965) considered the mobility of water molecules in the monolayer for Ca- and Na-montmorillonites, a silica gel and a window glass powder by comparing entropy data derived from water adsorption isotherms at different temperatures with the theoretical ones deduced from statistical relationships applied to different models of physically adsorbed layers. The results showed that monolayer water molecules on hydroxylic surfaces of the silica gel and glass powder are immobile, while on the oxygen surface of montmorillonite they are mobile. In addition, the observed differential heats of adsorption for the silica gel and glass powder were of the order of magnitude expected for hydrogen bonds. These facts may indicate that water molecules form hydrogen bonds with the hydroxylic surfaces of the silica gel and glass powder, but do not have strong interactions, including hydrogen bonding, with the oxygen surface of montmorillonite. Leonard (1970) studied partially deuterated water adsorbed on Na-, K-, Mg-, and Ca-montmorillonites using infrared spectroscopy. His conclusion was that there seems to be little tendency for water to form strong H-bonds with the silicate surface, and at low water contents, the exchangeable cations apparently exert the dominant force affecting water structure.

Iwata et al. (1989) measured differential heats of water vapor adsorption for Cs-saturated montmorillonite, kaolinite and allophane at low relative humidity, p/p_0 from 10^{-4} to 10^{-1}. Clay surface atom-water molecule and counter ion-water molecule interactions are involved simultaneously in the process. The amount of adsorbed water and differential heat due to the counter ion-water molecule interaction were estimated as a function of the water vapor pressure using equilibrium constants of $Cs^+ (H_2O)_{n-1} + H_2O = Cs(H_2O)_n$ (Dzidic and Kebarle, 1970). Water vapor adsorption for each clay surface could then be determined from the experimental results and the estimated values. It should be stressed that all the experimental results described above were obtained under conditions of very low water contents. The tentative conclusions were: 1) The clay surface atom-water molecule interactions in allophane and kaolinite were recognized to be dominant in a pressure range of p/p_0 below around 10^{-2}. The degree of surface coverage, θ, is smaller than about 1.0. 2) A dominant role of water adsorption due to C_s^+-water molecule interaction was found for montmorillonite in a pressure range from 10^{-3} to about 3×10^{-2} ($0.2 < \theta < 0.75$). 3) Affinity of water to the clay surface is strongest for allophane and weakest for montmorillonite. This finding seems to be consistent with the conclusion by Fripiat et al. (1965) relating to the mobility of water molecules.

Entropy

Entropies of water adsorbed on clay surfaces are generally measured either from adsorption isotherms at two different temperatures, or by combining calorimetric heat of immersion data with one adsorption isotherm at the same temperature.

Factors contributing to the generation of heat of immersion have been analyzed from various points of view (Iwata, 1972, 1976; Keren and Shainberg, 1975, 1980; Iwata et al., 1988), including intermolecular forces between a clay surface and a water molecule, hydration of counter ions, electric charges in or on the clay, chemical interaction between water and the surface, and redistribution of counter ions by addition of water. These factors, except the last one, generate heat. Counter ion redistribution induces an increment of electrical energy stored in the adsorbed phase, resulting in an endothermic reaction.

Therefore, the calorimetric method gives the entropy change for the whole system of clay, exchangeable ions and adsorbed water.

On the other hand, the entropy values obtained by the two-isotherm method may represent the entropy change of the water phase only. Iwata (1972) concluded that the partial entropy calculated from this method is approximately equal to that of the outermost phase, as the adsorbed water is regarded to be a heterogeneous system as described in Sec. 1.2.2. Regarding the adsorbed water to be a homogeneous system, it becomes possible to calculate the average molar entropy of the adsorbed water (integral entropy) from the results of the two-isotherm method (Hill, 1950; Hill et al., 1951).

There are, however, three weak points in the calculation. The equation used for the calculation is derived on the basis of an assumption that the adsorbed water shows a spreading pressure. Whether the adsorbed water, especially the first adsorbed water layer, can move freely in the directions parallel to the surface has not been clarified. Even if the adsorbed water molecules are recognized to be mobile, it is not technically easy to evaluate exactly the spreading pressure as a function of water vapor pressure from the experimental isotherm, especially in the low vapor pressure region. This function is indispensable in calculating entropy values. The last problem is in the differential operation generally used to calculate the entropy values. An integral operation is inclined to reduce the difference between measured values, while a differential operation has a tendency to enlarge the difference between measured values. The fact that entropy values for water adsorbed on clay surfaces found in literature are very different may result from these weaknesses.

Kohl et al. (1964) and Keren and Shainberg (1980) found negative changes of entropy relative to bulk water for water adsorbed by clays, which probably means the structure of the adsorbed water is more ordered than that of bulk water (Fig. 2.43). Conversely, Cary et al. (1964) and Sharma et al. (1969) found positive values for water adsorption on soils, and Mizota et al. (1989) on clays (Table 2.6). Keren and Shainberg (1980) concluded the large difference between Na- and Ca-montmorillonite suggested that the interaction is mainly with the adsorbed cations. In addition, they compared the entropy values estimated from the two-isotherm method with those from the calorimetric method for the same sample, and provide a good explanation of the difference in physical meaning of the results obtained by the two methods. The data in Table 2.6, calculated from results obtained using an apparatus newly developed by Mizota et al. (1989), indicate fairly large positive values. This may be due to comparatively high water contents of the samples. The entropy values of ice, liquid water and water vapor at 25°C are 38, 69.9, and 188.72 $J{\cdot}K^{-1}{\cdot}mol^{-1}$, respectively. The authors stated that each measured entropy value might be that of the outermost phase. These large values suggest the existence of the disordered zone in the three-layer model proposed by Drost-Hansen (1969), and described earlier.

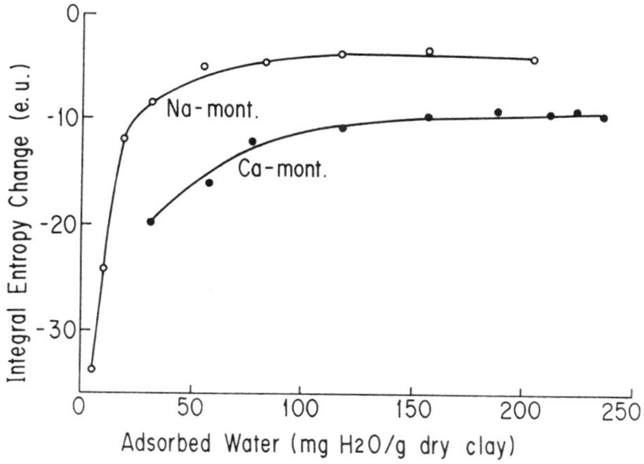

Fig. 2.43. Integral entropy changes of water adsorbed on Na- and Ca-montmorillonite. (From Keren and Shainberg, 1980)

Table 2.6. Entropy values of water adsorbed by montmorillonites. (From Mizota et al., 1989)

Desorbed H_2O, mmol g^{-1}	T, °C	S, JK^{-1} mol^{-1}
Matsuida		
1.2	30	70.2
1.4	30	62.1
1.6	30	76.5
1.8	40	74.5
Aterazawa		
0.60	30	83.9
0.70	30	78.8
0.80	30	76.6
0.90	30	76.3
1.00	40	75.0
1.10	40	82.1

Johnson et al. (1982) measured thermodynamic quantities for well-characterized specimens of analcime, a zeolite, and dehydrated analcime, over temperatures from 5 to 600°C, using adiabatic, solution, and drop-calorimetric techniques. The average molar entropies of water adsorbed by zeolite (zeolitic water) could be calculated from the differences in measured heat capacities between the analcime and dehydrated analcime. The zeolitic water was equivalent to water lost by heating a dry sample to 600°C for 24 hours.

Entropy values have been obtained for other zeolites using the same techniques (Johnson et al., 1983, 1985). The average standard molar entropies of zeolitic water are shown in Table 2.7, together with the amounts of zeolitic water, the cation exchange capacities estimated from the chemical compositions and the entropies of liquid water, ice and water vapor at 25°C. The cation exchange capacities are very large, and the zeolitic water contents are low. The entropy values of heulandite and analcime are considerably larger than those of natrolite and scolecite. Though the authors did not discuss the differences, the larger cation exchange capacities of the latter may be the reason. Furthermore, it is interesting that the entropy values of natrolite ($Na_2Al_2Si_3O_{10}{\cdot}2H_2O$) and scolecite ($CaAl_2Si_3O_{10}{\cdot}3H_2O$) are identical, while the counter ion is different. Zeolitic water is held in the very small channels of zeolite with a diameter of the order of several angstroms. This means that a zeolitic water molecule in a channel is under a large potential field due to intermolecular forces acting between the channel wall and the water molecule. In addition, the water molecule is adsorbed by a counter ion coexisting there, and may have a chemical interaction with an oxygen atom on the wall.

The method used by Johnson et al. (1982) is regarded as one of the most reliable methods to estimate the average molar entropy of adsorbed water. It would be desirable to use this method to measure average molar entropies of water adsorbed by sheet silicate minerals such as montmorillonite and kaolinite. Judging from the existence of very large potential fields, the very large cation exchange capacities and the very small contents of zeolitic water, it may be reasonable to consider that the average molar entropy of water in the first two water molecular layers adsorbed on a montmorillonite surface is larger than that of zeolitic water in heulandite and analcime.

Viscosity

On the theoretical side, there are at least two effects that result in an increase in viscosity of an electrolyte solution between two charged surfaces (Hunter, 1981). One is the electroviscous effect, which arises from the resistance to flow of the counter ions in the diffuse double layers. The apparent viscosity could be up to 30% higher than the bulk viscosity (Levin et al., 1975). The second effect, known as the viscoelectric effect, arises from a real increase in solvent viscosity

Table 2.7. Entropy values of zeolitic water.

Material	S, JK^{-1} mol^{-1}	Water content (%)	CEC[4] (meq/100 g)
Heulandite[1]	50.5	15.35	346
Natrolite[2]	32	9.2	581
Scolecite	32	13.8	591
Analcime[3]	55	8.20	477
Liquid water	69.9		
Ice	38		
Water vapor	188.72		

[1]Johnson et al. (1985), [2]Johnson et al. (1983), [3]Johnson et al. (1982), [4]Estimated by Iwata

in the presence of strong electric fields in the vicinity of charged surfaces. Israelachvili (1986) has pointed out that the magnitude of this effect rarely exceeds a few percent. In addition to the above effects, the structural change of water adjacent to the surface can be considered. The structural change has been regarded to be due to interaction between the surface atoms and water molecules (e.g. Derjaguin et al., 1967). The electric fields near the surfaces produced by surface charges and counter ions, whose intensities are very strong but whose directions are random, could also cause the structure change. However, this effect, contrary to the above effects, is likely to decrease the viscosity if the structure of water is broken by the electric fields (c.f. the flickering cluster model of Frank and Wen in Sec. 2.1.2).

Kemper et al. (1964) estimated mobility of water in the vicinity of Na- and Ca-bentonite surfaces from diffusion rates of deuterium hydroxide in oriented clay pastes at several moisture contents. Their results are: 1) In the first molecular layer of adsorbed water on Na- and Ca-bentonites, mobility of water molecules was reduced to 30% and 5% of that in bulk water. 2) In the sodium saturated system, a slight reduction in viscosity was recognized even at 40 to 50 Å from the surface. 3) In the stable Ca-bentonite hydrate with three molecular layers of water between bentonite platelets, the middle layer of water molecules clearly had much greater mobility than the layers directly adjacent to the surfaces. The authors explained this phenomenon by the positions of Ca^{2+} ions directly on the surfaces rather than midway between surfaces. 4) Water outside the first adsorbed molecular layer did not have viscosities more than 2.5 times the viscosity of bulk water.

Peschel and Adlfinger (1970) determined the viscosity of water in boundary layers at different shearing forces by means of a special device in which a spherical surface approaches a planar plate. Both surfaces were of fused silica and entirely covered with hydroxyl groups. The results are shown in Fig. 2.44, in which the relative surface zone viscosity, η_G/η, is expressed as a function of the plate distance h′ for three values of the shearing forces, K. The surface zone viscosity of water was found to be enhanced at distances larger than 1.6 x 10^{-5} cm, especially for small shearing forces. The increase of viscosity was not found at sufficiently high shearing forces. They concluded that this anomalous effect is due to the modified water formed under the influence of the surfaces.

Low (1976) calculated the viscosities of water in Na-montmorillonite systems at various water contents by different equations using data from experiments conducted by different investigators on viscous flow of water at different temperatures, self-diffusion of water, and neutron scattering by water. The results showed that viscosity increases exponentially with decreasing water content. The equation for the relative viscosity, η/η_o, at a water content, m_w/m_c, and a temperature, T, was given by

$$\eta/\eta_o = \exp\left[122.8/(m_w/m_c)T\right] \tag{2.50}$$

According to the above equation, $\eta/\eta_o \approx 1.5$ when $m_w/m_c = 1$ (corresponding to an interlayer distance of 26 Å) and T 25°C, and $\eta/\eta_o \approx 2.3$ when $m_w/m_c = 0.5$ and T 25°C.

The viscosities of water near solid surfaces reviewed above were estimated from indirect measurements. Israelachvili (1986) has developed a method that directly measures viscosity of liquids in thin films between two surfaces. His results (Table 2.8) show clearly that the viscosity of water between two mica surfaces is within about 10% of its bulk value (0.0095 at 23°C) in films whose thickness ranged from 265 to 18.5 Å, corresponding to three water molecular layers. In addition, the author has pointed out from his experimental results that the "plane of slip" is within a few angstroms of each interface, i.e. that at most, one layer of molecules is immobilized at each surface.

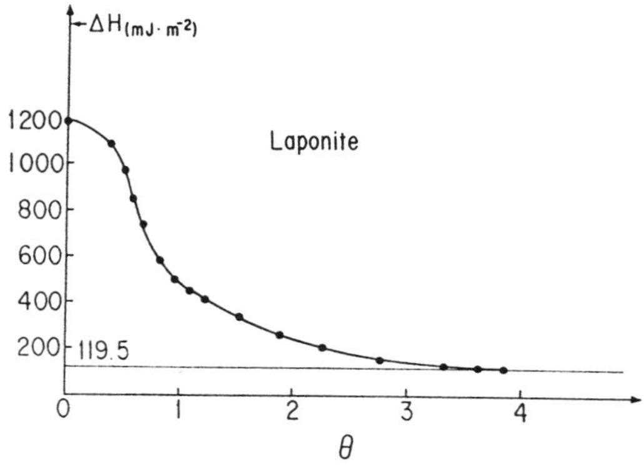

Fig. 2.44. Relative surface zone viscosity η_G/η vs. plate distance h′ for water at 20°C: (1) K = 253 dynes; (2) K = 523 dynes; (3) K = 657 dynes. (From Peschel and Adlfinger, 1970)

2.5.3 The Range of Influence for Surface Forces

There has been a long standing difference of interpretation about the range of action of the surface forces responsible for disruption of the structure of water molecules near clay surfaces. This issue has not been resolved.

Many spectroscopic data indicate that the range of surface effects on the vibrational, rotational, and translational motions of water molecules adsorbed by smectites is limited to a few molecular layers. For example, Prost (1981) concluded that the near-IR spectroscopic properties are not perturbed significantly

Table 2.8. Relationship between measured viscosity η, and average separation of mica surfaces, D. (From Israelachvili, 1986)

D (Å)	η(P)
265	0.0104
180	0.0102
70	0.0102
18.5	0.0098

beyond about two layers in Na-hectorite water systems. Woessner (1980) concluded from an NMR investigation that the hectorite surface influences and slows down the rotation of only the first layer or two of water molecules. Sposito and Prost (1982) believed the narrowness of the range was due to exchangeable cations. In addition, it should be borne in mind that the spectroscopic data were obtained at low water contents, and the physical and/or chemical states of water near the surfaces would change with increase in water content.

On the other hand, many researchers discussed the existence of ordered elements of water over a considerable distance from a solid surface. Three types of studies were used to verify the existence of such water: 1) measurements of macroscopic properties of adsorbed water, 2) measurements of disjoining pressure produced by the structural change of water held between two surfaces and temperature dependence of disjoining pressure, and 3) measurements of thermal anomalies in the properties of adsorbed water. Only the results from macroscopic studies will be introduced here. The review by Drost-Hansen (1969) has material related to thermal anomalies.

Experimental results showing viscosity of adsorbed water to be larger than that of the bulk have been reviewed in the preceding section. Etzler and Fagundus (1987) found the density of water in silicas of various pore diameters to be less than that of bulk water. Their density of 980 kg/m^3 at 4 nm from the surface agreed perfectly with the value of partial specific water volume near 4 nm for Na-montmorillonite obtained by Low (1979). From their results, and from those collected earlier by Low, they concluded water was modified to distances of 3-5 nm. Etzler and White (1987) measured the heat capacities of water confined in silica pores with radii from 1.2 to 12.1 nm, and concluded that the structural change of adsorbed water extends to distances of at least 6 nm.

The contradiction between the spectroscopic and the macroscopic data may be illuminated by results of two papers, one by Israelachvili (1986) and the other by Fripiat et al. (1982). Israelachvili (1986) measured viscosity of adsorbed water directly, while previous values were estimated from indirect measurements. Viscosity of adsorbed water on mica at distances from 265 to 18.5 Å was equal to or a little larger than the bulk value (Table 2.8). In addition, according to the theory of Levine et al. (1975), the change in viscosity due to the electroviscous effect in this case was calculated to be a 20% increase. It may be suggested from these results that the structure of water near the surface remains unchanged or has less structure than bulk water.

Fripiat et al. (1982) used both microscopic scale (NMR) and macroscopic scale (thermodynamic) measurements to determine the extent of the effect on water molecules near the surfaces. They used hectorite, laponite, and kaolinite clays. Nuclear magnetic resonance relaxation times of nuclei ^1H or ^2H were measured in H_2O or D_2O suspensions with increasing solids content. Heats of immersion were measured for clay surfaces onto which an increasing number of water layers had been preadsorbed. From a thermodynamic analysis it is expected that no surface effects are detectable when the heat of immersion is equal to the surface energy of pure liquid water. The result for heats of immersion for laponite are shown in Fig. 2.45. The "apparent θ" is the degree of coverage obtained by assuming that all the water molecules are absorbed on the external surfaces. This value is, of course, larger than the real degree of coverage. From the figure, the number of layers affected by the solid surface is about 4 at maximum. The average thickness of adsorbed water directly influenced by the external field obtained from the NMR data was about 10 Å, corresponding to 3.2 layers for the laponite. The values estimated from the two measurements agree well. Similar results were obtained for hectorite and kaolinite.

We cannot settle the question of range of action of surface forces responsible for disruption of water structure on clays from only the results in these two reports by Israelachvili (1986) and Fripiat et al. (1982). More measurements at macroscopic and microscopic scales, and their interpretation, will be needed.

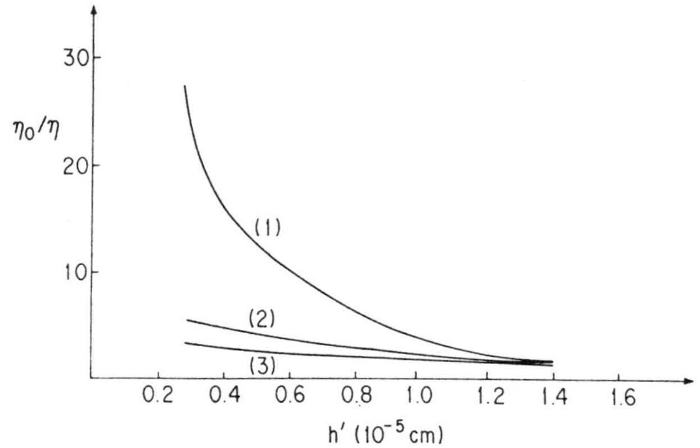

Fig. 2.45. Heat of immersion at 30°C vs. apparent water coverage θ for laponite. (From Fripiat et al., 1982)

ACKNOWLEDGEMENTS

Figure 2.1 reproduced from *Journal of Chemical Physics*. Figure 2.4 is reprinted by permission of Kyoritsu Shutsupan, Inc. Figures 2.13, 2.14, 2.16 and 2.17 are reproduced from *Higardia*. Figures 2.10, 2.18, 2.19, 2.44, and 2.45 are reprinted by permission of *Academic Press*. Figures 2.22, 2.23, 2.24, 2.25, 2.26, 2.27, 2.28, and 2.29 are reprinted from *Soil Science*, Copyright © by the Williams and Wilkins Co., Baltimore. Used with permission. Figure 2.33 is reproduced from Tapir Publishers. Figure 2.37 is reproduced from the *Soil Science Society of America Proceedings* by permission of the Soil Science Society of America. Figure 2.38 is reprinted by permission of Bolkema, Rotterdam. Figure 2.41 is reproduced from *Industrial and Engineering Chemistry*. Figure 2.42 is reprinted from *Journal of Chemical Physics*. Figure 2.43 is reproduced from *Clays and Clay Minerals*.

REFERENCES

Anderson, D. M., and P. Hoekstra, Migration of interlamellar water during freezing and thawing of Wyoming bentonite, *Soil Sci. Soc. Am. Proc.*, 29:498-504 (1965).

Anderson, D. M., and N. Morgenstern, Physical chemistry and mechanics of frozen ground, In *Permafrost*, the North American contribution to the 2nd Int. Conf., Washington, D.C., Natl. Acad. Sci. (1973).

Anderson, D. M., and G. Sposito, Heat of immersion of Arizona bentonite in water, *Soil Sci.* 97:214-219 (1964).

Arakawa, K., *Structures and Physical Properties of Water and Aqueous Solutions*, Hokukaido Daigaku Tosho Kankokai (1989). (Japanese)

Babcock, K. L., Theory of chemical properties of soil colloidal systems at equilibrium, *Hilgardia* 34:417-542 (1963).

Barak, P., Double layer theory prediction of Al-Ca exchange on clay and soil, *J. Colloid Interface Sci.* 113:479-490 (1989).

Barer, S. S., N. V. Churaev, B. V. Derjaguin, O. A. Kiseleva, and V. D. Sobolev, Viscosity of nonfreezing thin interlayers between the surfaces of ice and quartz, *J. Colloid Interface Sci.* 74:173-180 (1980).

Barouch, E., and S. Kulkarni, Exact solution of the Poisson-Boltzmann equation for two spheres with fixed surface potentials, *J. Colloid Interface Sci.* 112:396-402 (1986).

Benedict, W. S., N. Gailar, and E. K. Plyler, Rotation-vibration spectra of deuterated water vapor, *J. Chem. Phys.* 24:1139-1165 (1956).

Bentz, J., Electrostatic potential between concentric surfaces: spherical, cylindrical, and planar, *J. Colloid Interface Sci.* 90:164-181 (1982).

Bernal, J. D., and R. F. Fowler, A theory of water and ionic solution, with particular reference to hydrogen and hydroxyl ions, *J. Chem. Phys.* 1:515-548 (1933).

Beskow, G., Soil freezing and frost heaving with special application to roads and railroads, J. O. Osterberg, trans., 1947, The Technologies Institute, Northwestern Univ., Evanston, IL (1935).

Bjerrum, N., Structure and properties of ice, *Science* 115:385-390 (1952).

Black, P. B., and R. D. Miller, Hydraulic conductivity and unfrozen water content of air-free frozen silt, *Water Resour. Res.* 26:323-329 (1990).

Black, P. B., Interpreting uncofined unfrozen water content, In *Proc. 6th Intl. Symp. on Ground Freezing*, p. 3-6, Balkema, Rotterdam (1985).

Blanchard, D., and M. Fremond, Soils, frost heaving and thaw settlement, In *Proc. 4th Intl. Symp. on Ground Freezing*, p. 209-216, Balkema, Rotterdam (1985).

Bockris, J. O. M., and P. P. S. Saluja, Ionic solvation numbers from compressibilities and ionic vibration potential measurements, *J. Phys. Chem.* 76:2140-2151 (1972).

Bolt, G. H., and M. G. M. Bruggenwert, eds., *Soil Chemistry*, Elsevier, Amsterdam (1976).

Bolt, G. H., and R. D. Miller, Calculation of total and component potentials of water in soil, *Trans. Am. Geophys. Union* 39:917-928 (1958).

Brown, S. C., and D. Payne, Frost action in clay soils, 1. A temperature-step and equilibration differential scanning calorimeter technique for unfrozen water content determinations below 0°C, *J. Soil Sci.* 41:535-546 (1990).

Cary, J. W., R. A. Kohl, and S. A. Taylor, Water adsorption by dry soil and its thermodynamic functions, *Soil Sci. Soc. Amer. Proc.* 28:309-314 (1964).

Chan. D. Y. C., R. M. Pashley, and J. P. Quirk, Surface potentials derived from co-ion exclusion measurements on homoionic montmorillonite and illite, *Clays Clay Miner.* 32:131-138 (1984).

Del Bene, J., and J. A. Pople, Intermolecular energies of small water polymers, *Chem. Phys. Lett.* 4:426-428 (1969).

Derjaguin, B. V., Investigation of the forces of interaction of surfaces in different media and their application to the problem of colloid stability, *Disc. Faraday Soc.* 18:24-41 (1954).

Derjaguin, B. V., and N. V. Churaev, The theory of frost heaving, *J. Colloid Interface Sci.* 67:391-396 (1978).

Derjaguin, B. V., N. N. Fedyakin, and M. V. Talaev, Concerning the modified state and structural polymorphism of liquids condensed from their undersaturated vapors in quartz capillaries, *J. Colloid Interface Sci.* 24:132-133 (1967).

Dirksen, C., and R. D. Miller, Closed-system freezing of unsaturated soil, *Soil Sci. Soc. Am. Proc.* 30:169-173 (1966).

Dixon, J. B., and S. B. Weed, eds., *Minerals in Soil Environments*, Soil Science Society of America, Madison, WI (1977).

Drost-Hansen, W., Structure of water near solid surfaces, *Ind. Eng. Chem.* 61:10-47 (1969).

Drost-Hansen, W., Effects of vicinal water on colloid stability and sedimentation processes, *J. Colloid Interface Sci.* 58:251-262 (1977).

Ducros, P., and M. Dupont, Résonance magnetique nucléaire—Étude par résonance magnétique nucléaire des protons dans les argiles, *Compt. Rend.* 254:1409-1410 (1962).

Dzidic, I., and P. Kebarle, Hydration of the alkali ions in the gas phase. Enthalpies and entropies of reactions $M(H_2O)_{n-1} + H_2O = M(H_2O)_n$, *J. Phys. Chem.* 74:1466-1474 (1970).

Edlefsen, N. E., and B. C. Anderson, Thermodynamics of soil moisture, *Hilgardia* 15:31-298 (1943).

Edwards, D. G., and J. P. Quirk, Repulsion of chloride by montmorillonite, *J. Colloid Sci.* 19:872-882 (1962).

Edwards, D. G., and J. P. Quirk, Repulsion of chloride ions by negatively charged clay surfaces. I. Monovalent cation Fithian illite, *Trans. Faraday Soc.* 61:2808-2815 (1965a).

Edwards, D. G., and J. P. Quirk, Repulsion of chloride ions by negatively charged clay surfaces. II. Monovalent cation montmorillonites, *Trans. Faraday Soc.* 61:2816-2819 (1965b).

Edwards, D. G., and J. P. Quirk, Repulsion of chloride ions by negatively charged clay surfaces. III. Di- and tri-valent cation clays, *Trans. Faraday Soc.* 62:2820-2823 (1965c).

Egashira, K., Viscosities of allophane and imogolite clay suspensions, *Clay Sci.* 5:87-95 (1977).

Eisenberg, D., and W. Kauzmann, *The Structure and Properties of Water*, Clarendon Press, Oxford (1969).

Enüstün, B. V., H. S. Sentürk, and O. Yurdakul, Capillary freezing and melting, *J. Colloid Interface Sci.* 65:509-516 (1978).

Etzler, F. M., and D. M. Fagundus, The extent of vicinal water, *J. Colloid Interface Sci.* 115:513-519 (1987).

Etzler, F. M., and P. J. White, The heat capacity of water in silica pores, *J. Colloid Interface Sci.* 120:94-99 (1987).

Everett, D. H., The thermodynamics of frost damage to porous solids, *Trans. Faraday Soc.* 57:1541-1551 (1961).

Fletcher, N. H., *The Chemical Physics of Ice*, Cambridge University Press, London (1970).

Fletcher, N. H., in *Physics and Chemistry of Ice*, Royal Society of Canada, Ottawa (1973).

Førland, K. S., T. Førland, and S. K. Ratkje, Frost Heave, In *Proc. 5th Intl. Conf. on Permafrost*, p. 344-348, Tapir Publ. (1988).

Frahm, J., and S. Diekmann, Numerical calculation of diffuse double layer properties for spherical colloidal particles by means of a modified nonlinearlized Poisson-Boltzmann equation, *J. Colloid Interface Sci.* 70:440-447 (1979).

Frank, H. S., and W.-Y. Wen, Structural aspects of ion-solvent interaction in aqueous solutions: a suggested picture of water structure, *Discuss. Faraday Soc.* 24:133-140 (1957).

Fremond, M. and M. Mikkola, Thermodynamical modeling of freezing soil, In *Proc. Conf. on Ground Freezing*, p. 17-24, Balkema, Rotterdam (1991).

Frenkel, J., *Kinetic Theory of Liquids*, Oxford University Press, London (1946).

Fripiat, J. J., A. Jelli, G. Poncelet, and J. Andre, Thermodynamic properties of adsorbed water molecules and electrical conduction in montmorillonite and silicas, *J. Phys. Chem.* 69:2185-2197 (1965).

Fripiat, J. J., J. Cases, M. Francois, and M. Letellier, Thermodynamic and microdynamic behavior of water in clay suspensions and gels, *J. Colloid Interface Sci.* 89:378-400 (1982).

Fujii, K., and M. Nakano, Chemical potential of water adsorbed to bentonite, *Trans. Jpn. Soc. Irrig. Drain. Reclam. Eng.* 112:43-54 (1984).

Gayer, K. H., L. C. Thompson, and O. T. Zajicek, The solubility of aluminum hydroxide in acidic and basic media 25°C, *Can. J. Chem.* 36:1268-1271 (1958).

Gieseking, J. E., *Soil Components*, Vol. 2, *Inorganic Components*, Springer-Verlag, Berlin (1975).

Gilpin, R. R., A model of the "liquid-like" layer between ice and a substrate with applications to water regelation and particle migration, *J. Colloid Interface Sci.* 68:235-251 (1979).

Gilpin, R. R., A model for the prediction of ice lensing and frost heave in soils, *Water Resour. Res.* 16:918-930 (1980a).

Gilpin, R. R., Theoretical studies of particle engulfment, *J. Colloid Interface Sci.* 74:44-63 (1980b).

Gold, L. W., A possible force mechanism associated with the freezing of water in porous materials, *Highw. Res. Board Bull.* 168:65-72 (1957).

Groenevelt, P. H., and B. D. Kay, On the interaction of water and heat transport in frozen and unfrozen soils, *Soil Sci. Soc. Am. Proc.* 38:400-404 (1974).

Gur, Y., I. Ravina, and A. J. Babchin, On the electrical double layer theory: II. The Poisson-Boltzmann equation including hydration forces, *J. Colloid Interface Sci.* 64:333-341 (1978).

Halsey, G., Physical adsorption on nonuniform surfaces, *J. Chem. Phys.* 16:931-937 (1948).

Hankins, D., J. W. Moskowitz, and F. H. Stillinger, Water molecule interactions, *J. Chem. Phys.* 53:4544-4554 (1970).

Harada, Y., and K. Wada, Release and uptake of protons by allophanic soils in relation to their CEC and AEC, *Soil Sci. Plant Nutr. Tokyo* 19:73-82 (1973).

Harlan, R. L., Analysis of coupled heat-fluid transport in partially frozen soil, *Water Resour. Res.* 9:1314-1323 (1973).

Henmi, T., and K. Wada, Morphology and composition of allophane, *Am. Mineral.* 61:379-390 (1976).

Hill, T. L., Statistical mechanics of adsorption, IX. Adsorption thermodynamics and solution thermodynamics, *J. Chem. Phys.* 18:246-256 (1950).

Hill, T. L., Theory of physical adsorption, *Adv. Catal.* 4:212-258 (1952).

Hill, T. L., P. M. Emmett, and L. E. Joyner, Calculations of thermodynamic functions of adsorbed molecules from adsorption isotherm measurements. Nitrogen on graphon, *J. Amer. Chem. Soc.* 73:5102-5107 (1951).

Hoekstra, P., Water movement and freezing pressures, *Soil Sci. Soc. Am. Proc.* 33:512-518 (1969).

Hopke, S., A model for frost heave including overburden, *Cold Reg. Sci. Tech.* 14:147-153 (1980).

Horikawa, Y., R. S. Murray, and J. P. Quirk, The effect of electrolyte concentration on the zeta potentials of homoionic montmorillonite and illite, *Colloids and Surfaces* 32:181-195 (1988).

Hunter, R. J., *Zeta Potential in Colloid Science*, Academic Press, London (1981).

Iimura, K., Acidic properties and cation exchange of allophane and volcanic ash soils, *Bull. Nat. Inst. Agric. Sci.* B17:101-157 (1966). (Japanese with English summary)

Ishizaki, T., and N. Nishio, Experimental study of frost heaving of a saturated soil, In *Proc. 5th Intl. Symp. on Ground Freezing*, p. 65-72, Balkema, Rotterdam (1988).

Israelachvili, J. N., *Intermolecular and Surface Forces with Applications to Colloidal and Biological Systems*, Academic Press, New York (1985).

Israelachvili, J. N., Measurement of the viscosity of liquids in very fine films, *J. Colloidal Interface Sci.* 110:263-271 (1986).

Iwata, S., Thermodynamics of soil water. II, *Soil Sci.* 113:313-316 (1972).

Iwata, S., Thermodynamics of soil water. III, *Soil Sci.* 117:87-93 (1974a).

Iwata, S., Thermodynamics of soil water. IV, *Soil Sci.* 117:135-139 (1974b).

Iwata, S., A quantitative dependence of unfrozen water in a partially frozen soil on original water content, *Trans. 10th Int. Cong. Soil Sci.* 1:56-61 (1974c).

Iwata, S., Heat of immersion of clays in water, *Ceramic* 11:441-447 (1976). (Japanese)

Iwata, S., Driving force for water migration in frozen clayey soil, *Soil Sci. Plant Nutr. Tokyo* 26:215-227 (1980).

Iwata, S., A mechanism for the existence of an unfrozen liquid in the vicinity of a solid surface, In *Proc. 4th Intl. Symp. on Ground Freezing*, p. 25-31, Balkema, Boston (1985).

Iwata, S., F. Izumi, and A. Tsukamoto, Differential heat of water adsorption for montmorillonite, kaolinite and allophane, *Clay Miner.* 24:505-512 (1989).

Jackson, K. A., and B. Chalmers, Freezing of liquid in porous media with special reference to frost heave in soils, *J. Appl. Phys.* 29:1178-1181 (1958).

Jame, Y. W., and D. I. Norum, Heat and mass transfer in freezing unsaturated porous medium, *Water Resour. Res.* 16:811-819 (1980).

Jellinek, H. H. G., Liquid-like (transition) layer on ice, *J. Colloid Interface Sci.* 25:192-205 (1967).

Johnson, G. K., H. E. Flotow, P. A. G. O'Hare, and W. S. Wise, Thermodynamic studies of zeolites: analcime and dehydrated analcime, *Amer. Mineralog.* 67:736-748 (1982).

Johnson, G. K., H. E. Flotow, P. A. G. O'Hare, and W. S. Wise, Thermodynamic studies of zeolites: natrolite, mesolite, and scolecite, *Amer. Mineralog.* 68:1134-1145 (1983).

Johnson, G. K., H. E. Flotow, P. A. G. O'Hare, and W. S. Wise, Thermodynamic studies of zeolites: heulandite, *Amer. Mineralog.* 70:1065-1071 (1985).

Jones, G., and H. Dole, The viscosity of aqueous solutions of strong electrolyte with special reference to barium chloride, *J. Amer. Chem. Soc.* 51:2950-2964 (1929).

Jurinak, J. J., Multilayer adsorption of water by kaolinite, *Soil Sci. Soc. Am. Proc.* 27:269-272 (1963).

Kay, B. D., and E. Perfect, State of the art: Heat and mass transfer in freezing soils, In *Proc. 5th Intl. Symp. on Ground Freezing*, p. 3-21, Balkema, Rotterdam (1988).

Kemper, W. D., D. E. L. Maasland, and L. K. Porter, Mobility of water adjacent to mineral surfaces, *Soil Sci. Soc. Amer. Proc.* 28:164-167 (1964).

Keren, R., and I. Shainberg, Water vapor isotherms and heat of immersion of Na- and Ca-montmorillonite systems, I. Homoionic clay, *Clays Clay Miner.* 23:193-200 (1975).

Keren, R., and I. Shainberg, Water vapor isotherms and heat of immersion of Na- and Ca-montmorillonite systems, II. Mixed systems, *Clays Clay Miner.* 27:145-151 (1979).

Keren, R., and I. Shainberg, Water vapor isotherms and heat of immersion of Na- and Ca-montmorillonite systems, III. Thermodynamics, *Clays Clay Miner.* 28:204-210 (1980).

Kinoshita, S., Soil water movement and heat flux in freezing ground, *Proc. Conf. Soil Water Probl. Col Reg.*, Calgary, Alta. (1975).

Kitagawa, Y., The unit particle of allophane, *Am. Mineral.* 56:465-475 (1971).

Kitagawa, Y., A short discussion on the chemical composition of allophane based on the data of Yoshinaga in 1966, *Soil Sci. Plant Nutr. Tokyo* 19:321-341 (1973).

Kohl, R. A., J. W. Cary, and S. A. Taylor, On the interaction of water with a Li-kaolinite surface, *J. Colloid Sci.* 19:699-707 (1964).

Kollman, P. A., and L. C. Allen, Theory of the hydrogen bond: electronic structure and properties of the water dimer, *J. Chem. Phys.* 51:3286-3293 (1969).

Konrad, J. M., and N. R. Morgenstern, A mechanistic theory of ice lens formation in fine-grained soils, *Can. Geotech. J.* 17:473-486 (1980).

Konrad, J. M., and N. R. Morgenstern, The segregation potential of a freezing soil, *Can. Geotech. J.* 18:482-491 (1981).

Koopmans, R. W. K., and R. D. Miller, Soil freezing and soil water characteristic curves, *Soil Sci. Soc. Am. Proc.* 30:680-685 (1966).

Kuroda, T., Theoretical study of frost heaving—Kinetic process at water layer between ice lens and soil particles, In *Proc. 4th Intl. Symp. on Ground Freezing*, p. 39-45, Balkema, Rotterdam (1985).

Leonard, R. A., Infrared analysis of partially deuterated water adsorbed on clay, *Soil Sci. Soc. Amer. J.* 34:339-343 (1970).

Levine, S., and G. M. Bell, Theory of a modified Poisson-Boltzmann equation: the volume of hydrated ions, *J. Phys. Chem.* 64:1188-1195 (1960).

Levine, S., and G. M. Bell, In *Chemical Physics of Ionic Solutions*, B. E. Conway and R. G. Barrads, eds., Wiley, New York (1964).

Levine, S., J. R. Marriott, and K. Robinson, Theory of electrokinetic flow in a narrow parallel-plate charger, *J. Chem. Soc., Faraday Trans. II* 71:1-11 (1975).

Loch, J. P. G., Thermodynamic equilibrium between ice and water in porous media, *Soil Sci.* 126:77-80 (1978).

Loch, J. P. G., and B. D. Kay, Water redistribution in partially frozen, saturated silt under several temperature gradients and overburden loads, *Soil Sci. Soc. Am. J.* 42:400-406 (1978).

Loch, J. P. G., and R. D. Miller, Tests of the concept of secondary frost heaving, *Soil Sci. Soc. Am. Proc.* 39:1036-1041 (1975).

Loeb, A. L., P. H. Wiersema, and J. T. G. Overbeek, *The Electrical Double Layer Around a Spherical Colloid Particle*, MIT Press, Cambridge, MA (1961).

Low, P. F., Physical chemistry of clay-water interaction, *Advan. Agron.* 13:269-327 (1961).

Low, P. F., Viscosity of interlayer water in montmorillonite, *Soil Sci. Soc. Amer. J.* 40:500-505 (1976).

Low, P. F., Nature and properties of water in montmorillonite-water systems, *Soil Sci. Soc. Amer. J.* 43:651-658 (1979).

Low, P. F., D. M. Anderson, and P. Hoekstra, Some thermodynamic relationships for soils at or below the freezing point: 1, *Water Resour. Res.* 4:379-394 (1968).

Mamy, J., Recherches sur l'hydration de la Montmorillonite: Propriétés diélectriques et structure du film d'eau, *Ann. Agron.*, 19:175-246 (1968).

Marshall, C. E., *The Colloid Chemistry of the Silicate Minerals*, Academic Press, New York (1949).

Mayer, A. F., Effect of temperature on ground-water levels, *J. Geophys. Res.* 65:1747-1752 (1960).

Miller, R. D., Freezing phenomena in soils, In *Applications of Soil Physics*, D. Hillel, ed., Academic Press, New York, pp. 254-299 (1980).

Miller, R. D., J. H. Baker, and J. H. Kolaian, Particle size, overburden pressure, pore water pressure, and freezing temperature of ice lenses in soil, *Proc. Int. Cong. Soil Sci.* 122-129 (1960).

Miller, R. D., J. P. G. Loch, and E. Bresler, Transport of water and heat in a frozen permeameter, *Soil Sci. Soc. Am. Proc.* 39:1029-1036 (1975).

Miller, S. and P. F. Low, Characterization of the electrical double layer of montmorillonite, *Langmuir* 6:572-578 (1990).

Miyata, Y., A frost heave mechanism model based on energy equilibrium, In *Proc. 5th Intl. Symp. on Ground Freezing*, p. 91-98, Balkema, Rotterdam (1988).

Mizota, T., T. Goto, and H. Koshi, Development of an apparatus for measuring equilibrium water vapor pressure of hydrous silicate minerals and entropy of water molecule, *J. Clay Sci. Soc. Jap.* 29:151-158 (1989). (Japanese with English summary)

Moelwyne-Hughes, E. A., *Physical Chemistry*, 2nd ed. Macmillan, New York (1964).

Morgan, J., and B. E. Warren, X-ray analysis of the structure of water, *J. Chem. Phys.* 6:666-673 (1938).

Morokuma, K., and L. Pedersen, Molecular-orbital studies of hydrogen bonds, *J. Chem. Phys.* 48:3275-3282 (1968).

Mortland, M. M., J. J. Fripiat, J. Chaussidon, and J. Uytterhoeven, Interaction between ammonia and the expanding lattices of montmorillonite and vermiculite, *J. Phys. Chem.* 67:248-258 (1963).

Nakano, Y., Transport of water through frozen soils, In *Proc. Conf. on Ground Freezing*, p. 65-70, Balkema, Rotterdam (1991).

Nakano, Y., Mathematical model on the steady growth of an ice layer in freezing soils, In *Physics and Chemistry of Ice*, N. Maeno and T. Hondoh (eds.), p. 364-369, Hokkaido Univ. Press (1992).

Nakano, Y., and K. Horiguchi, Role of heat and water transport in frost heaving of fine-grained porous media under negligible overburden pressure, *Adv. Water Resour.* 7:93-102 (1984)

Narten, A. H., M. D. Danford, and H. A. Levy, X-ray diffraction study of liquid water in the temperature range 4-200°C, *Discuss. Faraday Soc.* 43:97-107 (1967).

Nersesova, Z. A., and N. A. Tsytovich, Unfrozen water in frozen soils, In *Permafrost, Proc. Int. Conf.*, Washington, D.C., Natl. Acad. Sci. pp. 230-234 (1966).

Nightingale, E. R., *Chemical Physics of Ionic Solutions*, Conway, B. E., and R. G. Barradas (eds.), Wiley, New York, p. 87 (1966).

Oliphant, J. L., and P. F. Low, The relative partial specific enthalpy of water in montmorillonite-water systems and its relation to the swelling of these systems, *J. Colloid Interface Sci.* 89:366-373 (1982).

O'Neill, K., and R. D. Miller, Exploration of a rigid ice model of frost heave, *Water Resour. Res.* 21:281-296 (1985).

Panjukov, V. V., The surface charge density/surface potential relationship and the Poisson-Boltzmann equation, *J. Colloid Interface Sci.* 110:556-560 (1986).

Parks, G. A., The isoelectric points of solid oxides, solid hydroxides and aqueous hydroxo complex systems, *Chem. Rev.* 65:177-198 (1965).

Penner, E., Heaving pressure in soils during unidirectional freezing, *Can. Geotech. J.* 4:398-408 (1967).

Penner, E., Grain size as a basis for frost heaving, In *Proc. 2nd Conf. Soil Water Prbl. Cold Reg.*, Edmonton, Alta., Am. Geophys. Union, 103-109 (1976).

Peschel, G., and K. H. Adlfinger, Viscosity anomalies in liquid surface zones. IV, The apparent viscosity of water in thin layers adjacent to hydroxylated fused silica surfaces, *J. Colloid Interface Sci.* 34:505-510 (1970).

Philip, J. R., and D. A. de Vries, Moisture movement in porous materials under temperature gradients, *Trans. Am. Geophys. Union* 38:222-232 (1957).

Philip, J. R., Diffuse double layer interactions in one-, two-, and three-dimensional particle swarms, *J. Chem. Phys.* 52:1387-1396 (1970).

Pickett, A. G., and M. M. Lemcoe, An investigation of shear strength of the clay-water system by radio-frequency spectroscopy, *J. Geophys. Res.* 64:1579-1586 (1959).

Pople, J. A., Molecular association in liquids: II. A theory of the structure of water, *Proc. Roy. Soc. London* A205:163-178 (1951).

Prost, R., Near infrared properties of water in sodium hectorite, *Proc. Int. Clay Conf.*, Bologna, Italy 187-195 (1981).

Quirk, J. P., Negative and positive adsorption of chloride by kaolinite, *Nature* 188:253-254 (1960).

Rand, B., and I. E. Melton, Particle interactions in aqueous kaolinite suspensions, 1. Effect of pH and electrolyte upon the mode of particle interaction in homoionic sodium kaolinite suspensions, *J. Colloid Interface Sci.* 60:308-320 (1977).

Ravina, I., and Y. Gur, Application of the electrical double layer theory to predict ion adsorption in mixed ionic systems, *Soil Sci.* 125:204-209 (1978).

Römkens, M. J. M., and R. D. Miller, Migration of mineral particles in ice with a temperature gradient, *J. Colloid Interface Sci.* 42:103-111 (1973).

Ross, C. S., and P. F. Kerr, Halloysite and allophane, *U.S. Geol. Surv. Prof. Pap.* 185G:135-148 (1934).

Rowlinson, J. S., The lattice energy of ice and the second virial coefficient of water vapour, *Trans. Faraday Soc.* 47:120-129 (1951).

Schuhmann, D., and B. D'Epenoux, The effect of electrical nonhomogeneities of particles on mean surface potential, *J. Colloid Interface Sci.* 116:159-167 (1987).

Secor, R. B., and C. J. Radke, Spillover of the diffuse double layer on montmorillonite particles, *J. Colloid Interface Sci.* 103:237-244 (1985).

Shainberg, I., and W. D. Kemper, Hydration status of adsorbed ions, *Soil Sci. Soc. Am. Proc.* 30:707-713 (1966).

Sharma, M. L., G. Uehara, and A. Mann, Thermodynamic properties of water adsorbed on dry soil surfaces, *Soil Sci.* 107:86-93 (1969).

Sposito, G., and R. Prost, Structure of water adsorbed on smectites, *Chem. Rev.* 82:554-573 (1982).

Squet, H., R. Prost, and H. Pézérat, Étude par spectroscopie infrarouge de l'eau adsorbée par la saponite-calcium, *Clay Miner.* 12:113-126 (1977).

Sridharan, A., and M. S. Jayadeva, Approximate potential-distance relationship, *Aust. J. Soil Res.* 18:461-466 (1980).

Strang, O., and O. J. Fix, *An Analysis of the Finite Element Method*, Prentice-Hall, Englewood Cliffs, NJ (1973).

Sutherland, H. B., and P. N. Gaskin, Pore water and heaving pressure developed in partially frozen soils, In *Permafrost, Proc. 1st Int. Conf.*, Washington, D.C. Natl. Res. Count., 409-419 (1973).

Suzuki, K., *Water and Aqueous Solution*, Kyoritsu Shupan, Inc., Tokyo (1980). (Japanese)

Taber, S., The mechanism of frost heaving, *J. Geol.* 38:303-317 (1930).

Takagi, S., Fundamentals of the theory of frost heaving, In *Permafrost, Proc. 1st Int. Conf.*, Washington, D.C. Natl. Res. Coun., 203-216 (1963).

Takagi, S., Principles of frost heaving, *Res. Rep. 140*, U.S. Army Cold Reg. Res. Eng. Lab., Hanover, NH (1965).

Takagi, S., An analysis of ice lens formation, *Water Resour. Res.* 6:736-749 (1970).

Takagi, S., Fundamentals of ice lens formation, Paper 28, *15th Nat. Heat Transfer Conf.*, San Francisco, CA (1975).

Takagi, S., Segregation-freezing temperature as the cause of suction force, In *Proc. Int. Symp. Frost Action Soils*, Div. Soil Mech., Univ. of Lulea, Sweden 1:59-66 (1977).

Takagi, S., Segregation freezing as the cause of suction force in ice lens formation, *Report 78-6*, U.S. Army Cold Reg. Res. Eng. Lab., Hanover, NH (1978).

Takagi, S., Segregation freezing as the cause of suction force for ice lens formation, *Eng. Geol.* 13:93-100 (1979).

Takagi, S., The adsorption force theory of frost heaving, *Cold Reg. Sci. and Tech.* 3:57-81 (1980).

Takashi, T., T. Ohrai, H. Yamamoto, and J. Okamoto, Upper limit of heaving pressure derived by pore water pressure measurements of partially frozen soil, *Proc. 2nd Int. Symp. Ground Freezing* 713-724, Trondheim (1980).

Taylor, A. W., Concentration of ions at the surface of clays, *J. Am. Ceram. Soc.* 42:182-184 (1959).

Taylor, G. S., and J. N. Luthin, A model for coupled heat and moisture transfer during soil freezing, *Can. Geotech. J.* 15:548-555 (1978).

Teubner, M., and J. Frahm, The electric double layer around axisymmetric particles for small potentials calculated by the method of images: 1. Charged prolate ellipsoids of revolution, *J. Colloid Interface Sci.* 82:560-568 (1981).

Touillaux, R., P. Salvador, C. Vandermeersche, and J. J. Fripiat, Study of water layers adsorbed on Na- and Ca-montmorillonite by the pulsed nuclear magnetic resonance technique, *Israel J. Chem.* 6:337-348 (1968).

Van Loon, W. K. P., I. A. van Haneghem, and H. P. A. Boshoven, Thermal and hydraulic conductivity of unsaturated frozen sands, In *Proc. 5th Intl. Symp. on Ground Freezing*, p. 81-89, Balkema, Rotterdam (1988).

Verwey, E. J. W., Charge distribution in the water molecule and the calculation of the intermolecular forces, *Rec. Trav. Chim. Pays-Bas* 60:887-896 (1941).

Verwey, E. J. W., and J. T. G. Overbeek, *Theory of the Stability of Lyophobic Colloids*, Elsevier, New York (1948).

Vignes, M., and K. M. Dijkema, A model for the freezing of water in a dispersed medium, *J. Colloid Interface Sci.* 49:165-172 (1974).

Vignes-Alder, M., On the origin of the water aspiration in a freezing dispersed medium, *J. Colloid Interface Sci.* 60:162-171 (1977).

Wada, K., The distinctive properties of Andosols, *Adv. Soil Sci.* 2:173-229, Springer-Verlag (1985).

Wada, K., and M. E. Harward, Amorphous clay constituents of soils, *Adv. Agron.* 26:211-260 (1974).

Wada, S., and K. Wada, Fine configuration and structure of allophane, *20th Discuss. Clay Sci. Jpn.* (1976). (Japanese)

Weissmann, M., and N. V. Cohan, Molecular orbital study of the hydrogen bond in ice, *J. Chem. Phys.* 43:119-123 (1965).

Williams, P. J., Properties and behavior of freezing soils, *Norwegian Geotech. Inst. Publ.*, no. 72 (1968).

Williams, P. J., Thermodynamic and mechanical conditions within frozen soils and their effects, In *Proc. 5th Int. Conf. on Permafrost*, p. 493-498, Tapir Publ. (1988).
Woessner, D. E., An NMR investigation into the range of the surface effect on the rotation of water molecules, *J. Magn. Reson.* 39:297-308 (1980).
Yong, R. N., Soil suction effects on partial freezing, *Highw. Res. Board, Res. Rec.* 68:31-42 (1965).
Yong, R. N., On the relationship between partial soil freezing and surface forces, In *Phys. Snow Ice, Proc. Int. Confr. Low Temp. Sci.*, 1375-1385 (1966).
Yoshida, T., and S. Iwata, Consideration on pH of soil solution, *Proc. Jap. Soil Sci. Soc. Conf.* (1970). (Japanese)
Yoshinaga, N., Chemical composition and some thermal data of eighteen allophanes from Ando soil and weathered pumices, *Soil Sci. Plant Nutr. Tokyo* 12:47-54 (1966).

3

INTERACTION BETWEEN PARTICLES THROUGH WATER

3.1 DLVO THEORY

The DLVO theory is a quantitative description of interaction of clay particles. The basic concepts are the mutual repulsion due to the interaction of overlapping electric double layers and the attraction due to the London-van der Waals forces. This theory was proposed independently by Derjaguin and Landau (1941) and by Verwey and Overbeek (1948). The Derjaguin and Landau method uses force considerations to obtain the repulsive force, whereas that of Verwey and Overbeek uses energy methods; but the results are essentially and physically identical. As the method employed by Derjaguin and Landau is generally understood more easily, we will use it to introduce the DLVO theory. In addition, we assume that colloidal particles are planar particles, because most clay minerals are plate-shaped and the application of the theory for planar surfaces can readily be understood. For details, the reader is referred to *Theory of the Stability of Lyophobic Colloids* by Verwey and Overbeek (1948).

3.1.1 Potential Energy Due to the Interaction of Two Double Layers

If two plate-shaped clay particles are brought so close together that the diffuse ion parts of their double layers overlap, a force of repulsion acts between these two plates. Consequently, work must be performed to bring the two particles together, and a potential energy exists due to the repulsive force. If the repulsive force as a function of the distance 2d between two plates is given, the potential energy, V_R (ergs/cm^2), can be obtained by the equation

$$V_R = -2 \int_{-\infty}^{d} P \, dz \tag{3.1}$$

where P denotes the pressure on the plane midway between the two plates or the external pressure added to the outside surfaces of the two plates.

Repulsive Force Acting Between Two Plates

Assume that two infinite parallel plates with negatively charged surfaces are placed in a solution facing each other as shown in Fig. 3.1. The two plates are balanced by an outside force, P_d (dyne/cm^2). We assume that the solution is ideal and contains a single symmetrical electrolyte, that the number of ions per cm^3 in the solution is $2n_0$, and that the electric potentials at the surfaces of the particles and at the midway plane between plates are ψ_0 and ψ_d, respectively. Then, according to Eq. (1.28), the chemical potential of water at point A in the outer solution (Fig. 3.1), μ_A, is expressed by

$$\mu_A = \mu_0 + Z_A g = -\frac{RT}{M}\sum x_i + Z_A g = -\frac{RT}{M}\left(\frac{2n_0}{1/M} \times \frac{1}{N}\right) + Z_A g \tag{3.2}$$

$$= -2kTn_0 + Z_A g$$

where μ_0 is the decrement in chemical potential due to ions, M the molecular weight of water, x_i the mole fraction of the ith ions, Z_A the distance from the standard height to A, and N is Avogadro's number.

The chemical potential at point B on the central plane between plates (Fig. 3.1), μ_B, is given by Eq. (1.66) (where $\bar{v}_w$ is the partial volume of water).

$$\mu_B = \mu_f + \mu_0 + \int_0^{P_{in}} \bar{v}_w \, dP + Z_B g$$

Assuming that μ_f is negligible and $\bar{v}_w$ is not dependent on pressure and is 1.0, the equation becomes

$$\mu_B = \mu_0 + P_{in} + Z_B g \tag{3.3}$$

As $P_{in} = P_d$ and the concentrations of cations and anions at point B, n^+ and n^-, are expressed by

$$n^+ = n_0 e^{-e\psi_d/kT} \qquad \textit{and} \qquad n^- = n_0 e^{+e\psi_d/kT}$$

Eq. (3.3) becomes

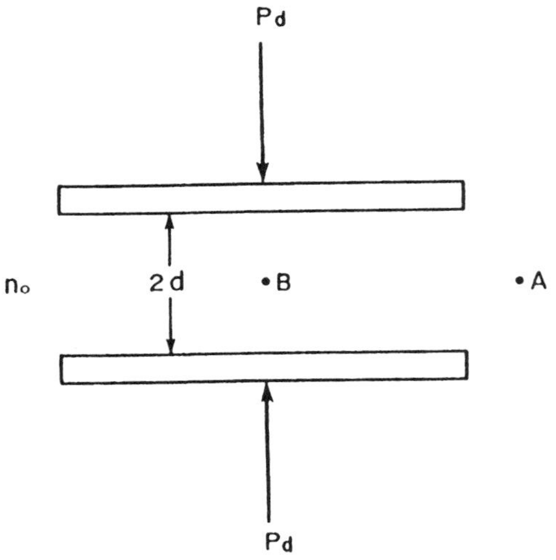

Fig. 3.1. Schematic diagram of the repulsive force acting between two charged plates.

$$\mu_B = -kT(n^+ + n^-) + P_d + Z_B g$$
$$= kTn_0\left(e^{-e\psi_d/kT} + e^{+e\psi_d/kT}\right) + P_d + Z_B g \qquad (3.4)$$

where k is the Boltzmann constant and T is the absolute temperature. When the system is at equilibrium, μ_A is equal to μ_B. Then

$$-2kT_0 + Z_A g = -kTn_0\left(e^{-e\psi_d/kT} + e^{+e\psi_d/kT}\right) + P_d + Z_B g \qquad (3.5)$$

Considering $Z_A = Z_B$, we arrive at

$$-2kT_0 = -kTn_0\left(e^{-e\psi_d/kT} + e^{+e\psi_d/kT}\right) + P_d \qquad (3.6)$$

Rearranging Eq. (3.6), we arrive at

$$P_d = n_0kT\left(e^{-e\psi_d/kT} + e^{+e\psi_d/kT} - 2\right) = 2n_0kT\left(\cosh\frac{e\psi_d}{kT} - 1\right) \qquad (3.7)$$

As P_d can be considered equal to the repulsive force between the plates, Eq. (3.7) shows the repulsive force when the distance between plates is 2d. P_d has been named the disjoining pressure (e.g. refer to Derjaguin et al, 1964; Derjaguin and Churaev, 1974, 1977, 1978). The disjoining pressure P_d is a function of d and defined by

$$P_d(d) = \frac{\mu_f - \mu_0}{\bar{v}}$$

Here μ_f is the chemical potential of water in the solution between the plates, assuming that P_d vanishes, but other state variables such as the concentrations of cations and anions remain unaltered; μ_0 is the chemical potential of the bulk liquid; and $\bar{v}$ is the partial volume of water in solution.

The concept of disjoining pressure has also been used to investigate the thickness of the wetting film formed when a gas bubble is pressed against a smooth surface. The disjoining pressure is defined in the way described above.

Potential Energy Due to the Repulsive Force

We can obtain the expression for the potential energy due to the repulsive force by substituting Eq. (3.7) into Eq. (3.1).

$$V_R = -2\int_{\infty}^{d} P\,dz = -2\int_{\infty}^{d} \left[2n_0kT\left(\cosh\frac{e\psi_d}{kT} - 1\right)\right] dz \tag{3.8}$$

In addition, it is important to understand whether the particles under consideration satisfy the condition of constant surface potential or constant surface charge before integrating Eq. (3.8). The former condition means that the electric potential at the surface of each particle is kept constant regardless of the distance between the plates, but the amount of charge at the surface depends on the distance. In the latter, the potential varies with the distance between the plates. The condition of constant surface charge is satisfied in surfaces having permanent charges, such as flat surfaces of montmorillonite. On the other hand, the assumption of constant surface potential has been considered to be valid for the interaction between particles whose surface charge density is defined by the concentration of a potential-determining ion in the equilibrium solution, for example, the edges of kaolinite or allophane particles. In this case, H^+ is the potential-determining ion.

Constant Potential When the interaction between the plates is small, we can assume, as a first approximation, that the electric potential between the plates, ψ, is simply the sum of the electric potentials due to the two unperturbed ion layers (Fig. 3.2). That is,

$$\psi(z) = \psi'(z) + \psi'(2d - z) \qquad (z \leqq d) \tag{3.9}$$

where 2d is the distance between plates, z the distance from one of the plates, and ψ' the electric potential when one of the plates does not exist. Consequently, denoting ψ at z = d as ψ_d, we have

$$\psi_d = 2\psi'(d)$$

If Eq. (2.21) is adopted as $\psi'(d)$, Eq. (3.8) becomes

$$V_R^{\psi} = -2\int_{\infty}^{d} P\, dz = -2\int_{\infty}^{d} \left\{ 2n_0 kT \left[\cosh \frac{e\psi_0^{\infty}\exp(-\kappa z)}{kT} - 1) \right] \right\} dz \tag{3.10}$$

where ψ_0^{∞} is the electric potential at the particle surface when the particle separation is infinite and κ is the Debye κ (see Sec. 2.3.2). Expanding $\cosh[e\psi_0^{\infty} \exp(-\kappa z)/kT]$ in a power series

$$\cosh(x) = 1 + \frac{x^2}{2!} + \frac{x^2}{4!} + \dots .$$

and considering

$$\frac{e\psi_0^{\infty}\exp(-\kappa z)}{kT} << 1$$

we obtain

$$V_R^{\psi} = -2\int_{\infty}^{d} 2n_0 kT \left(\frac{e\psi_0^{\infty}}{kT} e^{-\kappa z} \right)^2 dz = 4\, \frac{n_0 e^2 \left(\psi_0^{\infty}\right)^2}{kT} \int_{d}^{\infty} e^{-2\kappa z}\, dz$$

$$= 2\, \frac{n_0 e^2 \left(\psi_0^{\infty}\right)^2}{kT\kappa}\, e^{-2\kappa d} \tag{3.11}$$

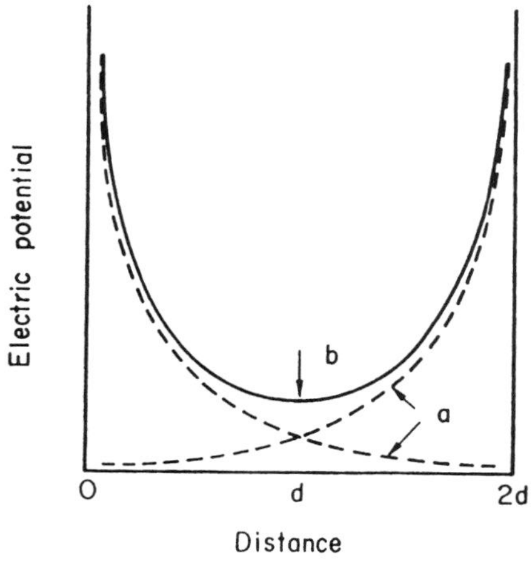

Fig. 3.2. Approximate electric potential between two interacting double layers: (a) single plate; (b) interacting plates.

This equation may be applied approximately to the repulsive interaction between the edges of kaolinite.

In addition, Derjaguin (1939) introduced an approximate formula for the potential energy of interaction of two identical spherical colloidal particles at constant potential assuming that ψ_0 is not too large,

$$V_R^{\psi} = \frac{\epsilon a(\psi_0)^2}{2} \ln[1 + \exp(-\kappa H)] \tag{3.12}$$

where ε is the dielectric constant of the solution, a the radius of the spherical particle, and H the minimum separation of the particle surfaces. Equation (3.12) may be used to calculate approximately the potential energy of interaction between the double layers of two allophane particles.

Constant Charge The electric potential at the surface of charged plates varies with distance between the plates under the condition of constant surface charge. Therefore, Eq. (3.1) cannot be integrated easily. Gregory (1973) has derived the following approximate expression for the repulsive energy, V_R^{σ}, between two flat, diffuse ion layers. In the case where $e\psi_d/kT$ is small,

$$V_R^{\sigma} = \frac{2n_0 kT}{\kappa} (y_0^{\infty})^2 \left(\coth \frac{\kappa D}{2} - 1\right) \tag{3.13}$$

where D is the distance between the plates. When $e\psi_d/kT$ is not small, V_R^σ is

$$V_R^\sigma = \frac{2n_0kT}{\kappa}\left\{2y_0^\infty \ln\left[\frac{B + y_0^\infty \coth(\kappa D/2)}{1 + y_0^\infty}\right] - \ln[(y_0^\infty)^2 + \cosh(\kappa D) + B\sinh(\kappa D) + \kappa D]\right\} \tag{3.14}$$

where $y_0^\infty = e\psi_0^\infty/kT$ and $B = \sqrt{1 + (y_0^\infty)^2 \operatorname{csch}^2(\kappa D/2)}$. Equation (3.14) can be applied approximately to the repulsive potential energy of interaction between flat surfaces of montmorillonite particles.

In addition, expressions for repulsive energy at constant potential or constant charge under various conditions have been given by many researchers (Verwey and Overbeek, 1948; Derjaguin, 1954; Hogg et al., 1966; Jones and Levine, 1969; McCartney and Levine, 1969; Bell et al., 1970; Honig and Mul, 1971; Usui, 1973; Barouch et al., 1978; McQuarrie et al., 1980; Smith and Deen, 1980; James and Williams, 1981, 1985; Ruckenstein, 1981; Wilemski, 1982; Bentz, 1982; Smith and Deen, 1983; Bratko, 1988; Ohshima and Kondo, 1988a,b; Ruckenstein and Krape, 1990; Sengupta and Papadopoulos, 1992).

Differences Between V_R^σ and V_R^ψ

Frens and Overbeek (1972) give an equation that relates the energy of repulsion between two flat plates at constant charge to that at constant potential.

$$V_R^\sigma = V_R^{y_0^D} + \frac{8n_0kT}{\kappa}\left[y_0^D - y_0^\infty \sinh\frac{y_0^\infty}{2} - 2\cosh\frac{y_0^D}{2} - \cosh\frac{y_0^\infty}{2}\right] \tag{3.15}$$

where y_0 represents $e\psi_0/kT$, ψ_0 is the surface electric potential of the plates, y_0^∞ the value of y_0 when the plate separation is infinite, y_0^D the corresponding value when the distance between the plates is D, and $V_R^{y_0^D}$ the repulsive energy at constant surface potential when the distance is D. This equation shows that for large particle distances the results are not much different for the two cases, but the constant charge repulsion energy becomes very much larger as the particles come closer together.

According to Wiese and Healy (1970), an approximate formula for potential energy of repulsion for interaction between two identical spherical particles is given by

$$V_R^\sigma = V_R^{\psi_0^\infty} - \frac{\epsilon a(\psi_0^\infty)^2}{2}\{\ln[1 - \exp(-2\kappa H)]\} \tag{3.16}$$

where $V_R^{\psi_0^\infty}$ is the repulsion energy between two particles of constant surface potential, ψ_0^∞, and is expressed by Eq. (3.12). Equation (3.16) indicates also that at close approach the repulsive potential energy at constant charge is much larger than that at constant potential. Figure 3.3 shows the change of V_R^σ and V_R^ψ as a function of D or H.

Factors Determining the Repulsive Potential Energy

As understood from Eqs. (3.11), (3.12), (3.13), and (3.14), V_R^σ and V_R^ψ are not only functions of D or H but of T, n_0 (the ion concentration in the equilibrium solution), κ, and ψ_0^∞. On the other hand, κ is a function of T, n_0, ε (the dielectric constant of water), and ν (the valence of the counter ions), and ψ_0^∞ is a function of T, n_0, and σ^∞ (the surface charge density when the particle separation is infinite), as found in Eq. (2.23). In addition, σ^∞ of the edges of kaolinite or allophane particles is defined by the pH in the equilibrium solution. Consequently, V_R^σ and V_R^ψ depend on T, n_0, ν, and σ^∞, ε being dependent only on T. Since V_R^σ and V_R^ψ depend in a qualitatively similar way on these factors, denoting V_R^σ or V_R^ψ as V_R, the dependences are summarized as follows:

1. $V_R(d,T_1) > V_R(d,T_2)$ when $T_1 > T_2$.
2. $V_R(d,n_0^1) > V_R(d,n_0^2)$ when $n_0^2 > n_0^1$.
3. $V_R(d,\nu_A) > V_R(d,\nu_B)$ when $\nu_B > \nu_A$.
4. $V_R(d,\sigma_1^\infty) > V_R(d,\sigma_2^\infty)$ when $\sigma_1^\infty > \sigma_2^\infty$.

Charge-Regulation Model for Particles with Variable Charges

Following Verwey and Overbeek (1948), the interaction between two particles with variable charge has been based on the condition of constant surface potential. This assumption has been shown not to be realistic for oxides. The behavior of many colloidal particles with variable charge falls between the extremes of constant surface charge and constant surface potential.

The condition of constant surface charge applies when the time scale of particle interaction is too short for the surface charge to adjust. If desorption of

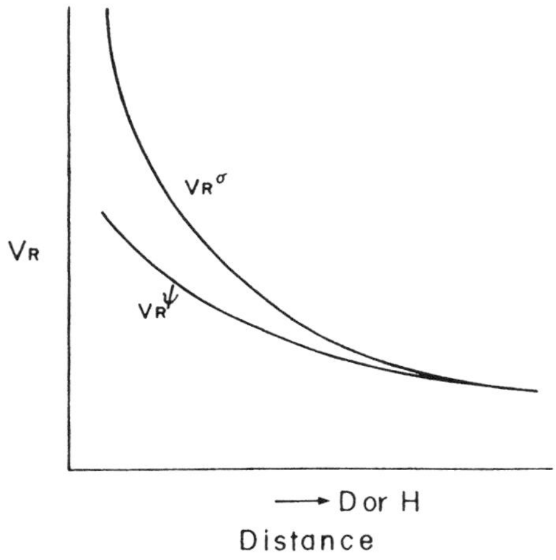

Fig. 3.3. Schematic representation of the relation between potentials at constant charge V_R^{σ} and at constant potential V_R^{Ψ}.

surface charge is fast compared to the time scale of particle interaction, equilibrium between the surface charge and ions in solution will be maintained. If the time scales of interaction and surface charge relaxation are of the same order, the potential is defined by a charge regulation mechanism. The double layer is always in equilibrium, and the surface charge as well as the surface potential adjusts during particle interaction. The mechanism may produce large differences in colloidal stability compared with the stability predicted by assuming either constant surface charge or constant surface potential (Prieve, 1977; Healy et al., 1980). Various regulation models have been proposed (e.g. Van Riemsdijk et al., 1988; Ohshima and Kondo, 1988a,b; Krozel and Saville, 1992).

As described in Section 2.3.2, Chan et al. (1984), Horikawa et al. (1988), and Miller and Low (1990) reported that the surfaces of homoionic montmorillonite appear to behave more like constant surface potential than constant surface charge models. According to the Gouy theory, the surface potential should be a strong function of electrolyte concentration. Further investigation is needed on the Stern-layer as a store for charges and on the position of the zeta potential plane in these regulation models.

3.1.2 Potential Energy Due to van der Waals Forces Between Two Particles

Attraction Energy Due to van der Waals Forces

According to London's theory, the potential of the van der Waals forces, ψ, between two atoms is given by

$$\psi = -\frac{C}{r^6} \tag{3.17}$$

where r is the distance between the atoms and C is expressed by

$$C_L = \frac{3}{2}\alpha_1\alpha_2\frac{h\lambda_1\lambda_2}{\lambda_1 + \lambda_2} = \frac{3}{2}\alpha_1\alpha_2\frac{I_1 I_2}{I_1 + I_2} \tag{3.18}$$

in which the subscripts represent atoms 1 and 2, α is the polarizability, h is Planck's constant, λ is the characteristic frequency, and I is the energy of ionization. Slater and Kirkwood (1931) introduced the following expression for C in Eq. (3.17):

$$C_{SK} = \frac{3ek}{4\pi\sqrt{m}}\frac{\alpha_1\alpha_2}{(\alpha_1/N_2)^{1/2} + (\alpha_2/N_2)^{1/2}} \tag{3.19}$$

where e is the elementary electric charge, m the mass of the electron, k the Boltzmann constant, and N the number of electrons in the outermost shell.

Differentiating Eq. (3.17) with r, we find for the force f,

$$f = \frac{6C}{r^7} \tag{3.20}$$

On the other hand, if the atoms are an appreciable distance apart, the time taken for the electric field from one instantaneous dipole to interact with a neighbor may be comparable with the lifetime of the dipole. In this case the interactions are out of phase. Casimir and Polder (1948) have indicated that the force is proportional to r^{-8} for such retarded van der Waals forces.

Flat Plates The attraction potential V_A between two plates is calculated from the summation of interatomic forces. For two plates of thickness δ at a distance 2d from each other, the force normal to the plates is

$$V_A = -\frac{A}{48\pi}\left[\frac{1}{d^2} + \frac{1}{(d+\delta)^2} - \frac{2}{(d+\delta/2)^2}\right] \tag{3.21}$$

where A is the Hamaker constant given by

$$A = \pi^2 q^2 C \tag{3.22}$$

where q is the number of atoms contained in the unit volume of the substance that makes up the particles. When $\delta >> d$, Eq. (3.21) becomes

$$V_A = -\frac{A}{48\pi}\frac{1}{d^2} = -\frac{A}{12\pi}\frac{1}{D^2} \tag{3.23}$$

where D is the distance between two plates ($D = 2d$).

Differentiating Eq. (3.21) with d and Eq. (3.23) with D, we obtain

$$F_A = \frac{A}{24\pi}\left[\frac{1}{d^3} + \frac{1}{(d+\delta)^3} - \frac{2}{(d+\delta/2)^3}\right] \tag{3.24}$$

$$F_A = \frac{A}{6\pi}\frac{1}{D^3} \tag{3.25}$$

where F_A represents the mean attractive force per unit area. In addition, when the force is retarded, the attractive force between two plates is expressed by

$$F_A = \frac{B}{D^4} \tag{3.26}$$

in which B is the appropriate constant for the given material (the retardation constant). Equations (3.24) and (3.25) may be applied when D or 2d is less than 100 to 200 Å and Eq. (3.26) is supposed to be valid at larger distances (Kitchener and Prosser, 1957; Derjaguin and Abrikosova, 1958; Tabor and Winterton, 1968).

Equal Spherical Particles Hamaker (1937) introduced an expression for potential energy between two equal spherical particles of the same material due to van der Waals forces:

$$V_A = -\frac{A}{6}\left(\frac{2}{S^2-4} + \frac{2}{S^2} + \ln\frac{S^2-4}{S^2}\right) \tag{3.27}$$

where $A = \pi^2q^2C$ and $S = R/a = 2 + H/a$; q is the number of atoms per unit volume in the particles, C is the same as C in Eq. (3.17), R is the distance between the centers of the spheres, a is their radius, and H is the shortest distance between their surfaces.

Schenkel and Kitchener (1960) derived two equations for V_A for equal spheres, corrected for the effects of retardation, which are valid at different separations. When H > 150 Å,

$$V_A = \frac{Aa}{\pi}\left(\frac{2.45\lambda}{120H^2} - \frac{\lambda^2}{1045H^3} + \frac{\lambda^3}{5.62 \times 10^4 H^4}\right) \tag{3.28}$$

where λ is the wavelength of the London frequency. This equation is accurate to within 5%. When H < 150 Å,

$$V_A = -\frac{Aa}{12H}\frac{\lambda}{\lambda + 3.54\pi H} \tag{3.29}$$

In addition, Wiese and Healy (1970) have introduced equations of potential energy for two spheres of unequal radii.

Hamaker Constant

If we can determine the value of the Hamaker constant for the system under consideration, it is possible to calculate V_A. This definition, $A = \pi^2q^2C$, is applicable only when the medium between the particles is a vacuum. For a substance other than a vacuum, A should be redefined as

$$A = A_{11} + A_{22} - 2A_{12} \tag{3.30}$$

where A_{11}, A_{22}, and A_{12} are the respective Hamaker constants for particle-particle, water-water, and particle-water interactions. In general, A_{12} is approximated by $(A_{11} \cdot A_{22})^{1/2}$.

Tabor and Winterton (1968, 1969) and Israelachvili and Tabor (1972) measured $(1.07 \pm 0.05) \times 10^{-12}$ erg and $(1.35 \pm 0.15) \times 10^{-12}$ erg, respectively, as the Hamaker constant for mica/air/mica. It may be reasonable to assume that A_{11} for clay is approximately equal to that for mica/air/mica. According to Verwey and Overbeek (1948), the Hamaker constant, A_{22}, for water-water interactions becomes 6×10^{-13} erg; Tabor and Winterton (1969) have obtained 3.4×10^{-13} erg as A_{22}. If A_{22} is 6×10^{-13} erg, the Hamaker constant, A, for clay/water/clay becomes about 1×10^{-13} erg. Assuming that A_{22} is 3.4×10^{-13} erg, A is about 2.0 $\times 10^{-13}$ erg. Israelachvili and Adams (1978) measured 2.2×10^{-13} erg as the

Hamaker constant for mica/aqueous electrolyte solution (10^{-1} to 10^{-3} mol KNO_3)/mica.

The value of the retardation constant, B, for clay/air/clay is considered to be about 10^{-19} erg cm from the experimental values shown in Table 3.1. In addition, appropriate formulas for retarded van der Waals interaction under various conditions have been proposed by several researchers (Czarnecki, 1979; Gregory, 1981; Pailthorpe and Russel, 1982).

3.1.3 Total Potential Energy

The total potential energy for two charged particles, V_t, is the sum of the repulsive potential energy V_R and the attractive potential energy V_A.

$$V_t = V_A + V_R \tag{3.31}$$

It should be noted that V_A is always negative and V_R is always positive.

V_t is also a function of d, n_0, ν, δ, and T, because V_R depends on d, n_0, ν, δ, and T, as described above. Usually, in studying dispersion and flocculation of colloidal particles, it is very important to know the change of V_t as a function of d corresponding to the magnitude of n_0. The values of V_t as a function of d corresponding to the magnitude of n_0 are shown schematically in Fig. 3.4. Curve (a) in Fig. 3.4 is produced when n_0 is very large. Then the diffuse ion layers are very thin and the magnitude of V_R is smaller than V_A for any value of d. Consequently, V_t is always negative. Curve (c) appears when n_0 is very small. Then the diffuse ion layers are thick and V_R becomes larger than V_A except in the region where d is very small.

In the intermediate region of n_0, a curve of type (b) is seen, with a maximum and two minima. In general, the primary minimum at a small separation is very deep, and the secondary minimum is shallow. The position and magnitude of the maximum and of the secondary minimum depend on the relative values of V_R and V_A. The difference between the maximum value and the secondary minimum value, V^*_{max}, is an energy barrier obstructing flocculation (coagulation), which occurs at the primary minimum at small separation.

Table 3.1. Values of the retardation constant.

Substance	B (erg cm)	Reference
Quartz	$1 - 2 \times 10^{-19}$	Derjaguin and Abrikosova (1958)
Glass	1.1×10^{-19}	Kitchener and Prosser (1957)
Mica	$(0.81 \pm 0.04) \times 10^{-19}$	Tabor and Winterton (1968, 1969)

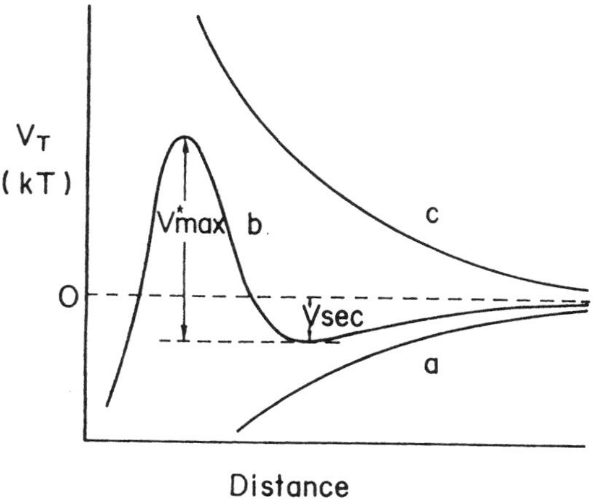

Fig. 3.4. Total potential energy for different component energy values.

Since force is the negative derivative of the potential, we can read from curve (a) in Fig. 3.4 that the force between particles is attractive for all interparticle distances. Accordingly, when two particles are brought together by Brownian motion, they form a lasting combination. The flocculation due to the potential of the type (a) curve is named fast flocculation. If the potential between two particles belongs to the type (c) curve, particles are permanently in a dispersed condition. When the potential curve belongs to type (b), a weak attractive force is active between the particles when the distance changes from infinite to the distance at the secondary minimum. Between the secondary minimum and the maximum, a repulsive force is found, and at still smaller distances, interparticle attraction occurs. In case (b) when one particle collides with another, a lasting combination between the particles is not necessarily formed. If the magnitude of potential energy at the secondary minimum V_{sec} is smaller than the average kinetic energy of molecules of the solvent, kT, and V^*_{max} is larger than kT, the probability of a lasting combination is small.

3.1.4 Other Forces of Interaction Between Charged Surfaces

Swelling Clays

Norrish and Quirk (1954) and Norrish (1954a,b) measured the dependence of distance between two sheets on concentration of the outer solution using montmorillonite saturated with various cations. Table 3.2 and Fig. 3.5 show part of their results.

Table 3.2. Distance (Å) between two sheets of montmorillonite in various salt solutions. (From Norrish, 1954b)

Concentration (normality)	NH_4Cl	KCl	NaCl	$MgCl_2$	$CaCl_2$
8.0				(6.6)[a]	5.7
4.0	(4.6)[a]	12.7	5.8		
2.0					9.1
1.0			9.1		
0.5			9.3		
0.31			30.4		
0.2	5.4[b]	5.4[b]			9.4
H_2O	5.4[b]	5.4[b]		9.6	9.4

[a]Apparent distance associated with an irrational (001) series
[b]Distance depends on history of the montmorillonite

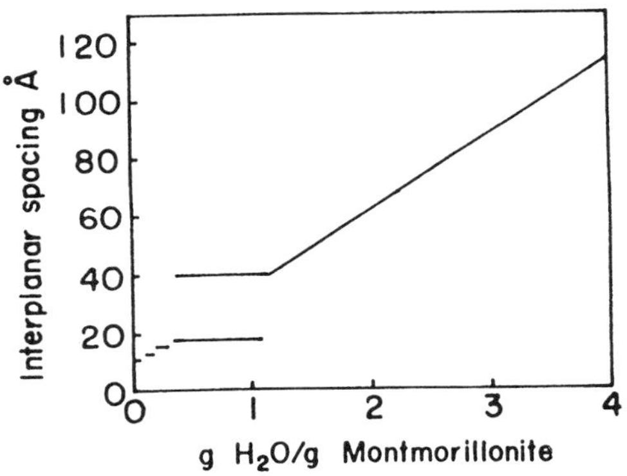

Fig. 3.5. Swelling of Na montmorillonite. Stage 2 swelling occurs above 100% water. (From Norrish, 1954b)

Following Norrish (1972), the swelling is divided into three stages. In the first stage the distance between two sheets ranges between 0.4 and about 10 Å, increasing stepwise as the clay takes up water. The interparticle distance at the end of this stage depends on the cation exchange capacity of the clay and the species of cation in the interlayer (see Section 3.3.1).

In the second stage, for monovalent cations with relatively small ion radius (Na, Li, and H), the distance increases spontaneously beyond 10 Å when the concentration of the outer solution decreases below a certain value. The distance for divalent or trivalent ions remains even if the sheets are immersed in water. In Li-vermiculite, the jump begins at a lower concentration than for Li-montmorillonite (0.15 N compared to 0.4 N) (Aylmore and Quirk, 1959). This may be due to its higher surface charge density (143 meq per 100 g). In this stage, if particles are in parallel arrangement, the repulsive force due to diffuse ion layers is dominant. When the concentration of the outer solution is near 0.2 to 0.3 N, van der Waals attractive forces have to be included.

Clay particle separation at the third stage is limited only by the volume of water available. The sheets may associate due to their thermal motion, forming edge-to-edge or edge-to-face bonds. Although these bonds are weak, they influence such properties as viscosity and thixotropy (Van Olphen, 1954).

The dependence of distance between two sheets on concentration of the outer solution can be described using the DLVO theory for the second stage, but not for the first stage. The distance between two sheets in the first stage changes by jumps. For example, the jump for Na-montmorillonite is from 5.8 to 9.1 Å, and for Ca-montmorillonite from 5.7 to 9.1 Å (Table 3.2). In addition, Pennino et al. (1981) found that the platelets can stack into stable aggregates with water interlayers of typical thickness 2.5 and 5.5 Å between them. It should be noted that the magnitude of the jumps is roughly equal to the diameter of a water molecule. The jump in distance is also observed in the change from the first stage to the second stage.

Discrete Charges

The jumps suggest the breakdown of continuum theories such as electrostatic double-layer forces and van der Waals forces. Ions and water molecules existing in the interlayer cannot be regarded as solution, they must be treated as discrete ions and water molecules with definite sizes.

The repulsive force between two double layers in the DLVO theory is calculated on the basis of ions regarded as point charges. The DLVO theory is, therefore, valid only where the distance between particles is so large that the magnitudes of the ions can be neglected, larger than 30 to 40 Å. The concept of surface charge density used in the DLVO theory was adopted from macroscopic electromagnetism. As described in Section 2.3.3, this concept is not valid for electric intensity or potential at a point where the distance from the charged surface is smaller than the average distance between adjacent charges on the surface. An example to clarify this limitation is shown in Fig. 3.6.

Assume that electrons are distributed on an infinite plane as shown in Fig. 2.21. Figure 3.6 shows a distribution of the electric intensities on the plane at distance 0.3a from the charged plane, where a is the distance between surface charges (Iwata, 1974). According to macroscopic electromagnetism, the intensity of the electric field in the space in contact with an infinite plane with uniform surface charge density σ is uniform and given as $2\pi\sigma/\varepsilon$. On the other hand, from Figs. 2.22 and 3.6 it is clear that the electric intensity within distance a from the charged surface is nonuniform vertically and horizontally. The repulsive force between two plates is calculated on the assumption that the electric intensity at the midplane between two plates is zero. Therefore, the horizontal nonuniformity in the electric intensity may prevent application of the DLVO theory to a system in which the distance between two plates is smaller than 2a. In addition, a repulsive force due to interaction between particles and interlayer water molecules must be taken into consideration. For example, this repulsive force would be present for montmorillonite when the distance between sheets is less than 100 Å.

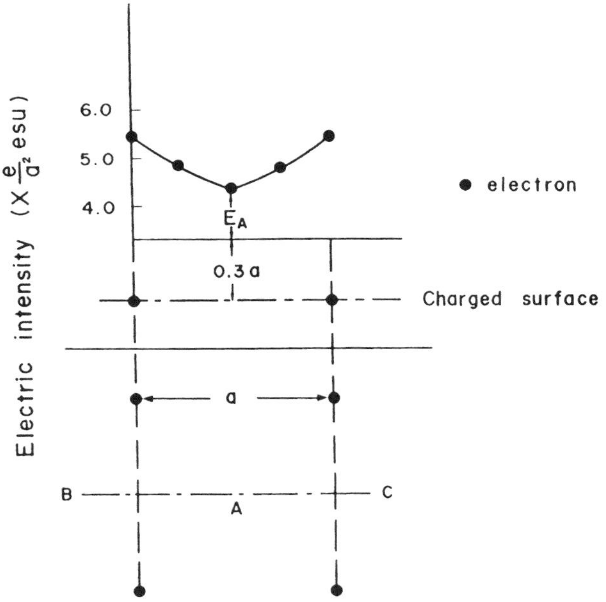

Fig. 3.6. Distribution of electric intensities along line BC on a plane at distance 0.3a from the charged plane. (From Iwata, 1974)

Forces Between Mica Surfaces

Israelachvili and Pashley (1983) measured forces at very small distances between two curved mica surfaces in 10^{-3} N KCl solution at pH 5.5 (Fig. 3.7). The ordinate is F/R, where F is the force and R is the radius of curvature. The dashed line in the main figure and the solid line in the insert indicate the theoretical electrostatic double layer force for a surface potential of -78 mV. Diffuse electric double-layer theory accurately predicts the repulsive force at separations >3 nm but not at closer distances. According to the DLVO theory, the attractive van der Waals force exceeds the repulsive double-layer force at separations smaller than a certain value. However, in Fig. 3.7, the net force is repulsive even at separations smaller than 1.5 nm. Such phenomena have been observed for various materials, e.g. Rabinovich et al. (1982) for a glass fiber; Horn et al. (1989) for silica; Pashley and Quirk (1989) for muscovite mica.

Israelachvili (1978) and Israelachvili and Adams (1978) found that four different forces operate from 200 nm down to contact separations. The first are van der Waals forces, which are nonretarded up to about 6.5 nm with a Hamaker constant of 2.2×10^{-13} erg and appear to be slightly stronger than expected from the Lifshitz theory. Second, there are repulsive double-layer forces. In 10^{-4} mol dm^{-3} 1:1, 2:1, and 2:2 electrolytes, repulsive forces are well accounted for by the theory, but at concentrations above 10^{-3} mol dm^{-3} there are discrepancies with the theory, especially for 2:1 electrolytes. The third are additional repulsive forces that decay exponentially with a characteristic decay length of about 1.0 nm. Their magnitude varies from mica to mica and is largely independent of ionic strength. They are negligible at separations above 7.5 nm. Fourth are complex adhesion forces that depend on a host of factors, such as pH, type of electrolyte cations, mica orientation, and so on, and are not simply given by extrapolating the long-range forces down to separations of the order of interatomic spacing. The additional repulsive forces have been named "hydration forces" (Israelachvili, 1978).

Pashley (1981a), who measured surface forces between mica sheets in Na^+ and K^+ aqueous solutions at various concentrations, suggested that when the counterions in the double layer next to the mica surface are hydrated at a certain separation and at a high surface charge, a hydration force would arise since further approach of surfaces would require removal of some fraction of the "hydration atmosphere" around these cations.

In Fig. 3.7, the force oscillates with mean periodicity 0.25 ± 0.03 nm. This periodicity is approximately equal to the diameter of the water molecules. In addition, there are three minima, at contact and at separations of 0.28 ± 0.03 and 0.56 ± 0.03 nm. These minima explain the existence of stable conditions in interlayer distances of around 0.28 and 0.56 nm described from swelling studies.

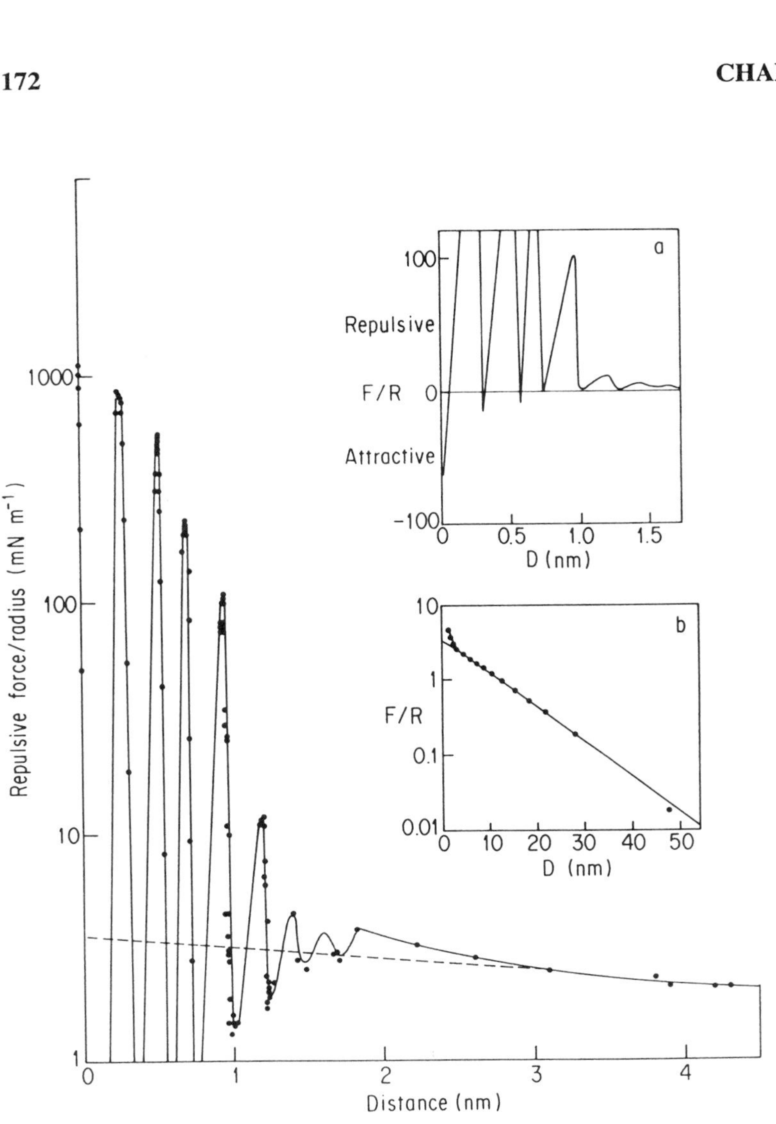

Fig. 3.7. Measured force between two curved mica surfaces of initial radius R = 1 cm as a function of distance D in 10^{-3} M KCl solution. (From Israelachvili and Pashley, 1983)

The hydration force, due to removal of some fraction of the "hydration atmosphere" around cations, would depend upon the specific ion on the surface. Pashley (1981a) found that replacement of surface K^+ ions with H^+ ions by lowering pH, completely removed the additional repulsive force (Fig. 3.8). The magnitude of repulsion increased with hydration of the bound cation, namely $Mg^{2+} > Ca^{2+} > Li^+ \sim Na^+ > K^+ > Cs^+$ (Israelachvili, 1985, 1991; Pashley and

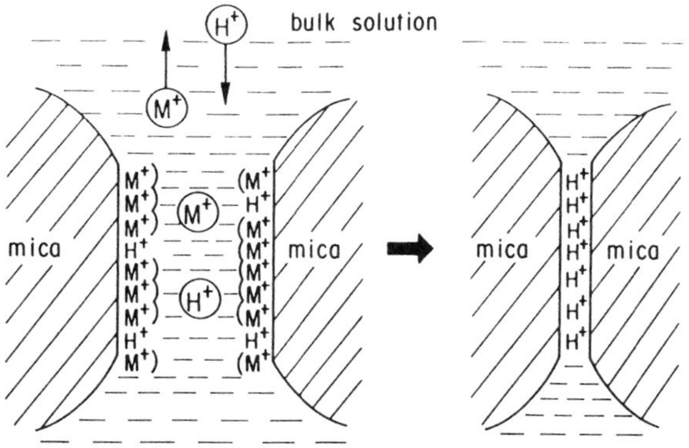

Fig. 3.8. Schematic diagram to illustrate the replacement of hydrated metal ions by hydrogen ions prior to the jump into contact in a primary minimum. (From Pashley, 1981a)

Israelachvili, 1984). No repulsive hydration forces were observed in acid solution, presumably because H^+ ions can penetrate into the mica lattice (Pashley, 1981b; Israelachvili, 1985). Israelachvili (1991) summarizes discussions on the magnitude of hydration forces and the role of protons in interparticle forces.

The swelling behavior (Table 3.2) suggests that the total potential energy between montmorillonite plates has a deep minimum (a strong attractive force) at a distance of 9.1 ~ 9.4 Å. The force measured for K-mica (Fig. 3.7) has a minimum at around 10 Å, while the force is repulsive. It is not clear whether this inconsistency is due to the difference in the counterions, the existence of an irreversibility (hysteresis) of the force which may appear in expanding and compressing processes, or other physical causes.

Other Forces

MacEwan (1948) assumed a repulsive force due to ion hydration rather than an osmotic repulsive force, and an electrostatic attractive force based on the model shown in Fig. 3.9 in addition to van der Waals forces. He considered that diffuse layers do not exist within surface separations of about 10 Å. The equations for the electrostatic attraction energy, E, which he introduced are

$$E = \frac{\sigma v e}{2d\epsilon} \quad erg\ cm^{-2} \qquad (3.32)$$

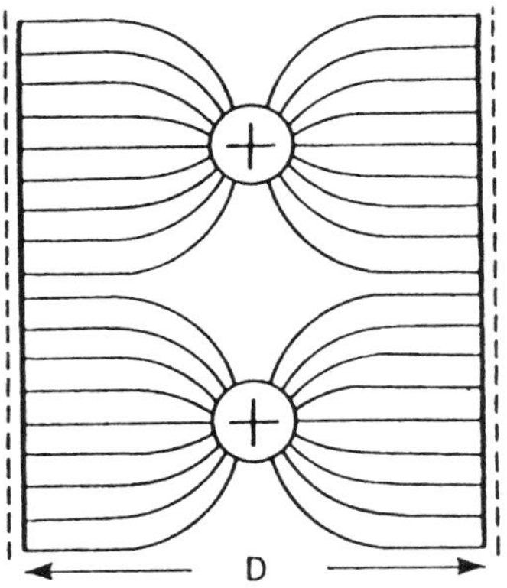

Fig. 3.9. Forces between cations and sheets of montmorillonite. (From Norrish, 1954b)

and

$$E = \frac{2\pi\sigma^2 d}{\epsilon} \qquad erg\ cm^{-2} \tag{3.33}$$

where d is a half distance between plates, σ the surface charge density, ν the valency of ions, e the elementary charge, and ε the dielectric constant of water. Pennino et al. (1981) investigated the dehydration of two montmorillonites saturated with 10 different cations at different conditions of temperature and relative humidity. They concluded that in a monoionic montmorillonite the cation can be treated as a hard sphere, and the interaction of the cation with water molecules and the silicate is mainly of an electrostatic nature.

Equation (3.32) is obtained by summation of the potential energy between each of the interlayer cations and each of the surface charges, assuming that they are point charges. Equation (3.33) is obtained by regarding the interlayer cations and the charged silicate as the plates of a condenser. The characteristics of this electrostatic force have been studied by several researchers (Jordine et al., 1962; Hurst and Jordine, 1964).

Kjellander and Marcelja (1986) and Kjellander et al. (1988) calculated double-layer interaction for electrolyte systems in clays using a model in which the ions are treated as charged, hard spheres and the solvent as a dielectric continuum. The method is based on integral equation theories of statistical mechanics, and makes possible the self-consistent, explicit calculation of anisotropic ion-ion correlation functions. The image charge interactions due to dielectric discontinuities at the surfaces are included. In contrast to predictions of the simple Poisson-Boltzmann theory, they found double-layer interaction for divalent ions were strongly attractive at around 1 nm (Fig. 3.10). The attraction is a consequence of the correlation between the ions, giving rise to an electrostatic fluctuation force. When the van der Waals force between the dielectric media is included, the attraction becomes even larger. Thus, the attractive force between particles saturated with a divalent ion gives rise to a stable state, a potential minimum at a distance of about 1 nm. Specific binding or hydration effects are not needed to explain the existence of the minimum.

Calculations by Kijlstra (1992) showed clearly how the electrostatic force decreases with polarizability of the particle, the decrease becoming larger with stronger double-layer overlap.

It is hoped that further investigation will establish a model that can consider all the forces acting between two clay plates at small distances.

3.2 TOTAL POTENTIAL ENERGY BETWEEN TWO CLAY PARTICLES

3.2.1 Montmorillonite

Taking a typical montmorillonite as an example, we will calculate total potential energy.

Repulsion Energy V_R

Repulsion energy due to the interaction of two diffuse ion layers can be obtained from the table, calculated without approximation, by Honig and Mul (1971). Repulsion energy at constant potential and constant charge are given as a function of κd and surface potential ψ_0^∞ when two plates are at infinite separation. κ is the Debye kappa, and d is the half distance between two plates. A part of the table is shown in Table 3.3. W^ψ and W^σ are given by the equation

$$W^\psi \times \frac{64 n_0 kT}{\kappa} = V_R^\psi$$

$$W^\sigma \times \frac{64 n_0 kT}{\kappa} = V_R^\sigma$$

where V_R^ψ and V_R^σ represent the repulsion energy at constant potential and constant charge, respectively.

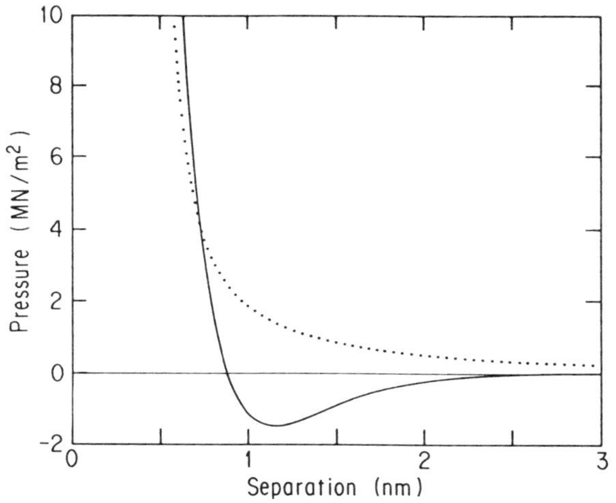

Fig. 3.10. The total pressure between two Ca-montmorillonite plates (one unit charge per 1.35 nm^2) immersed in pure water. (From Kjellander et al., 1988)

Table 3.3. Extract from the table of repulsion energy by Honig and Mul (1971).

	d	W^ψ	W^σ
$z = \frac{e\psi_0}{kT} = 2$	0.3	0.0873	0.1339
	0.6	0.0522	0.0641
	0.8	0.0365	0.0417
	1.0	0.0253	0.0277
	1.2	0.0174	0.0185
	1.4	0.0120	0.0125

From the surface charge density of Wyoming montmorillonite, 1.17 x 10^{-7} meq/cm^2, we can easily determine the concentration of the electrolyte for a given surface potential with Eq. (2.23). Consequently, we can also obtain the value of κ. Figure 3.11 shows the repulsion energy between two montmorillonite plates as a function of d at two concentrations of symmetrical electrolyte of valence 1.

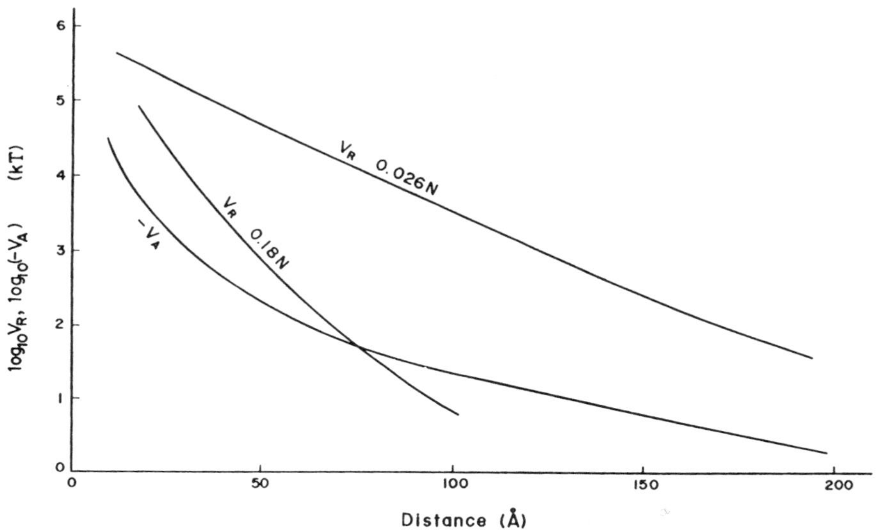

Fig. 3.11. Repulsion and attraction energy between two montmorillonite plates.

Attraction Energy V_A

The attraction energy due to the intermolecular interaction between the montmorillonite plates is calculated using Eq. (3.21) or (3.23), assuming that A is 1.5 x 10^{-13} erg (Fig. 3.11).

Total Potential Energy V_T

The total potential energy, V_T, is the sum of the repulsion energy, V_R, and attraction energy, V_A. Figure 3.12 shows the total potential energy for Wyoming montmorillonite as a function of interparticle distance for three concentrations of electrolyte, n_0. The values of total potential energy in Fig. 3.12 are expressed per surface area of a plate assuming that a single particle of Wyoming montmorillonite has a diameter of 8 x 10^{-5} cm (Van Olphen, 1956). There is a minimum potential energy for concentrations around 0.2 N. This minimum corresponds to the secondary minimum in Fig. 3.4.

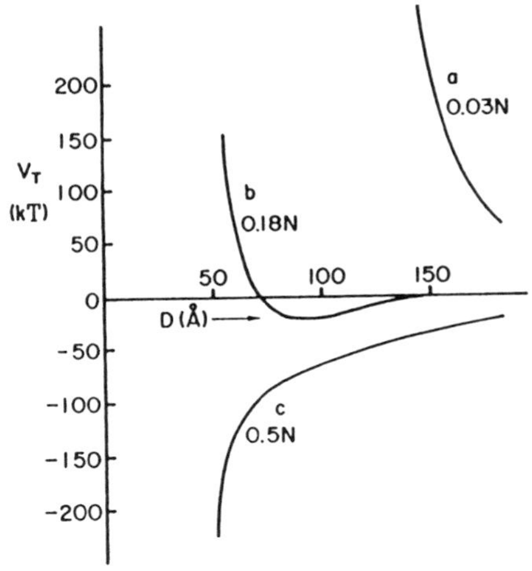

Fig. 3.12. Total potential energy between two montmorillonite plates, at three electrolyte concentrations.

Flocculation of Montmorillonite

Verwey and Overbeek (1948) have pointed out that an energy barrier of the order of 20 kT may be required for nearly permanent stability at a given concentration. Consequently, the depth of the minimum of curve (b) in Fig. 3.12 is enough to produce stable flocculation of montmorillonite. In other words, the DLVO theory predicts that the flocculation value of a single symmetrical electrolyte solution for Wyoming montmorillonite is about 0.2 N. Kahn (1958) measured 3 meq/liter as the flocculation value of NaCl solution for Na Wyoming montmorillonite. Van Olphen (1963) found a flocculation value of Na montmorillonite in NaCl solution of 12 to 16 meq/liter, and Aomine and Egashira (1968) measured 5.7 meq/liter at pH 6.5. Arora and Coleman (1979) obtained 7 to 20 meq/liter as the flocculation values of montmorillonites in NaCl solutions at pH 7.0.

Goldberg and Glaubig (1987) found that flocculation values of Na- and Ca-montmorillonite systems increase with increasing pH. Flocculation values of Na-montmorillonite were 14 meq/liter at pH 6.4 and 28 meq/liter at pH 9.4. Keren et al. (1988) also recognized that flocculation values of a Na-montmorillonite increase with pH, being 10, 13, 31, and 44 m mol/dm^3 of NaCl at pH 5, 7.5, 8.5 and 9.8, respectively. The dependence of flocculation of kaolinite on pH was observed to be greater than that of montmorillonite (Goldberg and Glaubig, 1987).

All of these values are much smaller than the predicted value. Van Olphen (1951) regarded the attraction of positive charges on the edges for negative charges on flat surfaces (edge-to-face flocculation) as the main cause of flocculation of montmorillonite. This explained the disagreement between experimental and predicted values based on a van der Waals potential.

Akae (1988) calculated the dependence of face-to-face, edge-to-face, and edge-to-edge interparticle linkages in a montmorillonite suspension on pH and salt concentration on the basis of both the double layer theory and the hetero-coagulation theory (Fig. 3.13). Each linkage can be formed only in the range below the solid line of 15 kT shown in the figure. Since the potential energy barrier for face-to-face linkage is 6,500 kT irrespective of pH, the linkage cannot occur. Edge-to-edge linkage is possible in the pH range between 6.8 and 9.2, and edge-to-face linkage can arise below pH 8.9.

Van Olphen (1951) analyzed modes of particle association when a suspension of platelike clay particles flocculates (edge-to-face, edge-to-edge, face-to-face). From measurements of the yield stress of gels, Van Olphen (1956) determined the force of a single edge-to-face bond as about 10^{-4} dyne. According to the calculation by Rausell-Colom (1958), the electrostatic attraction between the negative surface of a montmorillonite sheet and each charge, e, on an edge in contact with it, is about 7 x 10^{-5} dyne. Norrish and Rausell-Colom (1963) pointed out that the force of edge-to-face bonds for montmorillonite particles calculated from the estimates above would be sufficient to create forces of the magnitude required.

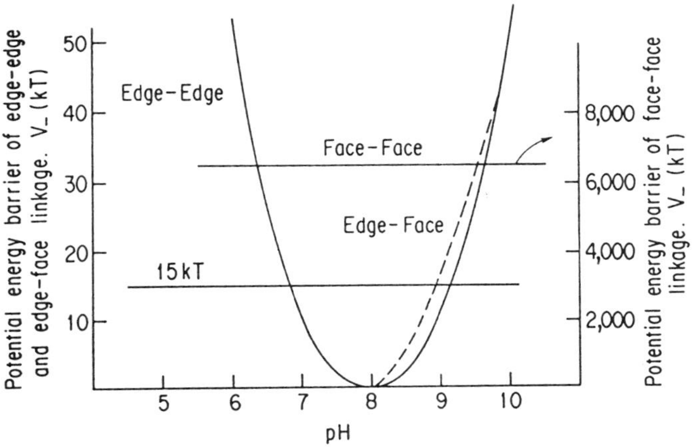

Fig. 3.13. Dependence of potential energy barrier per linkage on pH for monovalent ion system at salt concentration of 0.02 mol/dm^3. (From Aakae, 1988)

On the other hand, Van Olphen (1950) reported that with an optimum amount of Calgon (sodium polymetaphosphate) added, where very few edge-to-face bonds would be expected, Na-montmorillonite sols flocculated when the NaCl concentration reached about 0.3 N. This value can be considered to be a flocculation value under the condition where only face-to-face interaction exists. The predicted value is less than this value. Taking the hydration forces and/or the interaction between plates and water molecules into consideration as discussed before, it may be considered that the predicted value should become larger and could approach 0.3 N.

3.2.2 Allophane

To obtain a total potential energy for allophane, it is necessary to know the shape and size of the allophane particle, its surface charge density, and the Hamaker constant for allophane/water/allophane. The specific surface area of Chōyō allophane is 776 m^2/g (Aomine and Egashira, 1970), and the anion exchange capacity and cation exchange capacity at pH 6 are 16 and 10 meq per 100 g, respectively (Harada and Wada, 1973). Consequently, assuming that the net charge is effective, the surface charge density of Chōyō allophane is 2.2 x 10^3 esu/cm^2. In general, an allophane particle has been considered to be made up of many unit particles. Here we will assume the shape and size of the particle as a sphere with diameter of 300 Å. In addition, it is reasonable to assume that the Hamaker constant for allophane/water/allophane is approximately equal to that for mica/water/mica, 1.5 x 10^{-3} erg. We will calculate the potential energy between allophane particles at pH 6 using these values.

Repulsion Energy V_R and Attraction Energy V_A

We assume that the outer solution is a single symmetrical electrolyte. When the concentration of the outer solution is known, the surface potential ψ_0 is obtained from the following equation (Verwey and Overbeek, 1948):

$$\psi_0 = \frac{Q}{a\epsilon(1 + \kappa a)} \tag{3.34}$$

where a is the radius of the spherical particle, ε the dielectric constant of the medium, κ the Debye kappa, and Q the product of the surface charge density σ, and the surface area of the spherical particle, $4\pi a^2$. If the concentration of the outer solution is 20 mmol/liter, we obtain 6.59 x 10^{-5} esu as the value of ψ_0. As this value is small, Eq. (3.12) by Derjaguin is applicable for this system.

$$V_R^{\psi} = \frac{\epsilon a \psi_0^2}{2} \ln[1 + \exp(-\kappa H)]$$

where H is the minimum separation of the particle surfaces. The attraction energy between two particles as a function of minimum separation is given by Eq. (3.27).

Total Potential Energy

Figure 3.14 shows the calculated total potential energy for allophane at pH 6. When the concentration of a single symmetrical electrolyte solution is 10 mmol/liter, the curve of the total potential energy is always positive within the range of distances shown in Fig. 3.14. Flocculation does not occur at this concentration because the force acting between particles is always repulsive. When the concentration is 50 mmol/liter, the force acting between particles is always attractive, and consequently, allophane particles flocculate rapidly. When the concentration is 20 mmol/liter, the curve of the potential energy has one maximum and two minima. As the depth of the second minimum is very shallow, particles cannot be flocculated at this point. Particles can easily exceed the 1 kT maximum by their thermal motion. From these results, the flocculation value of a single symmetrical electrolyte solution for allophane at pH 6 is expected to be about 20 mmol/liter.

Flocculation of Allophane

Aomine and Egashira (1968) measured the flocculation values of various electrolyte solutions for allophane. Table 3.4 shows their results for Chōyō allophane, where single symmetric electrolyte solutions were used. The average of three flocculation values is about 21 mmol/liter. The agreement between the experimental and the theoretical results allows us to infer that the size of the allophane particle may be in the order of several hundred angstrom. Egashira (1977) reported that measured values of viscosities of allophane suspensions could be interpreted by assuming allophane particles to be platy particles with diameters of several hundred angstrom.

Karube (1982a,b; 1983) inferred from light scattering and membrane filtration measurements that the unit particles of allophane were connected like strings of beads several hundred nm in size. This was confirmed by electron microscopy (Fig. 3.15). This flocculation was explained by the DLVO theory.

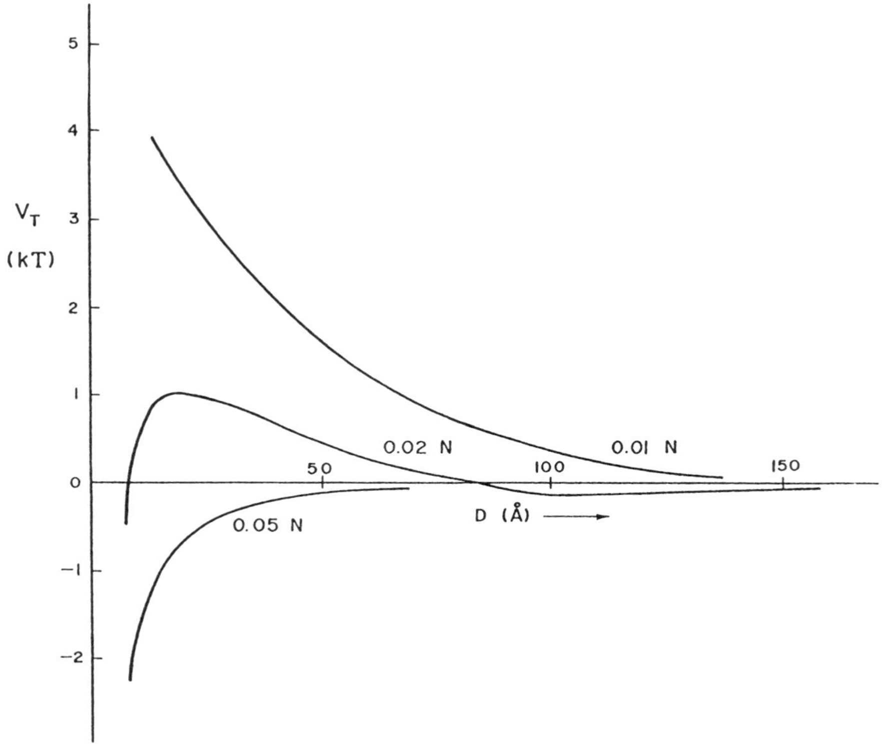

Fig. 3.14. Total potential energy for allophane, at pH 6, for different electrolyte concentrations.

Table 3.4. Flocculation values of allophane. (From Aomine and Egashira, 1968)

	pH	Flocculation value (mmol/liter)
NaCl	6.1	13.3
KCl	6.6	23.1
NH_4Cl	5.5	26.0

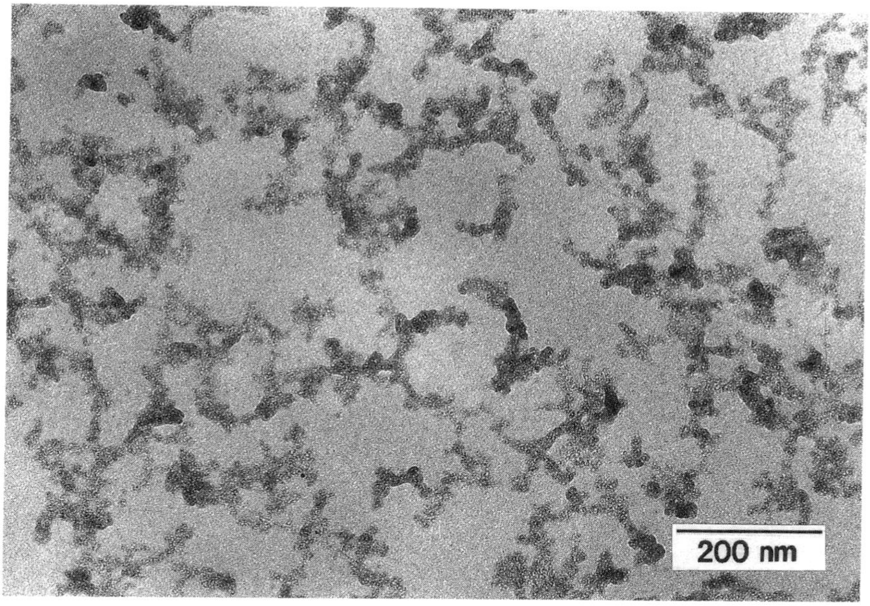

Fig. 3.15. Microphotograph of allophane. (From Karube, unpublished)

Thomas and McCorkle (1972) calculated the double-layer repulsive forces near the surfaces of two equally charged spherical particles in direct contact and seven spherical particles connected in a straight string of beads. They found in each case that a minimum in the repulsion energy barrier exists at the ends of the string. This means that floc growth by attachment of new particles at the ends of the floc is favored in stable sols.

Drying to water contents below pF 4.2 makes volcanic ash soils less dispersible through irreversible aggregation (Kubota, 1976). This effect of drying is not recognized in other mineral soils. As the concentration of a soil solution is increased by drying, the irreversible aggregation may result because the characteristic flocculation of allophane (low flocculation value and small maximum) is different from that of other clay minerals.

The surface charge density of allophane and the charge density of edges of montmorillonite and kaolinite are dependent on the pH of the outer electrolyte solutions, as discussed in Sec. 2.1. The net positive charge of allophane becomes large at low pH. This means that the repulsive force due to the interaction of the double layers increases, and the flocculation value for allophane should become larger. The results, shown in Fig. 3.16, confirm this.

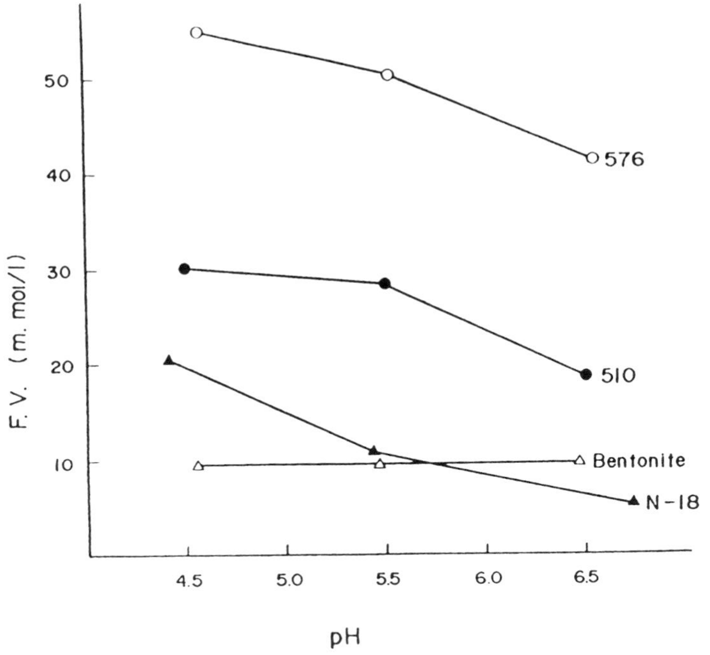

Fig. 3.16. Effect of pH on the flocculation value of three allophane samples and of bentonite. (From Aomine and Egashira, 1968)

3.3 SWELLING OF CLAYS

When ambient humidity or water content of a clay soil is increased, or concentration of the electrolyte solution is decreased, the soil volume increases. This phenomenon is termed swelling. An additional external pressure is needed to prevent volume increase due to swelling. We call this pressure the swelling pressure. Swelling occurs for particles with surface charges.

3.3.1 Crystalline Swelling

Clay swelling is broadly of two kinds. Crystalline swelling results from hydration of exchangeable cations; macroscopic swelling proceeds beyond crystalline swelling and can be analyzed by the DLVO theory. Crystalline swelling has been studied by many researchers (e.g. Kraehenbuehl et al., 1987;

Madsen and Müller-Vonmoos, 1989; Slade and Quirk, 1991; Slade et al., 1991; Sato et al., 1992).

Crystalline swelling occurs in steps, corresponding with one to four layers of water between the silicate layers (Norrish, 1954a,b; Norrish and Quirk, 1954). Kraehenbuehl et al. (1987) also found the stepwise nature of the process in water vapor adsorption isotherms. The pressure of crystalline swelling can be very large, reaching several megapascals (MPa), while pressure from macroscopic swelling is much smaller, but acts over greater distances (Slade and Quirk, 1991).

Slade and Quirk studied the limited crystalline swelling at basal spacings from ~15.5 to ~19.0 Å for some well-defined Mg-, Ca-, or La-saturated montmorillonites equilibrated with corresponding salt concentrations (Table 3.5). They concluded that 1) the higher its charge, the lower the salt concentration at which the clay swells, 2) Ca-saturated smectites with high surface charges, developed principally from Al^{3+} for Si^{4+} substitution in tetrahedral sites, have restricted swelling in water, 3) swelling involves a balance between cation hydration forces and interlayer electrostatic forces, and 4) the transition from 15.5 to 19.0 Å spacings can be considered an osmotic process.

The total, and the tetrahedrally derived charges, were 1.04 and 0.06 for Otay montmorillonite and 0.74 and 0.08 for Wyoming bentonite. The swelling of these clays with low and high charge, arising principally in octahedral sites, is contrasted in Table 3.5. The osmotic pressure of the outer solution at which the transition is half complete is lower for Otay montmorillonite with the higher surface charge. They explained this behavior, following Norrish (1954b), based on attractive and repulsive potentials in Eqs. (3.33) and (3.35).

$$V_R = \frac{2U\sigma}{ZeN} \tag{3.35}$$

where σ is the surface charge density, U is the hydration energy of an ion, Z is the valency, e is the elementary electric charge, and N is Avogadro's number.

The attractive potential from Eq. (3.33) varies as the square of the surface charge density, while the repulsive potential from Eq. (3.35) increases linearly with charge density. As a result, the transitions of the basal spacing from ~15.5 to ~19.0 Å for clays with higher charge densities require greater differences in osmotic pressure between solutions in the interlamellar space and enveloping solutions to overcome the larger attractive potentials. In other words, the concentration (osmotic pressure) of the outer solution for Otay has to be lower than that for Wyoming bentonite.

Sato et al. (1992) investigated swelling of ten homoionic smectites that differed in amount and location of layer charge, by X-ray diffraction analysis at various relative humidities, or after glycerol or ethylene glycol solvation. The

Table 3.5. Crystalline swelling of Wyoming and Otay smectites in relation to osmotic pressure, relative vapor pressure, and exchangeable cation. (From Slade and Quirk, 1991)

Ion	Pressure midway between ~15.5 Å and final spacing (MPa)	Relative vapor pressure (p/p_o)	Final spacing (Å)
Wyoming smectite			
Na	8.2	(0.943)	18.3
Mg	15.2	(0.894)	18.6
Ca	11.4	(0.920)	18.5
Otay smectite			
Na	4.0	(0.974)	18.8
Mg	12.5	(0.912)	18.0
	2.1	(0.983)	19.0
Ca	10.8	(0.923)	17.7
	3.2	(0.980)	18.5

results showed that the spacings are larger when the layer charge is located in octahedral sites than when it is in tetrahedral sites. The expansion is due to the combined effects of charge location and amount. The effects of layer charge magnitude and location are represented by an energy change (expansion energy) during the hydration and solvation processes.

It is therefore necessary that the total amount of charge and also its distribution in tetrahedral and octahedral sites be taken into consideration to clarify the interactions between two clay particles at small distances.

3.3.2 Mechanism of Macroscopic Swelling

Chemical Potential of Water

Swelling takes place when the chemical potential of water in the outer solution becomes larger than that of water in spaces between particles. The magnitude of swelling pressure depends only on this difference in chemical potential of water. This will be explained with a simple model. Suppose that a solution contains two charged plates which are kept at a distance 2d from each other (Fig. 3.17). Since equilibrium is maintained,

$$\mu_0 = \mu_{in} \tag{3.36}$$

where the chemical potential of water in the interlayer and that of water in the outer solution are μ_{in} and μ_0, respectively. In addition, taking the external pressure exerted on the plates, P_{ex}, as the standard, μ_0 can be given by

$$\mu_0 = \frac{-RT\ \Sigma\ n_i\pi_i}{1000}\,\bar{v}_w \tag{3.37}$$

where n_i is the molarity of solute i, π_i the osmotic coefficient of solute i, and $\bar{v}_w$ the specific volume of water.

When the concentration of the outer solution is changed by addition of water, the chemical potential μ_0' is

$$\mu_0' = \frac{-RT\ \Sigma\ n_i'\pi_i'}{1000}\,\bar{v}_w \tag{3.38}$$

where the prime indicates the condition after the addition of water. Therefore, the difference between μ_0' and μ_{in}, $\Delta\mu$, is

$$\Delta\mu = \mu_0' - \mu_{in} = \mu_0' - \mu_0 = \frac{RT\ \Sigma\ \left(n_i\pi_i - n_i'\pi_i'\right)\bar{v}_w}{1000} \tag{3.39}$$

Since before the addition of water the sum of the forces acting between the two plates was zero, $\Delta\mu/\bar{v}_w$ represents the magnitude of the swelling pressure produced by the addition of water. That is,

$$P_s = \frac{\Delta\mu}{\bar{v}_w} \tag{3.40}$$

where P_s is the swelling pressure.

Equations (3.39) and (3.40) mean that if the change in concentration of the outer solution is known, it is possible to calculate the swelling pressure. However, these equations cannot predict the change of distance between the two plates due to addition of water. In practice, we would require measurement of the volume change of a soil with change of solution concentration or change of external pressure. To predict the amount of volume change, it is necessary to know not only the change of concentration but the distance between the plates.

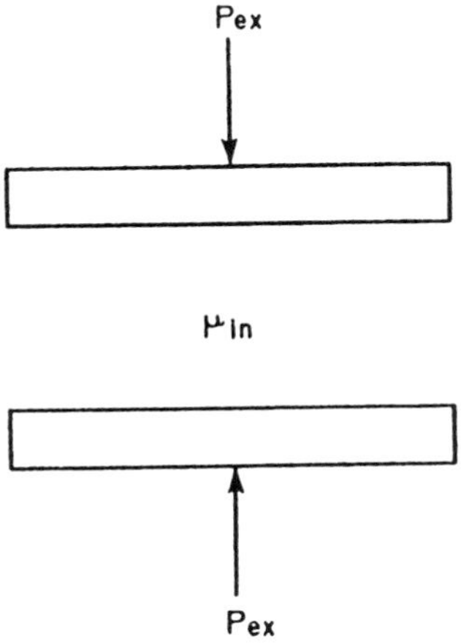

Fig. 3.17. Schematic representation of swelling pressure.

Swelling Under Unsaturated Conditions

Let us suppose a microscopic clay-water system as shown in Fig. 3.18. The distance between the plates is assumed to be larger than that at the first stage of swelling. As the system is at equilibrium, the chemical potential of water is the same at every point. Forces acting between the two plates are a repulsive force due to the interaction of the double layers, f_R, van der Waals attractive forces, f_A, and capillary forces due to existence of concave water surfaces, f_C. The capillary force is considered to act as an attractive force. As these forces are balanced,

$$f_A + f_R + f_C = 0 \tag{3.41}$$

In addition, if the attractive forces are expressed as positive values, f_C is given by

$$f_C = \frac{-2\sigma}{r} \tag{3.42}$$

where r is the radius of curvature of the concave surface (negative) and σ is the specific surface energy of water (surface tension). As f_C is attractive, we obtain

$$f_A + f_R < 0 \tag{3.43}$$

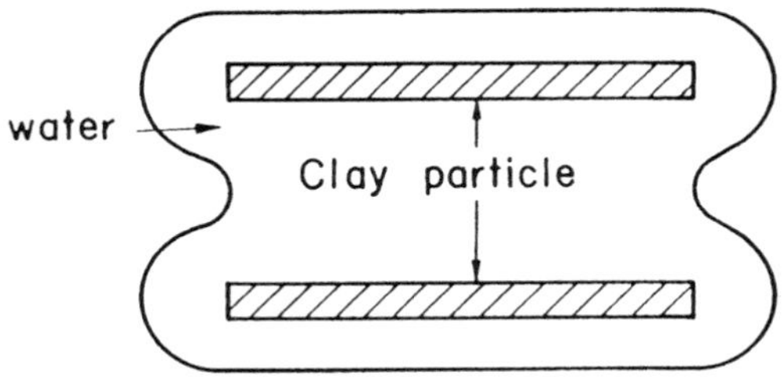

Fig. 3.18. Schematic diagram of an unsaturated microscopic clay-water system.

Let us now add pure water to the system. As the chemical potential of water in the system is evidently smaller than that of the added water, the water begins to move into the interlayer, and swelling results. When enough water is added, the concave surfaces become planar. Then the capillary force disappears, and a repulsive force corresponding to $(f_A + f_R)$ remains. This repulsive force is the swelling pressure. As the osmotic pressure of the added water is zero, swelling pressure P_w is given by

$$P_w = \pi_{in} \tag{3.44}$$

where π_{in} is the osmotic pressure at the midpoint between the plates.

3.3.3 Application of the DLVO Theory to Macroscopic Swelling

In macroscopic swelling, if particles of crystalline clay minerals are arranged parallel to each other, the DLVO theory can be used to predict the relationship between plate separation (the volume) and applied external pressure. When the concentration of the outer solution is low, the van der Waals attractive force is usually neglected, and the DLVO theory becomes the Gouy theory.

Various researchers have applied the DLVO or Gouy theory to explain swelling of Na-montmorillonite (Warkentin et al., 1956; Warkentin and Schofield, 1958, 1962; Yong et al., 1963; Norrish and Rausell-Colom, 1963; Shainberg et al., 1971; Madsen, 1989). In most of these reports, considerable agreement was shown between predicted and observed values. Figure 3.19 shows an example from among these results. Theoretical values agree quantitatively with measured swelling pressure for decompression and recompression cycles following the first compression. The authors explained this phenomenon as due to the random orientation of clay particles before the first compression and parallel orientation after the first compression above 20 atm.

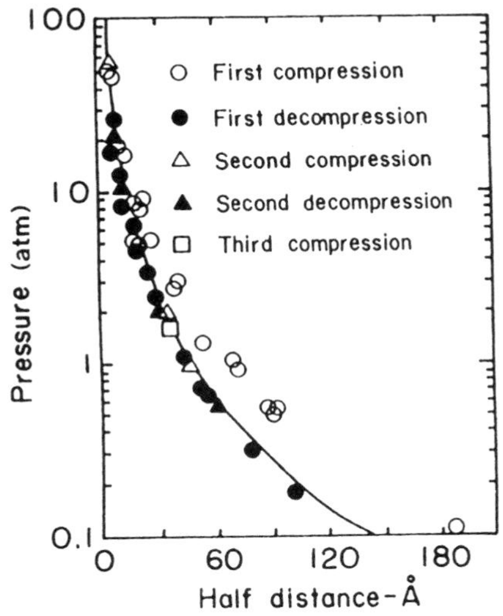

Fig. 3.19. Swelling pressure of Na montmorillonite in 10^{-4} M NaCl as a function of the half-distance separating the clay plates. Continuous line denotes calculated swelling pressure from DLVO theory. (From Warkentin et al., 1957)

However, the DLVO theory has not predicted measured swelling of Ca-montmorillonite and illite (Bolt and Miller, 1955; Warkentin et al., 1957). This disagreement has been explained using a model of "tactoids," "domains," or "hinge structure," in which dead volume is assumed as shown in Fig. 3.20 (Aylmore and Quirk, 1959; Greacen, 1959; Blackmore and Warkentin, 1960; Quirk and Aylmore, 1960; Blackmore and Miller, 1961; Mungan and Jessen, 1963). Such a special structure may result from stronger edge-to-face bonding because sheet separation of Ca-montmorillonite does not increase beyond 9.4 Å (Norrish and Quirk, 1954).

Figure 3.21 shows the dependence of plate separation on applied pressure for vermiculite in 0.03 N LiCl solution. Very good agreement is recognized between observed and calculated swelling pressures. This agreement may be due to the fact that the effect of edge-to-edge bonding is negligible or even absent because the plates of vermiculite are much larger than those of montmorillonite.

Total potential energy for clay particles in an electrolyte solution depends on the surface charge density of the particle, the concentration of the solution, and valence of the counterions, as discussed above. Consequently, these factors influence swelling behavior of clay soils.

Fig. 3.20. Schematic diagram illustrating a clay structure with dead volume. (From Greacen, 1959)

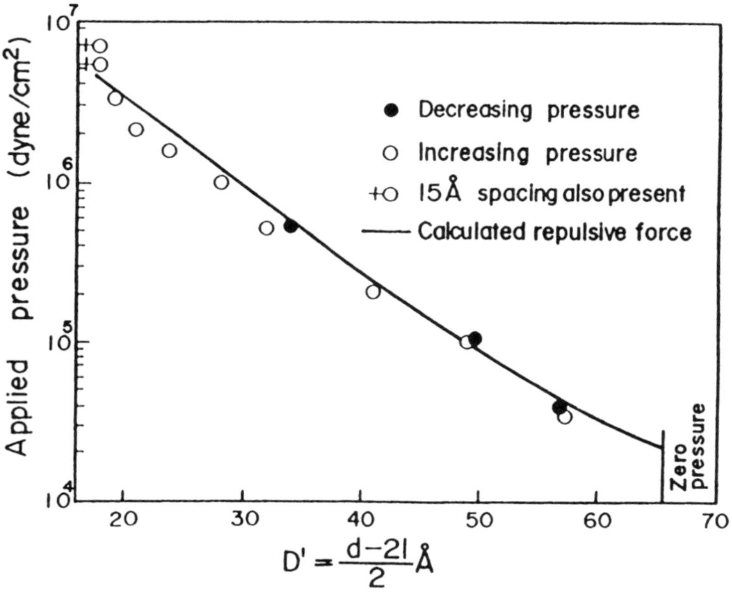

Fig. 3.21. Swelling of Li vermiculite in 0.03 N LiCl. (From Norrish and Rausell-Colom, 1963)

In addition, the pH of the outer solution is an important factor affecting swelling. As the pH decreases, the charge at the edges becomes positive and edge-face bonding increases. This means that swelling will decrease. If the charge at the edges is negative or if the edges lose their charge, the effects of edge-to-face bonding will be reduced. When the pH is high, the repulsive force due to the interaction between double layers becomes more effective and swelling will increase.

3.3.4 Swelling Theory Based on Structural Forces

Low (1980, 1991) has proposed a different theory for swelling of clays, especially for montmorillonite. The dominant repulsive force between clay layers in macroscopic swelling is not the double-layer force but a structural force due to the existence of structurally modified water in the vicinity of a clay surface (see Section 2.5). Low (1980) has pointed out the independence of repulsive pressure and zeta potential on surface charge density and the negligibly small values of repulsive double-layer force calculated from the measured zeta potential values in various montmorillonites as the main reasons for the failure of the electric double-layer theory in macroscopic swelling.

Viani et al. (1983) studied the relationships between repulsive pressure, π, and interlayer separation, λ, under an outer solution concentration of 10^{-4} N, using eight montmorillonites with surface charge densities, σ, ranging from 2.77 x 10^4 to 5.60 x 10^4 esu/cm^2 (Fig. 3.22). λ was measured directly by X-ray diffraction. The double-layer theory predicts a dependence of π on σ while the results show π independent of σ.

Miller and Low (1990) found zeta potentials, ψ_δ, for 34 different smectites normally distributed around a modal value of ~ -55 mV and essentially the same for all Na-smectites. In addition, they found that σ is independent of electrolyte concentration and, hence, smectites would be colloids with constant surface potential rather than constant charge. Table 3.6 shows the values of the charge density at the clay-water interface, σ_0, the zeta potential, ψ_δ, the surface charge density on the plane of the zeta potential, σ_δ, which is estimated from ψ_δ, and the fraction of exchangeable cations in the diffuse layer for Na-Upton montmorillonite in 10^{-4} N NaCl solution. They assumed that the diffuse layer originates in the plane of the zeta potential, namely, the plane is the boundary between the Stern layer and the range of the diffuse double layer. Since σ_δ/σ_0 is only 0.015, the calculated repulsive double-layer force becomes negligible compared with the measured repulsive pressure. Thus, Low has concluded that a structural force plays the most decisive role in macroscopic swelling, and the independence of π on σ can be understood only on the basis of that fact.

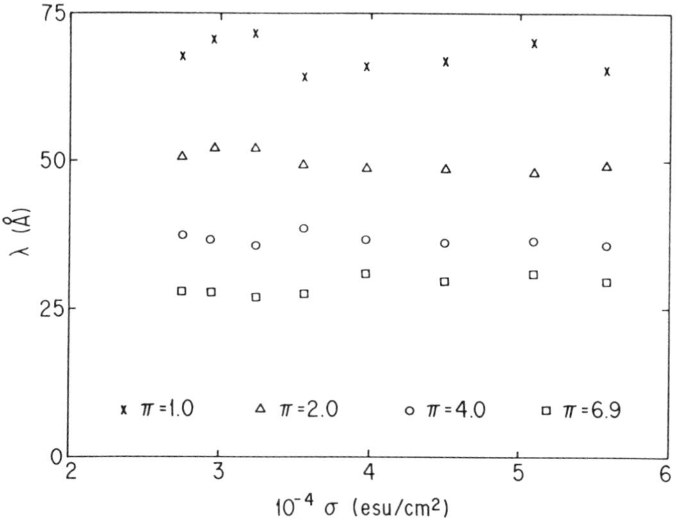

Fig. 3.22. The relation between interlayer separation, λ, and surface charge density, σ, at four values of repulsion pressure as indicated by the data for eight different montmorillonites. (From Viani et al., 1983)

On the other hand, many reports have confirmed the validity of the DLVO theory under certain conditions (e.g. Warkentin et al., 1957; Norrish and Rausell-Colom, 1963; Israelachvili and Pashley, 1983). In order to distinguish between the structural force theory and the DLVO theory, more theoretical and quantitative research related to macroscopic swelling forces seems to be required together with studies of molecular structure of water adsorbed by clays.

Table 3.6. Values of σ_o, the charge density at the clay-water interface, ψ_δ, the electric potential in the plane where the diffuse layer originates, σ_δ, the charge density in this plane, and σ_δ/σ_o, the fraction of exchangeable cations in the diffuse layer, for Na-Upton montmorillonite in 10^{-4} M NaCl. (From Miller and Low, 1990)

σ_o (esu/cm^2)	ψ_δ (mv)	σ_δ (esu/cm^2)	σ_δ/σ_o
32,500	-57.6	486	0.015

3.3.5 Effect of Clay Dispersion and Swelling on Permeability to Water

Permeability of a soil generally decreases with time when water is added by irrigation or rainfall. This may be due to swelling and/or dispersion from the decrease in concentration of the soil solution. Dispersion cannot occur unless the concentration of soil solution is smaller than the flocculation value. Clogging of conducting pores due to swelling and/or migration of dispersed clay particles causes a remarkable variation in soil hydraulic conductivity (HC) (McNeal et al., 1966).

The counterions in a soil also control the decrement in HC. For sodium, swelling proceeds to the second stage (see Section 3.3.2) and soil particles become dispersed. The HC then decreases greatly. The decrement of the HC is not as large for calcium or magnesium, since most of the clay platelets remain at the first stage of swelling. The pH of the soil solution also determines soil permeability through changes in amount and sign of variable charges on the edge surfaces of clay particles, and hence flocculation.

Effect of Electrolyte Concentration

The HC of a soil decreases with increasing exchangeable sodium percentage (ESP) and decreasing electrolyte concentration (e.g. Quirk and Schofield, 1955; McNeal et al., 1966; Rolf and Aylmore, 1977; Dane and Klute, 1977; Pupisky and Shainberg, 1979; Keren and O'Connor, 1982; Keren and Singer, 1988; Baudracco and Tardy, 1988). Quirk and Schofield (1955) found: 1) A 0.5 N percolating NaCl solution does not decrease HC but it becomes almost zero at 0.01 N. 2) For $CaCl_2$, even at 0.001 N, the decrease in HC is very small, and the percentage decrease is only around 30% even for pure water.

Pupisky and Shainberg (1979) suggested that at salt concentrations >10 mol_c/m^3, clay swelling is the main mechanism responsible for the HC decrease, whereas at salt concentrations below the flocculation value, dispersion and clay migration into the conducting pores occur. Keren and Singer (1988) studied the effect of electrolyte concentration in the percolating solutions on HC of Na/Ca-montmorillonite-sand mixtures at ESP 5, 10, and 20 (Fig. 3.23). HC_i and HC_0 denote the HC of the mixture leached with solution at a given electrolyte concentration and that for a concentration of 500 mol_c/m^3 having the same sodium adsorption ratio (SAR). The curve for SAR 10 was obtained by leaching consecutively with 0.5 L of 500, 50, and 10 mol_c/m^3 and then with distilled water. The curve shows: 1) SAR 10 is insufficient to reduce the HC of the mixture in equilibrium with 50 mol_c/m^3 solution. 2) The HC decreases by 60% when the solution of 50 mol_c/m^3 is replaced by a solution of 10 mol_c/m^3. Since

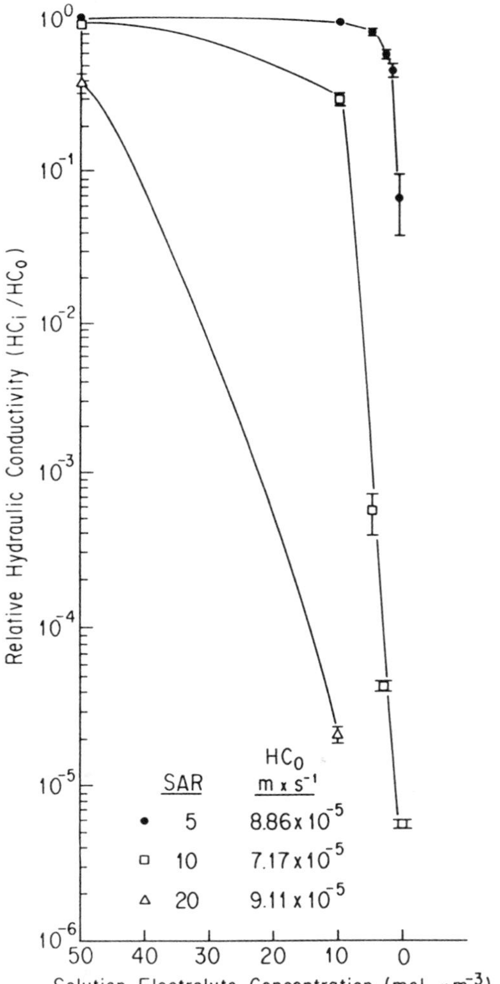

Fig. 3.23. Effect of electrolyte concentration and SAR in precolating solution on the hydraulic conductivity of Na/Ca-montmorillonite-sand mixture. (From Keren and Singer, 1988)

the flocculation value of Na/Ca-montmorillonite at ESP 10 is 5 mol_c/m^3 (Oster et al., 1980), the reduction in HC for the concentration of 10 mol_c/m^3 is probably due to swelling. 3) When the 10 mol_c/m^3 solution is replaced by distilled water, the HC of the mixture drops sharply to a very low value followed by a sharp increase. The presence of clay in the effluent indicates that the clay has dispersed.

Baudracco and Tardy (1988) studied the dependence of soil permeability on solution concentration and temperature using samples of unconsolidated Triassic sandstone. The effect of temperature is different at different ionic strengths of the percolating solution: at a low ionic strength of 0.01 the permeability coefficient decreases with increasing temperature; but at high ionic strength of 2 the permeability coefficient increases with temperature. They explained the phenomenon on the basis of double-layer theory.

Effect of pH

The flocculation values for montmorillonite and kaolinite increase with rise of pH of the outer solutions (see Section 3.2.1). Hesterberg and Page (1990) found over the pH range 6 to 10 that flocculation values increased from 6 to 58 mol Na/m^3 for Na-illite and from 3 to 14 mol K/m^3 for K-illite. These pH dependencies are due to the variable charges on the edges of these minerals. Since the increase in flocculation value with pH means an increased dispersion concentration range, the hydraulic conductivity (HC) may decrease at higher pH depending upon the concentration of the leaching solution. Suarez et al. (1984) found that HC values at pH 9 were lower than at pH 6 for montmorillonitic and kaolinitic soils.

The effect of pH on HC in a soil containing amorphous materials with pH-dependent charge may be much greater than for crystalline clay minerals. Nakagawa and Ishiguro (1994) studied the effect of solution pH on HC for an allophanic Andisol (Typic Hydrudand) containing 45% allophane and amorphous materials (Fig. 3.24). The ordinate indicates the dispersion ratio, defined by

$$\textit{Dispersion ratio} = \frac{\textit{absorbance after } 12 \textit{ hours settling}}{\textit{absorbance just after shaking}}$$

The larger the dispersion ratio, the more the soil disperses. The relative HC (Fig. 3.25), defined as the ratio of HC at a given pore volume to the initial HC, was measured at a constant hydraulic gradient in packed soil columns. The HC values decrease greatly when the influent solutions are pH 3 or pH 11. Observations on the soil structure at pH 3 and pH 11 showed that aggregates just below the surface (about 1 mm thickness) were totally collapsed, i.e. a crust was formed, while those in the lower soil remained without any collapse. The authors concluded that the decreases in HC were caused mainly by clogging of soil pores in the crust due to dispersed clay particles.

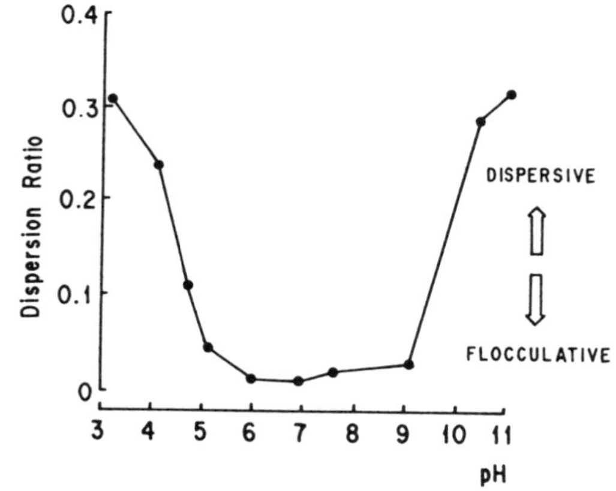

Fig. 3.24. Effect of pH on dispersion of an allophanic soil. (From Nakagawa and Ishiguro, 1994)

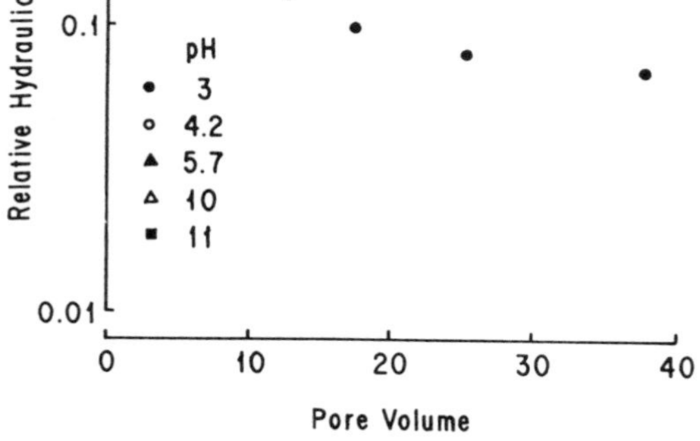

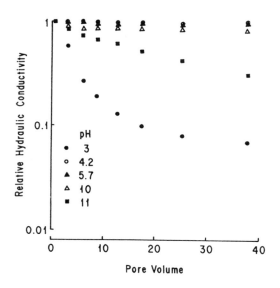

Fig. 3.25. Effect of influent pH on relative hydraulic conductivity of allophanic soil. (From Nakagawa and Ishiguro, 1993)

3.4 FLOW OF SOLUTIONS THROUGH CLAY LAYERS

The electric double layers formed at charged surfaces lead to interesting phenomena in the flow of solutions through charged membranes such as clay layers, phenomena that are not observed with uncharged membranes such as a sand layer. If the framework of the membranes is not rigid, as is the case for clay layers, DLVO theory is needed in addition to the Gouy theory to analyze the phenomena quantitatively. The theoretical analysis for nonrigid membranes has scarcely been begun.

3.4.1 Characteristics of Solution Flow Through Clay Layers

We will assume thin liquid films between clay platelets, in equilibrium with a symmetric electrolyte solution. The platelets are close enough together so that their potential fields overlap. As clay generally has negative charges, some anions are excluded from the films, and the number of cations in the films exceeds that of anions. This means that the concentration of free electrolyte in the films is always lower than that of the external solution. This characteristic induces remarkable phenomena.

Before beginning the discussion, we will show a schematic representation of the distributions of ion concentration and electric potential in the system under consideration (Fig. 3.26). C denotes the concentration of the salt, P the pressure, and π the osmotic pressure. The subscripts A and B refer, respectively, to the inflow and outflow solutions. The upper solid curve and the dashed curve, in the membrane, are the average concentration profiles of the cation and the salt (anion) in the membrane, respectively. $\bar{C}_A$ and $\bar{C}_B$ are the average concentrations of the salt on the plane in contact with solutions A and B, respectively. ΔE is the membrane potential. ΔE_A and ΔE_B are the two phase boundary potential differences, each of which is the difference in electric potential between the clay membrane and the adjacent solution phase. $\Delta\psi$ is the difference in electric potential between two sides of the membrane, which may be considered to be the sum of the streaming potential and the diffusion potential produced within the membrane. $\Delta\psi$ is called the liquid junction potential. The membrane potential, ΔE, can be measured with electrodes reversible to the cation or the anion and is given by

$$\Delta E = \Delta\psi + \Delta E_A + \Delta E_B = \Delta\psi - \frac{kT}{\nu e} \ln \frac{a_A}{a_B} \tag{3.45}$$

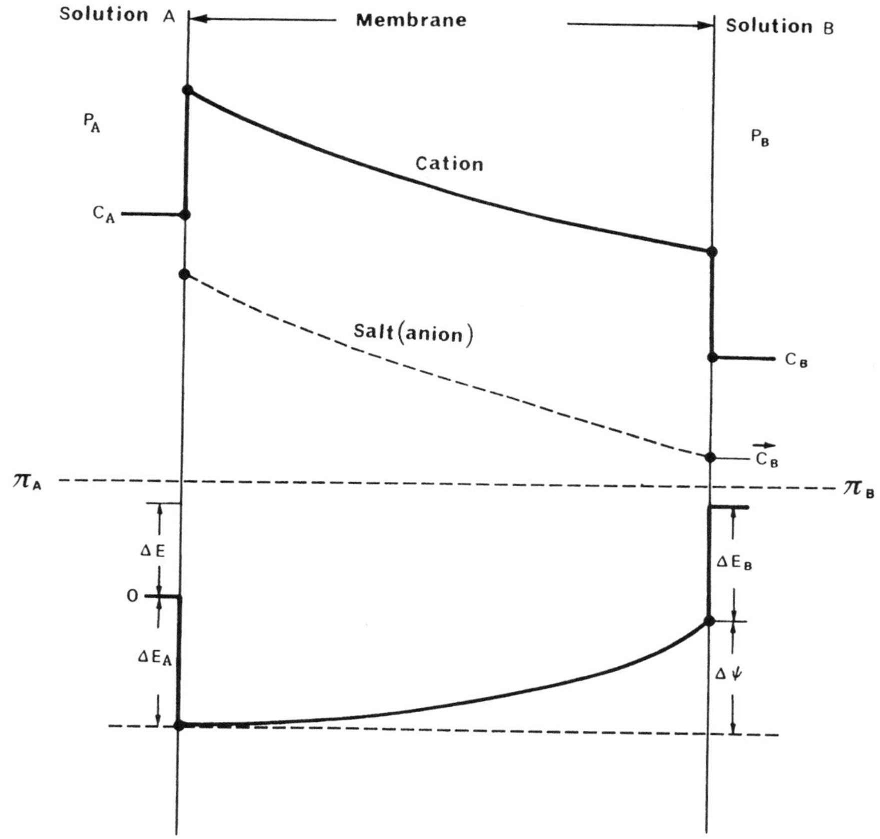

Fig. 3.26. Schematic representation of the distribution of ion concentrations and electric potential in a clay membrane and solutions on each side of the membrane when water flows from left to right.

where ν is the valence of the cation or the anion, k the Boltzmann constant, T the absolute temperature, e the elementary charge, and a_A and a_B the activities of the salt in solutions A and B, respectively.

Salt Sieving and the Salt Sieving or Rejection Coefficient

If a pressure gradient is applied to force salt solution through the pores between clay platelets, there results a net transport of charge and a buildup of the streaming potential because of an excess of cations in the film solution. The potential becomes increasingly positive at the outflow and increasingly negative at the inflow end. This electric field parallel to the surfaces brings about an increased transport of anions and a decreased transport of cations until an equal steady transport of positive and negative ions occurs. This is true provided that the feed and effluent solutions are electrically insulated from each other so that there is no current flow. In the steady state, although salt and water are transported through the pores, charges are not carried. The net effect of the anion exclusion (lower concentration of salt in the films than in the external solution) and the electric field induce the ratio of the molar salt flux to the volume flux of water to be less than the molar concentration on the high-pressure side. In other words, the pores between clay platelets tend to reject salt. This phenomenon is called salt sieving, and the value of $(C_A - \vec{C}_B)/C_A$ is defined as the salt sieving coefficient or rejection coefficient, R_j, where C_A is the salt concentration in the feed solution and $\vec{C}_B$ is the concentration coming through the clay. Clearly, R_j is unity for perfect osmotic membranes.

One of the first observations of salt sieving across a clay layer was made by Kemper (1960). A 0.01 N NaCl solution was poured on top of a Na-bentonite layer, and a pressure of 10 bars was applied to make the solution flow through the clay. The solution coming through the clay at first had a concentration of about 0.007 N. Similarly, 0.1 N and 1.0 N NaCl solutions decreased in concentration by about 10% and 3%, respectively, when forced through the clay. Clearly, R_j decreases as the concentration of feed solution, C_A, increases. This may be due to the fact that the thickness of the double layers in the film solution decreases as C_A increases, and consequently, the ratio of the average concentration of the free salt in the film solution to C_A increases. The inference above may be valid for rigid as well as for nonrigid membranes. In addition, it should be noted that R_j is a function of the hydraulic gradient because of its dependence on the streaming potential. This is discussed later.

Sands and Reid (1980) reported that osmotic potential measurements of soil solution extruded under pressure through membranes were unsatisfactory, probably due to salt sieving in the soil and/or at the membrane. A very large excess of counterions in thin films between platelets in equilibrium with dilute and medium concentrated solutions gives high values for the transport number of the cations. Shainberg and Kemper (1972) obtained 0.97 as the value of the transport number of Na ions for a compacted Na-clay layer in equilibrium with 0.0095 N NaCl solution. They measured the conductance across the clay layer. The transport number of the Na ion, of course, decreased as the concentration of the equilibrium solution increased.

Kharaka and Smalley (1976) recognized that the salt filtration efficiency for Na and Ca tends to increase with decreasing flow rate. Haydon and Graf (1986) found that the efficiency rises with increasing temperature.

Demir (1988) studied the interaction of hydraulic and electrical conductivities, streaming potential and salt filtration (salt sieving) during the flow of chloride brines through a smectite layer at elevated pressures. Solutions of NaCl or NaCl-$CaCl_2$ were forced through a clay plug made by compacting the 0.2-0.5 μm size fraction of Cheto montmorillonite. The initial concentration of NaCl solution was 1.10 molal (m), and the NaCl-$CaCl_2$ solution was 0.92 m in NaCl and 0.075 m $CaCl_2$. The thickness of the clay plug was 0.5 cm and the cross-sectional area is 65 cm^2. It took six to seven weeks to achieve constant effluent chemical composition and constant streaming potential. The compaction pressure was 34.5 MPa, and the fluid pressures at the upstream and the downstream end were 22.27 and 8.97 MPa. Hydraulic flow rate, streaming potential, and brine chemical composition were measured periodically until steady state. Electrical conductance of the clay plug was measured in the beginning and at the end of each experimental run. The main results are summarized in Table 3.7. Salt filtration efficiencies of 48% shown in the table compare reasonably well with published data (e.g. Hitchon and Friedman, 1969).

Reflection Coefficient or Osmotic Efficiency Coefficient

If a membrane restricts solutes completely, osmotic pressure and hydraulic pressure differences between the solutions on both sides of the membrane are equally effective in moving water through the membrane (Low, 1955; Kemper and Evans, 1963). On the other hand, when solutes are not completely restricted by a membrane, an osmotic pressure gradient is less effective than an equal hydraulic pressure gradient. Two concepts have been proposed to define this characteristic, the reflection coefficient (Kedem and Katchalsky, 1963), and the osmotic efficiency coefficient (Kemper and Rollins, 1966). The reflection coefficient σ is defined by

$$\sigma = \left(\frac{\Delta P}{\Delta \pi}\right)_{J^V=0,I=0} \tag{3.46}$$

where ΔP is the hydraulic pressure difference (P_A - P_B), Δπ the osmotic pressure difference (π_A - π_B), J^V the volume flux for the entire solution, and I the electric current. The osmotic efficiency coefficient σ′ is

$$\sigma' = \frac{(J^V)_{\Delta P=0,I=0/\Delta\pi}}{(J^V)_{\Delta\pi=0,I=0/\Delta P}} \tag{3.47}$$

Although the former is defined using forces and the latter is defined using fluxes, σ and σ′ are identical not only in physical meaning but also in numerical value. This fact is easily proved by the use of thermodynamics of irreversible processes

Table 3.7. Experimental results from flow of chloride brines through a smectite layer at elevated pressures. (From Demir, 1988)

Run no.	Fluid throughput rate ($cm^3s^{-1} \times 10^5$)*	Hydraulic conductivity ($cm^3s^{-1} \times 10^5$)	Electrical conductivity ($1/ohm\text{-}cm \times 10^2$) Beginning of run	Electrical conductivity End of run	Total measured potential (mV) Beginning of run	Total measured potential (mV) End of run	Asymmetry potential (mV)** Beginning of run	Asymmetry potential (mV)** End of run
1	1.19	0.69	0.152	0.164	33.7	33.0	17.8	19.0
2	0.88	0.51	0.169	0.178	30.5	30.0	12.1	19.0

Run no.	Streaming potential (mV)† Beginning of run	Streaming potential (mV)† End of run	Final Na molality Upstream	Final Na molality Downstream	Final Ca molality Upstream	Final Ca molality Downstream	Salt filtration efficiency (%) Na	Salt filtration efficiency (%) Ca	Salt filtration efficiency (%) Total cation
1	15.9	14.0	1.190	0.620	--	--	48	--	48
2	18.4	11.0	1.220	0.660	0.058	0.008	46	86	48

*Average of several readings at steady state.
**Measured at 15.9 MPa mean fluid pressure and zero differential hydraulic pressure.
†Total measured potential minus asymmetry potential.

(Bolt, 1979). However, it should be noted that σ and σ' are identical only when they are measured under the minimum condition that the concentrations of the equilibrium solutions are kept equal. This is discussed in the following section.

The physical meaning of the reflection coefficient can be understood more easily by considering the reflection coefficient of polymers by an uncharged membrane. Then J^V is given by

$$J^V = K\left(\frac{\Delta P}{\Delta X} + \sigma\frac{\Delta\pi}{\Delta X}\right) \tag{3.48}$$

where K is the hydraulic conductivity of the membrane. The equation shows that σ is the coefficient expressing the degree of effectiveness of an osmotic pressure difference for solution flow. The physical meaning of σ for a charged membrane does not change fundamentally.

Osmotic efficiency coefficients are increased by saturating a clay with monovalent rather than divalent cations, by using divalent rather than monovalent anions, by decreasing water content of the clay, and by decreasing the average concentration of the outside solution (Kemper and Rollins, 1966). An example is shown in Fig. 3.27. Bresler (1973, 1978) stated that salt concentration gradients (osmotic gradients) may become a major factor in movement of soil solution after infiltration. He proposed and solved unsaturated flow equations including the osmotic efficiency coefficient as a function of water content, the composition of both anions and cations, and the concentration of ions.

Characteristics of Solution Flow Through a Clay Layer

Soil physicists have traditionally dealt with linear flux laws such as Darcy's law or Ohm's law, where one driving force determines the flow. However, in solution flow through a clay layer there are three driving forces: the hydraulic pressure, the osmotic pressure, and the electric potential differences across the clay layer. The electric potential difference $\Delta\psi$, which may be the sum of a diffusion potential and a streaming potential, is a function of the hydraulic pressure difference ΔP and the osmotic pressure difference $\Delta\pi$ when the feed and effluent solutions are insulated from each other. So $\Delta\psi$ is necessarily present when flow is produced by ΔP and/or $\Delta\pi$; we cannot have flow due only to ΔP or $\Delta\pi$.

The ratio of the diffusion potential to the streaming potential depends on the concentrations of the outer solutions, $\Delta\pi$ and ΔP. For example, Kemper and Van Schaik (1966) reported that, at concentrations larger than 0.05 N, the calculated potential differences across a layer of Na montmorillonite under the condition where $\Delta P = 0$ are due mainly to the greater mobility of the anion as compared

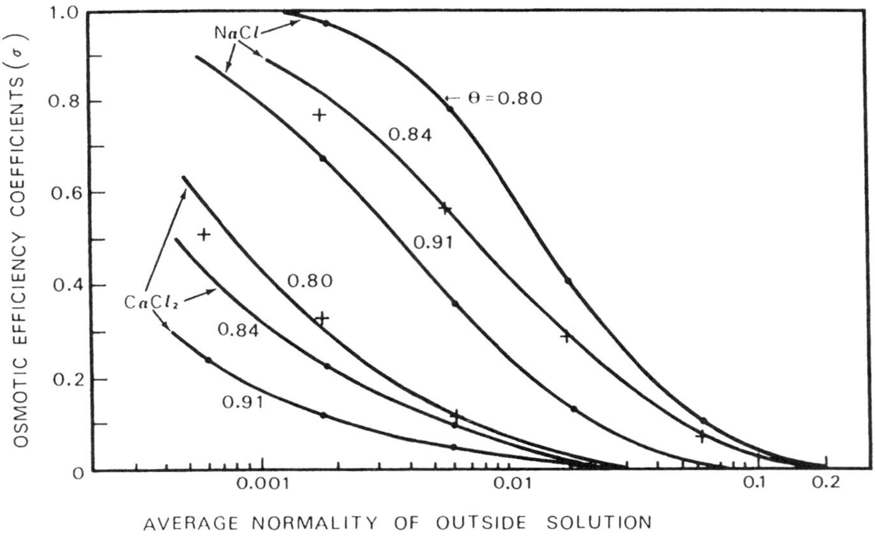

Fig. 3.27. Osmotic efficiency coefficients of clay at various moisture contents and NaCl and $CaCl_2$ solutions of various average concentrations. (From Kemper and Rollins, 1966)

with the cation. At lower concentrations, the streaming potential due to osmotic flow causes most of the potential differences.

The thickness of the double layers in the pores within the clay layer changes with changes in concentration of the outer solutions (solutions A and B in Fig. 3.26). This means that the properties of the clay layer in regard to salt transport are changed even if the clay layer is rigidly restricted.

If the clay layer is not perfectly rigid, the distance between the clay platelets varies with the concentrations of the outer solutions. Both salt transport and water movement can then be changed. The distance between the clay platelets in the clay layer as a function of salt concentration can be predicted from the DLVO theory. Studies on predicting changes in salt and water fluxes have scarcely begun. But it is known that coefficients relating to solution flow through a clay layer, such as σ and phenomenological coefficients described later, are all functions of, at least, the concentrations of the outer solutions.

3.4.2 Theoretical Analyses of Solution Flow Through Clay Layers

Rejection Coefficient (Salt Sieving Coefficient)

Kemper proposed a theory to calculate the salt sieving coefficient due to flow in films between two charged surfaces. The theory is introduced in detail in a textbook (Kemper, 1972). The salt solution between two charged surfaces is divided into homogeneous layers and the solution velocity in each layer is calculated using a difference equation derived from a simplified Navier-Stokes equation. The Gouy theory is used to estimate the electric force acting on the solution in each layer. Although the theory has several limitations, the method developed is typical of theories used to analyze heterogeneous systems such as solutions in the vicinity of a clay surface.

Jacazio et al. (1972) used a capillary membrane model to develop a theory for salt rejection characteristics of a charged membrane whose pore size is large compared with molecular dimensions. They considered reverse osmosis conditions, where a saline solution is moved through the membrane pores by an applied pressure gradient. They expressed the salt rejection coefficient, R, as a function of three parameters: the ratio of the Debye length (reciprocal of the Debye κ) to the effective pore radius, λ; a dimensionless wall potential, $\bar{\psi}_w$, related to the ζ potential; and a Peclet number, Pe, an index to indicate the ratio of convective to diffusive flow rates, given by VL/D, where V is the mean flow velocity, L the membrane thickness (pore length), and D the diffusion coefficient of the salt in water.

They started from the Nernst-Planck equations for ion fluxes. Adopting a cylindrical coordinate system (r, x) with x positive in the direction of flow and r the radial coordinate with origin at the axis of symmetry, the equations may be written

$$j_{\pm} = uC_{\pm} - D \exp(\mp\bar{\psi}) \frac{dC}{dx} \mp DC_{\pm} \frac{d\bar{\phi}}{dx} \tag{3.49}$$

where j denotes the ion flux per unit area, the subscript + refers to cations, and the subscript - refers to anions, $u = 2V(1 - \bar{r}^2)$ (where V is the mean flow velocity, $\bar{r} = r/a$, where a is the radius of the capillary), C+ and C- are the concentrations of the cation and the anion, respectively, D is the diffusion coefficient to be taken equal for the positive and negative ions, $\bar{\psi} = \nu F\psi(r,x)/RT$ [where ν is the absolute valency value of charge, to be taken equal for positive and negative ions, F the Faraday constant, R the gas constant, and $\psi(r,x)$ the radial component of the electric potential obtained by the Gouy theory (the potential due to the charged surface)], C is the salt concentration in the core of the capillary where ψ is zero, and $\bar{\phi} = \nu F\phi(x)/RT$ [where $\phi(x)$ is the electric potential corresponding to the axial component of the electrical field]. Assuming that the electric field has a negligible effect on the mean velocity, u, they obtained the total flux by integrating Eq. (3.49) over the pore cross-sectional area, and derived relationships between R and the three parameters. The salt rejection coefficient R is a function of ΔP, as V, which is included in Pe, is clearly dependent on ΔP (Fig. 3.28).

Neogi and Ruckenstein (1981) solved the equations of motion and conservation, which include the effect of the electric field neglected by Jacazio et al. (1972), together with the Poisson equation for a single charged cylindrical pore, over a wide range of conditions to compute the flow rate of water as well as the fluxes of salt and the rejection coefficient. The calculations were carried out for both constant surface potential and constant surface charge. The rejection coefficient is overpredicted by the equation of Jacazio et al. (1972) by 15% or less, due to the neglect of the electrical effects on fluid flow.

Streaming Potential

A streaming potential is induced whenever flow occurs through a charged membrane under the condition of zero electric current. The magnitude of the streaming potential depends on the charge density of the membrane; the electrokinetic radius λ, defined as the ratio of Debye length to the effective pore radius of the membrane; and the applied pressure gradient. Burgreen and Nakache (1964) presented a theory of electrokinetic flow for uni-univalent electrolyte through a channel between two charged plates in which the double layers overlap and the plate potential can be varied. Under steady flow conditions, the sum of pressure forces, viscous forces, and electric body forces generated by an axial electric field is zero, so we obtain

$$-\frac{dP}{dx} + \eta\frac{d^2V}{dy^2} - \rho Y = \frac{dP}{dx} + \eta\frac{d^2V}{dy^2} + \frac{\epsilon Y}{4\pi}\frac{d^2\psi}{dy^2} = 0 \qquad (3.50)$$

where P is the pressure in the fluid, x the axial distance, η the viscosity of the fluid, V the velocity at a given point and is equal to the sum of the electro-osmotic velocity V_r and the pressure-induced velocity V_p, y the distance measured from the charged surface, ρ the volume charge density, Y the axial electric field strength, ε the dielectric constant of the fluid, and ψ the electric potential due to the charged surface.

They obtained streaming potentials as a function of these parameters by solving Eq. (3.50) with the help of the Gouy theory and a numerical integration technique (Fig. 3.29). In Fig. 3.29, Y_s represents the streaming potential, M denotes $\psi_0\varepsilon/4\pi\eta$ (where ψ_0 is the surface potential), K_0 is the specific conductivi-

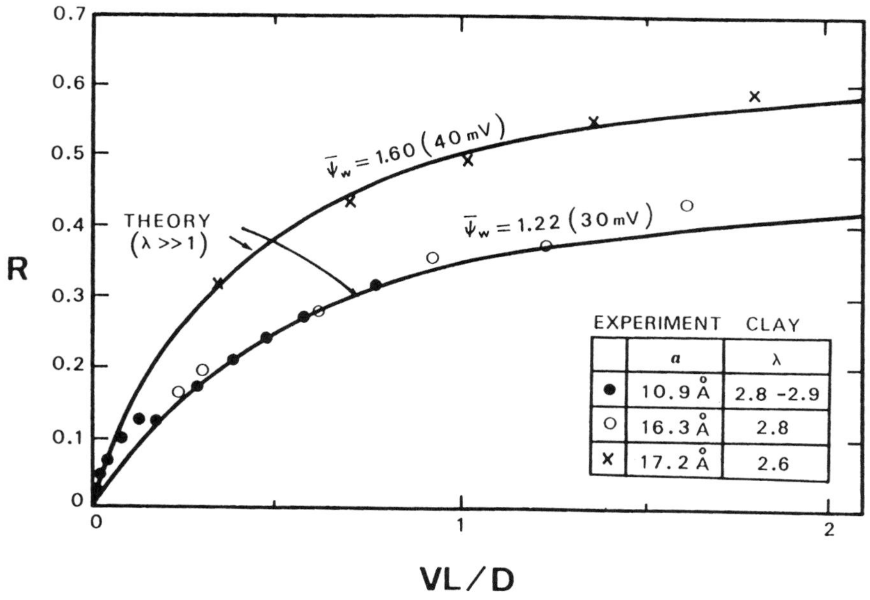

Fig. 3.28. Salt rejection coefficient of a cylindrical pore with a constant surface potential for large Debye ratio as a function of Peclet number, and comparison with experiments on compacted clay. (From Jacazio et al., 1972)

ty of the fluid, α is the surface ionic energy parameter and equal to $e\nu\psi_0/kT$ (where e is the elementary electric charge, ν the valence of the ions, ψ_0 the potential at the surface, k the Boltzmann constant, and T the absolute temperature), and B denotes $M^2\kappa^2\eta/K_0$ (where κ is the Debye κ). It can be seen from the figure that for a given value of ψ_0, the relative magnitude of streaming potential to the hydraulic pressure gradient increases as λ increases (the thickness of the electric double layer increases). We discuss this point again later.

Levine et al. (1975) developed the work of Burgreen and Nakache, considering that the conductivity within the double layer near the surfaces exceeds that of bulk electrolyte. They adopted an analytical method instead of the numerical integration used by Burgreen and Nakache to find electroosmotic velocities.

Ohshima and Kondo (1990) developed a theory for electrokinetic flow between two parallel similar plates (at separation h) covered by ion-penetrable surface charge layers (of thickness d) in which fixed charges are distributed at a uniform density. They derived simple approximate analytic formulas for the

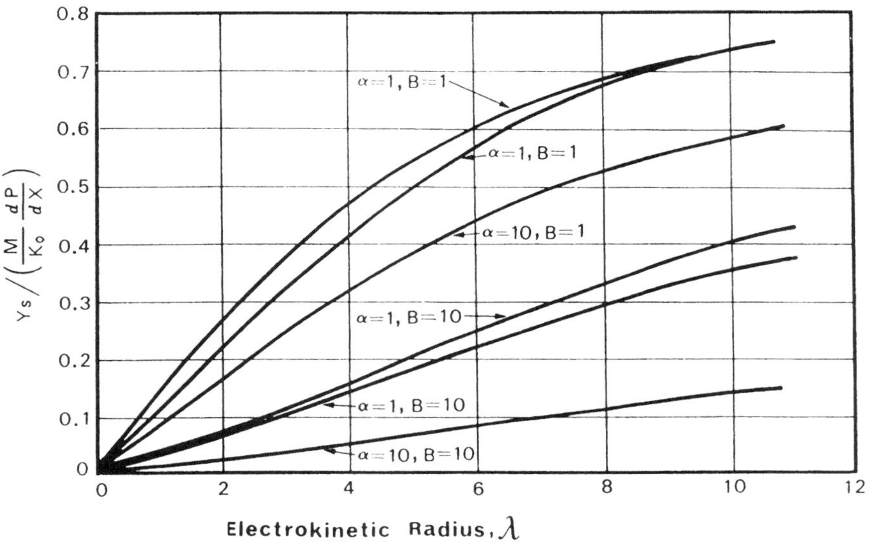

Fig. 3.29. Variation of streaming potential with electrokinetic radius for several values of α and B. (From Burgreen and Nakache, 1964)

electro-osmotic velocity, the volume flow, the electric current, and the streaming potential. Their formulas are regarded to be applicable when h >> d $\gtrsim$ 1/κ, 1/λ, where κ is the Debye-Hückel parameter and $\lambda = (\gamma/\eta)^{1/2}$, γ being the frictional coefficient of the surface charge layer and η the viscosity. Cohen and Radke (1991) presented numerical solutions for the general equations for the streaming potential for a system in which the surfaces of a slit are nonuniformly charged.

Jin and Sharma (1991) proposed a theoretical analysis of transport phenomena in inhomogeneous charged porous media using a two-dimensional network model composed of tubes with different radii. The network allows us to calculate the streaming potential and modified potential. In one of the results (Fig. 3.30) the network model is constructed of equal tubes with a radius (R) of 0.01 μm. Smoluchowski had shown that the streaming potential (Sp) is proportional to surface potential on the wall. However, from the figure, the relationship between surface potential and streaming potential for brine concentration of 0.05 is not linear in a high surface potential range.

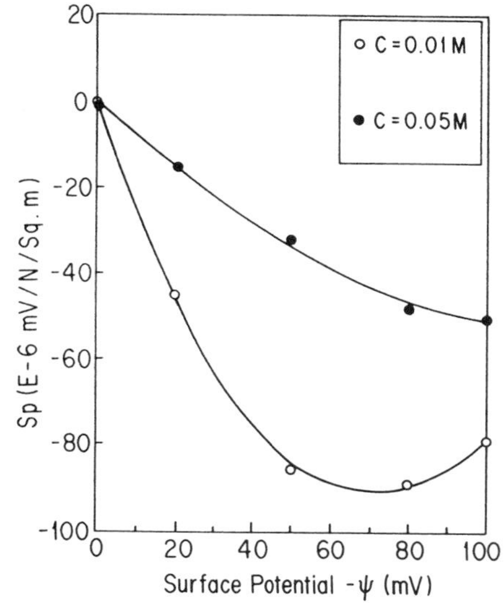

Fig. 3.30. Streaming potential versus surface potential for R = 0.01 μm. (From Jin and Sharma, 1991)

3.4.3 Relationships Between Factors Influencing Solution Flow Through Clay Layers

Streaming Potential and Hydraulic Pressure Gradient

The streaming potential accompanying solution flow through clay layers depends on the applied pressure (Gairon and Swartzendruber, 1975). According to Burgreen and Nakache (1964), the relationship between the streaming potential Y_s and the pressure gradient dP/dx, is given by

$$Y_s = \frac{(M/k)\{dP[1 - G(\alpha,\lambda)]/dx\}}{1 - (M/kh)\int_0^h (1 - \psi/\psi_0)\, dh} \tag{3.51}$$

where G(α,λ) is a parameter, h is the half distance between plates, and other symbols are defined in Sec. 3.4.2.

This equation shows that Y_s is proportional to dP/dx if other conditions are kept constant. Rolfe and Aylmore (1981) recognized that a linear relationship between streaming potential and flow pressure is maintained for Willalooka illite cores as shown in Fig. 3.31.

Demir (1988) also found that the streaming potentials are proportional to the applied differential hydraulic pressure. These facts indicate that the second postulate of nonequilibrium thermodynamics, that there is a linear relation between the force and the fluxes at or near steady-state, is obeyed.

Pressure-Induced and Electrokinetic Flows

The existence of the electrical potential (streaming potential) gradient across a membrane has a large effect on water flow through the membrane. In Fig. 3.32, the ratio of the retarding velocity due to the electrokinetic counterflow

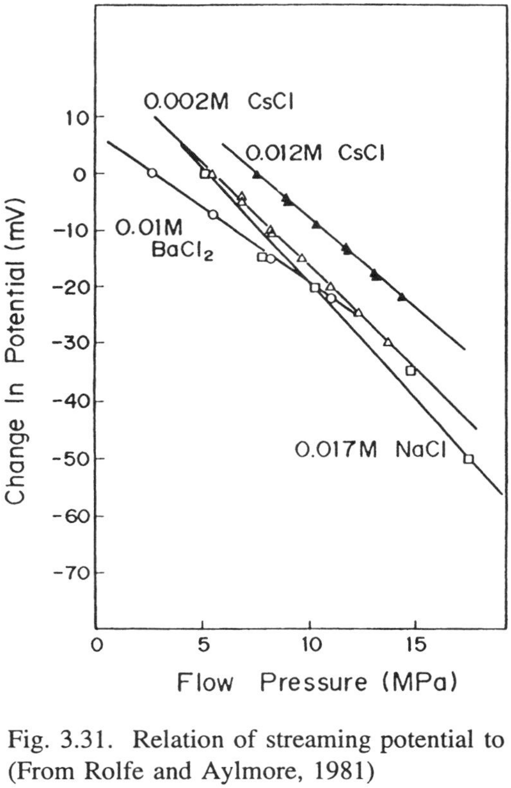

Fig. 3.31. Relation of streaming potential to flow pressure across a clay plug. (From Rolfe and Aylmore, 1981)

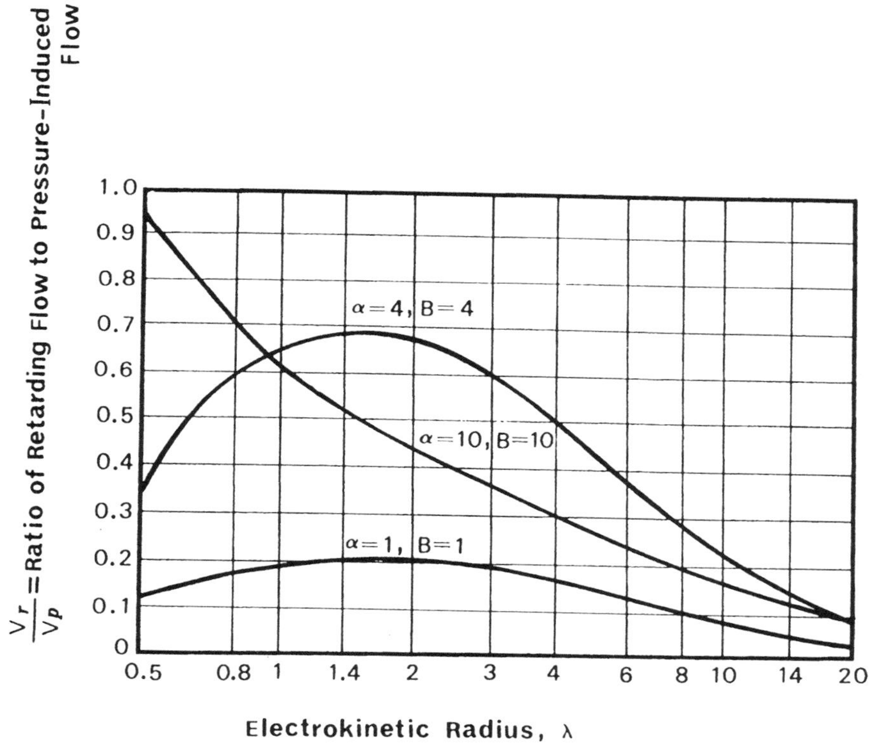

Fig. 3.32. Variation of retarding flow with electrokinetic radius for several values of α and B. (From Burgreen and Nakache, 1964)

component, V_r, to the pressure-induced velocity, V_p, is shown as a function of electrokinetic radius, λ, for several values of α and B (Burgreen and Nakache, 1964). The definitions of α and B are given in Sec. 3.4.2 (Fig. 3.29). The retarding flow component can be as much as 68% of pressure-induced flow for α = 4 and B = 4, which corresponds roughly to the α and B of distilled water in equilibrium with the CO_2 of the air. This occurs when the electrokinetic radius is 1.6.

Neogi and Ruckenstein (1981) also reported that the flow of water can be greatly decreased by the electric field. For example, the decrease at constant surface potential can be as large as 90% for a surface potential of 75 mV. At constant surface charge, the decrease is about 20% for a surface charge of 10^4 esu/cm^2. At constant potential, the electroviscous effect decreases as the flow pressure increases, while it remains approximately constant with flow pressure at constant charge.

Dependence of Rejection Coefficient on Applied Pressure Gradient and Surface Charge Density

McKelvey and Milne (1962) and Kemper and Maasland (1964) observed an increasing rejection coefficient R_j with increasing flow pressure for compacted Wyoming montmorillonite. Jacazio et al. (1972) found that R_j asymptotically reached a maximum value after increasing with flow pressure (Fig. 3.28). This dependence of R_j on the Peclet number (flow pressure) was predicted from theoretical models (Kemper and Maasland, 1964; Jacazio et al., 1972; Neogi and Ruckenstein, 1981). On the other hand, Rolfe and Aylmore found maximum rejection coefficients at the lowest flow pressures, and salt rejection decreased rapidly with increasing flow pressure for Willalooka illite cores. More resreach will be required to resolve this. The rejection coefficient of a membrane, R_j, becomes larger when the surface potential or the surface charge density is higher (Figs. 3.28 and 3.33).

3.4.4 Application of Irreversible Thermodynamics to Solution Flow Through Clay Layers

Thermodynamics of irreversible processes provides a suitable framework to study coupled phenomena such as material transport through a clay layer. Several fine books and reviews of the nonequilibrium thermodynamic treatment of water and salt transport through charged membranes are available (e.g. Kedem and Katchalsky, 1963; Katchalsky and Curran, 1965; Groenevelt and Bolt, 1969; Bolt, 1979). We will refer only to the work of Groenevelt and his co-workers, in which a typical theoretical analysis is applied to their experimental values (Elrick et al., 1976; Groenevelt and Elrick, 1976; Groenevelt et al., 1978).

They measured the effects of salt concentration differences across a thin layer of Na montmorillonite (10^{-3} and 10^{-4} M NaCl). The water pressure, concentration, and voltage differences, measured with electrodes reversible to the anion (Cl^-), were observed as functions of time (Fig. 3.34). After 20 hours the reversible electrodes were short-circuited.

They adopted the following phenomenological equations to describe their experimental results, and performed model calculations to obtain the phenomenological coefficients.

$$j^V = L_V F_V + L_{VD} F_D + L_{VE} F_E \tag{3.52a}$$

$$j^D = L_{DV} F_V + L_D F_D + L_{DE} F_E \tag{3.52b}$$

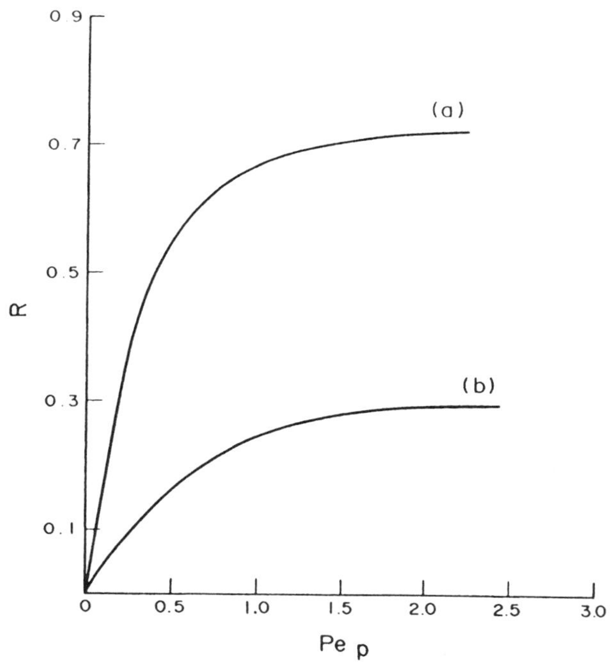

Fig. 3.33. Rejection coefficient R for KCl as a function of Peclet number Pe_p for two values of surface charge density, (a) 10^4 esu/cm^2 and (b) 2×10^3 esu/cm^2. (From Neogi and Ruckenstein, 1981)

$$I = L_{EV}F_V + L_{ED}F_D + L_E F_E \tag{3.52c}$$

where j^V is the volume flux, j^D the diffusion flux for the neutral salt relative to the solvent, I the electric current, L the phenomenological coefficients, and

$$F_V = -\frac{\Delta P}{\Delta X} \qquad F_D = -\frac{\Delta \pi}{\Delta X} \qquad F_E = -\frac{\Delta E}{\Delta X}$$

where ΔX is the thickness of the thin clay layer, ΔP the pressure difference across the clay layer, $\Delta\pi$ the osmotic pressure difference across the clay layer, and ΔE the electric potential difference measured with electrodes reversible to the anion. Under the assumption of parallel arrangement of clay platelets, model calculations were performed for the following conditions:

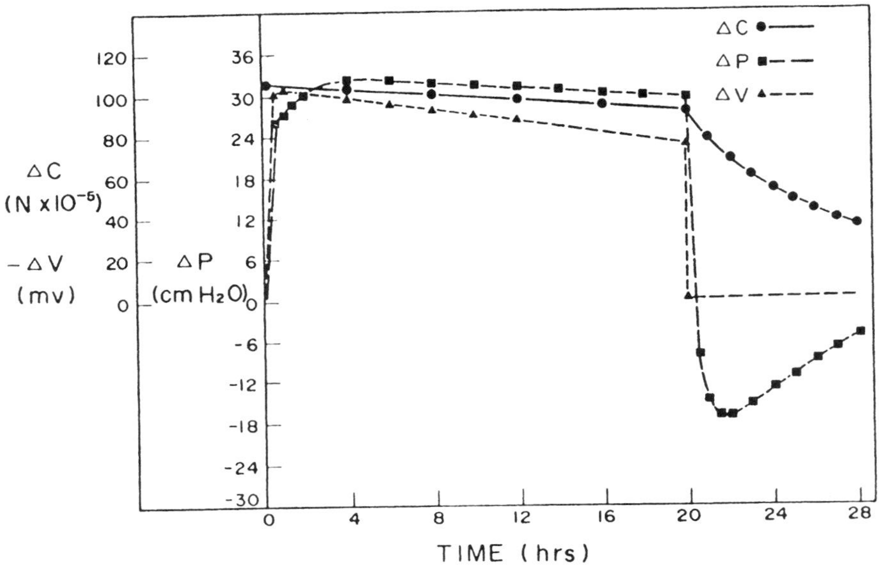

Fig. 3.34. Measured and calculated values of concentration, pressure, and voltage for flow across a clay membrane. (From Groenevelt et al., 1978)

1. The concentration distributions of the cations and the anions in the electric double layer are taken as logarithmically symmetric (Boltzmann's law of energy distribution), such that

$$C^+(y) = C_0 u(y) \qquad C^-(y) = \frac{C_0}{u(y)} \tag{3.53}$$

where C_0 is the equilibrium salt concentration of the solution, y the distance from the clay surface, and u the Boltzmann factor for the cations, $F\nu\psi/RT$.

2. All fluxes in Eqs. (3.52) are expressed in terms of integrals of the product of the local velocity V(y) and the local concentration $C^*(y)$ according to

$$j \equiv \frac{\theta}{\lambda}\frac{1}{d}\int_0^d V(y)C^*(y)\,dy \tag{3.54}$$

where θ is the volumetric liquid content of the system, λ the tortuosity factor, and d half the distance between two clay platelets. Of course, V(y) and $C^*(y)$ are physically different, corresponding to the flux under consideration. For example, for j^V, V(y) and $C^*(y)$ show the local velocity given by Eq. (3.56) and concentration of the liquid which is equal to unity, as described later.

3. Each phenomenological coefficient is obtained by dividing the flux by the driving force related to the coefficient under the condition where the other driving forces are kept at zero:

$$L_{VE} = \left[\frac{j^V}{-(\Delta E/\Delta X)} \right]_{P,\pi=0} \tag{3.55}$$

Here we will introduce their model calculation for L_{VE} as an example. To calculate L_{VE}, $[j^V]_{P,\pi}$ has to be obtained. It can be calculated from Eq. (3.54) if V(y) and $C^*(y)$ corresponding to $[j^V]_{P,\pi}$ are known. The local liquid velocity V_ℓ in the film adjacent to parallel clay platelets is found from

$$V_\ell = \int_0^{y''} \frac{1}{\eta} \int_{y'}^{d} f_\ell \, dy \, dy' \tag{3.56}$$

where η is the viscosity of the liquid, which is generally a function of y, and f_ℓ is the driving force acting on the liquid expressed per unit volume. f_ℓ corresponding to $[j^V]_{P,\pi}$ is given by

$$f_\ell = FC_0\left(u - \frac{1}{u}\right) \frac{-\Delta E}{\Delta X} \tag{3.57}$$

where F is the Faraday constant, and $C_0(u - 1/u)$ represents the volume charge density. f_ℓ in Eq. (3.57) is the force caused by the presence of a net charge in the liquid. C^*, the liquid concentration in the liquid, is equal to one. Thus we have

$$L_{VE} = \left(\frac{j^V}{-(\Delta E/\Delta X)} \right)_{P,\pi} = \frac{\theta}{\lambda d} \int_0^d \int_0^{y''} \frac{1}{\eta} \int_{y'}^{d} FC_0\left(u - \frac{1}{u}\right) dy \, dy' \, dy'' \tag{3.58}$$

In addition, by this method the phenomenological coefficients are equal, $L_{ik} = L_{ki}$. The value of u(y) in the integral expressions of phenomenological coefficients such as Eq. (3.58) can be obtained using the Gouy theory if boundary conditions are known. Assuming that the viscosity of the liquid and the electric mobilities of the ions are position independent, and the boundary conditions are known, one can find the analytical expressions for the coefficients.

Groenevelt et al. (1978) obtained the values of the phenomenological coefficients. Under their experimental conditions, at the first stage, $j^V = 0$ and $I = 0$. That is,

$$0 = L_V \, \Delta P + L_{VD} \, \Delta\pi + L_{VE} \, \Delta E \tag{3.59a}$$

$$-j^D \quad \Delta X = L_{DV} \, \Delta P + L_D \, \Delta\pi + L_{DE} \, \Delta E \tag{3.59b}$$

$$0 = L_{EV} \, \Delta P + L_{ED} \, \Delta\pi + L_E \, \Delta E \tag{3.59c}$$

To calculate the coefficients, three unknown constants—d, C_0, and λ/θ—in the analytical expressions [e.g. the right side of Eq. (3.58)] must be determined. The reflection coefficient σ and the coefficient τ, which is related to what may be called the "charge fluidity," are defined, respectively, by

$$\sigma = \left(\frac{\Delta P}{\Delta \pi} \right)_{j^V=0, I=0} = \frac{L_{ED}L_{VE} - L_{VD}L_E}{L_E L_V - L_{EV}L_{VE}} \tag{3.60}$$

$$\tau = -\left(\frac{\Delta E}{\Delta \pi} \right)_{j^V=0, I=0} = \frac{L_{ED}L_V - L_{VD}L_{EV}}{L_E L_V - L_{EV}L_{VE}} \tag{3.61}$$

Equations (3.60) and (3.61) are obtained from Eqs. (3.59a) and (3.59b). Since all analytical expressions for the phenomenological coefficients include the term $\theta/\lambda d$ as shown in the right side of Eq. (3.58), the right sides of Eqs. (3.60) and (3.61) are composed of only the integral parts in the analytical expressions. Consequently, the right sides of Eqs. (3.60) and (3.61) become functions only of d and C_0, respectively [see Eq. (3.58)]. The values of two coefficients, σ and τ, are calculated from the experimental values of the three forces at a time immediately after the stabilization period (i.e. at t = 4 hr). The values of d and C_0 can be obtained by finding values to satisfy Eqs. (3.60) and (3.61) by an iterative method (calculations of various combinations of values of C_0 and d). Then the value of λ/θ is calculated from the value of $-j^D \, \Delta X$ in Eq. (3.59b) obtained from the experimental data.

Finally, they analyzed their experimental data using Eqs. (3.52). If the numerical values of all the phenomenological coefficients are known, the response of the system to an imposed $\Delta\pi$ can be predicted from Eqs. (3.59). At all times, except for the initial stabilization period, the values of ΔP and ΔE, corresponding to the value of $\Delta\pi$, can be calculated from Eqs. (3.59a) and (3.59b). This, in turn, predicts the value of $\Delta\pi$ at the next state in time, and the procedure is then repeated.

The numerical values of d, C_0, λ/θ, and the nine phenomenological coefficients are now known. For d = 350 Å,

$$C_0 = 5.06 \times 10^{-4}\ mol/l \quad \left.\begin{aligned} L_{VD} &= 1.09 \times 10^{-10} \\ L_{VD} = L_{DV} &= 6.20 \times 10^{-10} \\ L_D &= 2.39 \times 10^{-8} \end{aligned}\right\} cm^2\ sec^{-1}/(dyne\ cm^{-2})$$

$$\lambda/\theta = 4.2\ (\theta = 0.96) \quad \left.\begin{aligned} L_{VE} = L_{EV} &= 0.105 \\ L_{DE} = L_{ED} &= 3.70 \end{aligned}\right\} cm^2\ sec^{-1}/(esu\ of\ potential)$$

$$L_E = 5.76 \times 10^8\ (esu\ of\ charge)\ cm^{-1}\ sec^{-1}/(esu\ of\ potential)$$

The measured and calculated (solid lines) changes in concentration, pressure, and voltage with time are shown in Fig. 3.33. The calculated values of ΔP and ΔC before shorting the electrodes coincide precisely with the measured values. The calculated values of ΔV (= ΔE x 300) before shorting the electrodes indicate that some type of drift probably occurred during the ΔV measurement. In addition, they concluded from the experimental data that shorting of the clay-water system is not perfect. According to the results of Demir (1988) (see Section 3.4.1), the values of L_V were 0.2132 Ohm for Run 1 and 0.2315 Ohm for Run 2, and the values of L_{EV} are 0.216 mA/MPa for Run 1 and 0.185 mA/MPa for Run 2.

3.4.5 Future Research Effort

In solution flow through a clay layer, coefficients such as R, and phenomenological coefficients such as hydraulic conductivity L_v, depend on concentrations of the outer solutions even if the clay layer is rigidly restricted. In addition, the electric potential difference, which is one of three driving forces for solution flow, is influenced not only by the concentration of the outer solution but also by the values of ΔP and/or $\Delta\pi$. Quantitative studies of these relationships have scarcely begun. Therefore, at the present stage, the application of irreversible thermodynamics to solution flow through a rigidly restricted clay layer under simple conditions $\Delta P = 0$ or $\Delta\pi = 0$ and/or $\Delta E = 0$ is very important to advance research on flow through clay layers. The dependence of the phenomenological coefficients and the streaming potential on concentrations of the outer solutions and dependence of the streaming potential on ΔP and/or $\Delta\pi$ must be measured. The dependence of L_v on concentration of the outer solution and that of streaming potential on ΔP at each concentration of the outer solution must be studied under conditions where $\Delta\pi = 0$. In addition to experimental measurements, the theoretical analysis of streaming potential has to be made using measured results and a model for solution flow through a clay layer. A dimensionless parameter such as the Peclet number may be useful in this analysis. It will be possible to analyze solution flows through clay layers under natural conditions, where $\Delta\pi \neq 0$, $\Delta P \neq 0$, and the clay layers are not rigidly restricted, only after these dependencies are clarified experimentally and theoretically.

ACKNOWLEDGEMENTS

Figures 3.5 and 3.9 are from *Discussions of the Faraday Society*, © The Royal Society of Chemistry. Figure 3.6 is reprinted from *Soil Science*, © by the Williams and Wilkins Co., Baltimore. Used with permission. Figures 3.7 and 3.18 are reprinted by permission from *Nature*, © Macmillan Journals Limited. Figures 3.8, 3.10, 3.29, 3.30 and 3.32 are reprinted by permission of Academic Press, Inc. Figure 3.13 is reproduced from the *Transactions of Irrigation, Drainage and Reclamation Engineering*, © The Japanese Society of Irrigation, Drainage and Reclamation Engineering. Figure 3.15 is reproduced from *Soil Science and Plant Nutrition*, © The Japanese Society of Soil Science and Plant Nutrition. Figures 3.18, 3.22, 3.26, 3.28, and 3.33 are reproduced from the *Soil Science Society of America Proceedings* by permission of the Soil Science Society of America. Figures 3.23 and 3.24 are reproduced from the *Journal of Environmental Quality*, The American Society of Agronomy. Figures 3.27, 3.28, and 3.31 are reprinted with permission from the *Journal of Physical Chemistry*, © America Chemical Society.

REFERENCES

Akae, T., Estimation of existing particle linkage modes in clay suspensions. Dispersion-flocculation behavior and rheological properties of bentonite-water systems (1), *Trans. Jpn. Soc. Irrig. Drain. Reclam. Eng.* 133:37-42 (1988); 148:75-80 (1990). (Japanese with English summary)

Aomine, S., and K. Egashira, Flocculation of allophane clays by electrolytes, *Soil Sci. Plant Nutr. Tokyo* 14:94-98 (1968).

Arora, H. S., and N. T. Coleman, The influence of electrolyte concentration on flocculation of clay suspensions, *Soil Sci.* 127:134-139 (1979).

Aylmore, L. A. G., and J. P. Quirk, Swelling of clay-water systems, *Nature* 189:1752-1753 (1959).

Barouch, E., E. Matijevic, T. A. Ring, and J. M. Finlan, Heterocoagulation: II. Interaction energy of two unequal spheres, *J. Colloid Interface Sci.* 53:1-9 (1978).

Baudracco, J., and Y. Tardy, Dispersion and flocculation of clays in unconsolidated sandstone reservoirs subjected to percolation with NaCl and $CaCl_2$ solutions at different temperatures, *Appl. Clay Sci.* 3:347-360 (1988).

Bell, G. M., S. Levine, and L. N. McCartney, Approximate methods of determining the double-layer free energy of interaction between two charged colloidal spheres, *J. Colloid Interface Sci.* 33:335-359 (1970).

Bentz, J., Electrostatic potential between concentric surfaces: spherical, cylindrical, and planar, *J. Colloid Interface Sci.* 90:164-182 (1982).

Blackmore, A. V., and R. D. Miller, Tactoid size and osmotic swelling of montmorillonite, *Soil Sci. Soc. Am. Proc.* 25:169-173 (1961).

Blackmore, A. V., and B. P. Warkentin, Swelling of calcium montmorillonite, *Nature* 186:823-824 (1960).

Bolt, G. H., *Soil Chemistry*, Chap. 11 Part B, *Physico-Chemical Models*, Elsevier, New York (1979).

Bolt, G. H., and R. D. Miller, Compression studies of illite suspensions, *Soil Sci. Soc. Am. Proc.* 19:285-288 (1955).

Bratko, D., A. Luzar, and S. H. Chen, Electrostatic model for protein/reverse micelle complexation, *J. Chem. Phys.* 89:545-550 (1988).

Bresler, E., Anion exclusion and coupling effects in nonsteady transport through unsaturated soils: 1. Theory, *Soil Sci. Soc. Am. Proc.* 37:663-669 (1973).

Bresler, E., Theoretical modeling of mixed-electrolyte solution flows for unsaturated soils, *Soil Sci.* 125:196-203 (1978).

Burgreen, D., and F. R. Nakache, Electrokinetic flow in ultrafine capillary slits, *J. Phys. Chem.* 68:1084-1091 (1964).

Casimir, H. B. G., and D. Polder, The influence of retardation on the London-van der Waals forces, *Phys. Rev.* 73:360-372 (1948).

Chan, D. Y. C., R. M. Pashley, and J. P. Quirk, Surface potentials derived from ion exclusion measurements on homoionic montmorillonite and illite, *Clays Clay Minerals* 32:131-138 (1984).

Cohen, R. R., and C. J. Radke, Streaming potentials of nonuniformly charged surfaces, *J. Colloid Interface Sci.* 141:338-347 (1991).

Czarnecki, J., van der Waals attraction energy between sphere and half-space, *J. Colloid Interface Sci.* 72:361-362 (1979).

Dane, J. H., and A. Klute, Salt effect on the hydraulic properties of a swelling soil, *Soil Sci. Soc. Am. J.* 41:1043-1049 (1977).

Demir, I., The interaction of hydraulic and electrical conductivities, streaming potential, and salt filtration during the flow of chloride brines through a smectite layer at elevated pressures, *J. Hydrol.* 98:31-52 (1988).

Derjaguin, B. V., A theory of interaction of particles in the presence of electric double layers and the stability of lyophobic colloids and disperse systems, *Acta Physicochim. URSS* 10:333-346 (1939).

Derjaguin, B. V., A theory of the heterocoagulation, interaction and adhesion of dissimilar particles in solutions of electrolytes, *Discuss. Faraday Soc.* 18:85-98 (1954).

Derjaguin, B. V., and I. I. Abrikosova, Direct measurements of molecular attraction of solids, *J. Phys. Chem. Solids* 5:1-10 (1958).

Derjaguin, B. V., and N. V. Churaev, Structural component of disjoining pressure, *J. Colloid Interface Sci.* 49:249-255 (1974).

Derjaguin, B. V., and N. V. Churaev, Disjoining pressure of thin layers of binary solutions, *J. Colloid Interface Sci.* 62:369-380 (1977).

Derjaguin, B. V., and N. V. Churaev, On the question of determining the concept of disjoining pressure and its role in the equilibrium and flow of thin films, *J. Colloid Interface Sci.* 66:389-398 (1978).

Derjaguin, B. V., and L. Landau, Theory of the stability of strongly charged lyophobic sols and of the adhesion of strongly charged particles in solutions of electrolytes, *Acta Physicochim. URSS* 14:633-662 (1941).

Derjaguin, B. V., T. N. Voropayeva, B. N. Kabanov, and A. S. Titiyevskaya, Surface forces and the stability of colloids and disperse systems, *J. Colloid Sci.* 19:113-135 (1964).

Egashira, K., Viscosities of allophane and imogolite clay suspensions, *Clay Sci.* 5:87-95 (1977).

Elrick, D. E., D. E. Smiles, N. Baumgartner, and P. H. Groenevelt, Coupling phenomena in homoionic montmorillonite: I. Experimental, *Soil Sci. Soc. Am. J.* 40:490-491 (1976).

Frens, G., and J. T. G. Overbeek, Repeptization and the theory of electrocratic colloids, *J. Colloid Interface Sci.* 38:376-387 (1972).

Gairon, S., and D. Swartzendruber, Water flux and electrical potentials in water-saturated bentonite, *Soil Sci. Soc. Am. Proc.* 39:811-817(1975).

Goldberg, S., and R. A. Glaubig, Effect of saturating cation, pH, and aluminum and iron oxide on the flocculation of kaolinite and montmorillonite, *Clays Clay Minerals* 35:220-227 (1987).

Greacen, E. L., Swelling forces in straining clays, *Nature* 184:1695-1697 (1959).

Gregory, J., Approximate expression for the interaction of diffuse electric double layers at constant charge, *J. Chem. Soc., Faraday Trans., ser. 2* 69:1723-1728 (1973).

Gregory, J., Approximate expression for retarded van der Waals interaction *J. Colloid Interface Sci.* 83:138-145 (1981).

Groenevelt, P. H., and G. H. Bolt, Non-equilibrium thermodynamics of the soil water system, *J. Hydrol.* 7:358-388 (1969).

Groenevelt, P. H., and D. E. Elrick, Coupling phenomena in saturated homoinic montmorillonite: II. Theoretical, *Soil Sci. Soc. Am. J.* 40:820-823 (1976).

Groenevelt, P. H., D. E. Elrick, and T. J. M. Blom, Coupling phenomena in saturated homoinic montmorillonite: III. Analysis, *Soil Sci. Soc. Am. J.* 42:671-674 (1978).

Hamaker, H. C., The London-van der Waals attraction between spherical particles, *Physica* 4:1058-1072 (1937).

Harada, Y., and K. Wada, Release and uptake of protons by allophanic soils in relation to their CEC and AEC, *Soil Sci. Plant Nutr. Tokyo* 19:73-82 (1973).

Haydon, P. R., and D. L. Graf, Studies of smectite membrane behavior. Temperature dependence, 20-180°C, *Geochim. Cosmochim. Acta* 50:115-121 (1986).

Healy, T. W., D. Chan, and L. R. White, Colloidal behavior of materials with ionizable group surfaces, *Pure Appl. Chem.* 52:1207-1219 (1980).

Hesterberg, D., and A. L. Page, Critical coagulation concentration of sodium and potassium illite as affected by pH, *Soil Sci. Soc. Am J.* 54:735-739 (1990).

Hitchon, B., and I. Friedman, Geochemistry and origin of formation water in western Canada sedimentary basin, I. Stable isotopes of hydrogen and oxygen, *Geochim. Cosmochim. Acta* 33:1321-1349 (1969).

Hogg, R., T. W. Healy, and D. W. Fuerstenau, Mutual coagulation of colloidal dispersions, *Trans. Faraday Soc.* 62:1638-1651 (1966).

Honig, E. P., and P. M. Mul, Tables and equations of the double layer repulsion at constant potential and at constant charge, *J. Colloid Interface Sci.* 36:258-272 (1971).

Horikawa, Y., R. S. Murray, and J. P. Quirk, The effect of electrolyte concentration on the zeta potentials of homoionic montmorillonite and illite, *Colloids and Surfaces* 32:181-195 (1988).

Horn, R. G., D. T. Smith, and W. Haller, Surface forces and viscosity of water measured between silica sheets, *Chem. Phys. Lett.* 162:404-408 (1989).

Hurst, C. A., and E. S. A. Jordine, Role of electrostatic energy barriers in the expansion of lamellar crystals, *J. Chem. Phys.* 41:2735-2745 (1964).

Israelachvili, J. N., Measurement of forces between immersed surfaces in electrolyte solutions, *Discuss. Faraday Soc.* 65:20-24 (1978).

Israelachvili, J. N., Measurements of hydration forces between macroscopic surfaces, *Chemica Scripta* 25:7-14 (1985).

Israelachvili, J. N., *Intermolecular and Surface Forces*, 2nd Ed., Academic Press (1991).

Israelachvili, J. N., and G. E. Adams, Measurement of forces between two mica surfaces in aqueous solutions in the range 0-100 nm, *J. Chem. Soc. Faraday Trans.* ser. 1, 74:975-1001 (1978).

Israelachvili, J. N., and R. M. Pashley, Molecular layering of water at surfaces and origin of repulsive hydration forces, *Nature* 306:249-250 (1983).

Israelachvili, J. N., and D. Tabor, The treatment of van der Waals dispersion forces in the range 1.5 to 130 nm, *Proc. Roy. Soc. London* A331:19-38 (1972).

Iwata, S., Thermodynamics of soil water: IV. Chemical potential of soil water, *Soil Sci.* 117:135-139 (1974).

Jacazio, G., R. F. Probstein, A. A. Sonin, and D. Young, Electrokinetic salt rejection in hyperfiltration through porous materials, theory and experiment, *J. Phys. Chem.* 76:4015-4023 (1972).

James, A. E., and D. J. A. Williams, Electrical double-layer interaction energy for two cylinders, *J. Colloid Interface Sci.* 79:33-46 (1981).

James, A. E., and D. J. A. Williams, Numerical solution of the Poisson-Boltzmann equation, *J. Colloid Interface Sci.* 107:44-59 (1985).

Jin, M., and M. M. Sharma, A model for electrochemical and electrokinetic coupling in inhomogeneous porous media, *J. Colloid Interface Sci.* 142:61-73 (1991).

Jones, J. E., and S. Levine, A comparison of the conditions of constant surface potential and constant charge in the DLVO theory of colloid stability, *J. Colloid Interface Sci.* 30:241-246 (1969).

Jordine, E. S. A., G. B. Bodman, and A. H. Gold, Effect of surface ions on the mutual interaction of montmorillonite particles, *Soil Sci.* 94:371-378 (1962).

Kahn, A., The flocculation of sodium montmorillonite by electrolytes, *J. Colloid Sci.* 13:51-60 (1958).

Karube, J., Microstructure of allophane in disperse system by light scattering method, Studies on the structure formation of volcanic ash soil, I. *Trans. Jpn. Soc. Irrig. Drain. Reclam. Eng.* 98:7-14 (1982a). (Japanese with English summary)

Karube, J., Domain size measurement of dispersed allophane with membrane filter and a discussion on the allophane disperse system, Studies on the structure formation of volcanic ash soil, II. *Trans. Jpn. Soc. Irrig. Drain. Reclam. Eng.* 99:17-23 (1982b). (Japanese with English summary)

Karube, J., Physical properties and microstructure of allophane, Studies on the structure formation of volcanic ash soil, III. *Trans. Jpn. Soc. Irrig. Drain. Reclam. Eng.* 107:47-53 (1983). (Japanese with English summary)

Katchalsky, A., and P. R. Curran, *Non-equilibrium Thermodynamics in Biophysics*, Harvard University Press, Cambridge, Mass. (1965).

Kedem, O., and A. Katchalsky, Permeability of composite membranes, *Trans. Faraday Soc.* 59:1918-1953 (1963).

Kemper, W. D., Water and ion movement in thin films as influenced by the electrostatic charge and diffuse layer of cations associated with clay mineral surfaces, *Soil Sci. Soc. Am. Proc.* 24:10-16 (1960).

Kemper, W. E., in Chap. 6 *Soil Water*, D. R. Nielsen, R. D. Jackson, J. W. Cary, and D. D. Evans, eds., American Society of Agronomy, Madison, Wis. (1972).

Kemper, W. D., and N. A. Evans, Movement of water as affected by free energy and pressure gradients: II. Restriction of solutes by membranes, *Soil Sci. Soc. Am. Proc.* 27:485-490 (1963).

Kemper, W. D., and D. E. L. Maasland, Reduction in salt content of solution on passing through thin films adjacent to charged surfaces, *Soil Sci. Soc. Am. Proc.* 28:318-323 (1964).

Kemper, W. D., and J. B. Rollins, Osmotic efficiency coefficient across compacted clays, *Soil Sci. Soc. Am. Proc.* 30:529-534 (1966).

Kemper, W. D., and J. C. Van Schaik, Diffusion of salts in clay-water systems, *Soil Sci. Soc. Am. Proc.* 30:534-540 (1966).

Keren, R., and G. A. O'Connor, Gypsum dissolution and sodic soil reclamation as affected by water flow velocity, *Soil Sci. Soc. Am. J.* 46:726-732 (1982).

Keren, R., I. Shainberg, and E. Klein, Setting and flocculation value of sodium-montmorillonite particles in aqueous media, *Soil Sci. Soc. Am. J.* 52:76-88 (1988).

Keren, R., and M. J. Singer, Effect of low electrolyte concentration on hydraulic conductivity of sodium/calcium-montmorillonite-sand system, *Soil Sci. Soc. Am. J.* 52:368-373 (1988).

Kharaka, Y. K., and W. C. Smalley, Flow of water and solutes through compacted clays, *Am. Assoc. Pet. Geol. Bull.* 60:973-980 (1976).

Kijlstra, J., Polarizability effects in the electrostatic repulsion between charged colloidal particles, *J. Colloid Interface Sci.* 153:30-36 (1992).

Kitchener, J. A., and A. P. Prosser, Direct measurement of the long-range van der Waals forces, *Proc. Roy. Soc. London* A242:403-409 (1957).

Kjellander, R., and S. Marcelja, Double layer interaction in the primitive model and the corresponding Poisson-Boltzmann description, *J. Phys. Chem.* 90:1230-1232 (1986).

Kjellander, R., S. Marcelja, and J. P. Quirk, Attractive double-layer interactions between calcium clay particles, *J. Colloid Interface Sci.* 126:194-222 (1988).

Kraehenbuehl, F., J. F. Stoeckli, F. Brunner, G. Kahr, and M. Muller-Vonmoos, Study of the water-bentonite-system by vapor adsorption, immersion calorimetry and X-ray techniques, I. Micropore volumes and internal surface areas, following Dubinin's theory, *Clays Clay Minerals* 22:1-9 (1987).

Krozel, J. W., and D. A. Saville, Electrostatic interactions between two spheres: Solution of the Debye-Hückel equation with a charge regulation boundary condition, *J. Colloid Interface Sci.* 150:365-373 (1992).

Kubota, T., Surface chemical properties of volcanic ash soil, *Nat. Inst. Agric. Bull.* 28:1-74 (1976).

Levine, S., J. R. Marriott, and K. Robinson, Theory of electrokinetic flow in a narrow parallel-plate channel, *J. Chem. Soc. Faraday Trans.* ser. 1, 71:1-11 (1975).

Low, P. F., The effect of osmotic pressure on the diffusion rate of water, *Soil Sci.* 80:95-100 (1955).

Low, P. F., The swelling of clay: II. Montmorillonites, *Soil Sci. Soc. Am. J.* 44:667-676 (1980).

Low, P. F., Structural and other forces involved in the swelling of clays, NATO Advanced Workshop on Clay Swelling and Expansive Soils, Cornell Univ., Ithaca, NY, Aug. 1991, in press.

MacEwan, D. M. C., Complexes of clays with organic compounds: I. Complex formation between montmorillonite and halloysite and certain organic liquids, *Trans. Faraday Soc.* 44:349-367 (1948).

Madsen, F., and M. Müller-Vonmoos, The swelling behavior of clays, *Appl. Clay Sci.* 4:143-156 (1989).

McCartney, L. N., and S. Levine, An improvement on Derjaguin's expression of small potentials for the double layer interaction energy of two spherical colloidal particles, *J. Colloid Interface Sci.* 30:345-354 (1969).

McKelvey, J. G., and J. H. Milne, The flow of salt solutions through compacted clay, *Clays Clay Minerals* 9:248-250 (1962).

McNeal, B. L., W. A. Norvell, and N. T. Coleman, Effect of solution composition on the swelling of extracted soil clays, *Soil Sci. Soc. Am. Proc.* 30:313-317 (1966).

McQuarrie, D. A., W. Olivers, D. Henderson, and L. Blum, On Derjaguin's formula for the force between planar double layers, *J. Colloid Interface Sci.* 77:272-273 (1980).

Miller, S., and P. F. Low, Characterization of the electrical double layer of montmorillonite, *Langmuir* 6:572-578 (1990).

Mungan, N., and F. W. Jessen, Studies in fractionated montmorillonite suspensions, *Proc. 11th Nat. Conf. Clays Clay Minerals*, 282-294 (1963).

Nakagawa, T., and M. Ishiguro, Hydraulic conductivity of an allophanic Andisol as affected by solution pH, *J. Environ. Qual.* 23:208-210 (1994).

Neogi, P., and E. Ruckenstein, Viscoelectric effects in reverse osmosis, *J. Colloid Interface Sci.* 79:159-169 (1981).

Norrish, K., Manner of swelling of montmorillonite, *Nature* 173:256-257 (1954a).

Norrish, K., The swelling of montmorillonite, *Discuss. Faraday Soc.* 18:120-134 (1954b).

Norrish, K., Forces between clay particles, *Proc. Int. Clay Conf.*, Madrid, 375-383 (1972).

Norrish, K., and J. P. Quirk, Use of electrolytes to control swelling, *Nature* 173:255-256 (1954).

Norrish, K., and J. A. Rausell-Colom, Low-angle X-ray diffraction studies of the swelling of montmorillonite and vermiculite, *Clays Clay Miner.* 10:123-149 (1963).

Ohshima, H., and T. Kondo, Comparison of three models on double layer interaction, *J. Colloid Interface Sci.* 126:382-383 (1988a).

Ohshima, H., and T. Kondo, Approximate analytic expression for double-layer interaction at moderate potentials, *J. Colloid Interface Sci.* 122:591-592 (1988b).

Ohshima, H., and T. Kondo, Electrokinetic flow between two parallel plates with surface charge layers: Electro-osmosis and streaming potential, *J. Colloid Interface Sci.* 135:443-448 (1990).

Oster, J. D., I. Shainberg, and J. D. Wood, Flocculation value and gel structure of Na/Ca montmorillonite and illite suspensions, *Soil Sci. Soc. Am. J.* 44:955-959 (1980).

Pailthorpe, B. A., and W. B. Russel, The retarded van der Waals interaction between spheres, *J. Colloid Interface Sci.* 89:563-566 (1982).

Pashley, R. M., Hydration forces between mica surfaces in aqueous electrolyte solutions, *J. Colloid Interface Sci.* 80:153-162 (1981a).

Pashley, R. M., DLVO and hydration forces between mica surfaces in Li^+, Na^+, K^+, and Cs^+ electrolyte solutions: A correlation of double-layer and hydration forces with surface cation exchange properties, *J. Colloid Interface Sci.* 83:531-546 (1981b).

Pashley, R. M., and J. N. Israelachvili, DLVO and hydration forces between mica surfaces in Mg^{2+}, Ca^{2+}, Sr^{2+}, and Ba^{2+} chloride solutions, *J. Colloid Interface Sci.* 97:446-455 (1984).

Pashley, R. M., and J. P. Quirk, Ion exchange and interparticle forces between clay surfaces, *Soil Sci. Soc. Am. J.* 53:1660-1667 (1989).

Pennino, U. D., E. Mazzega, S. Valeri, A. Alietti, M. F. Brigatti, and L. Poppi, Interlayer water and swelling properties of monoionic montmorillonites, *J. Colloid Interface Sci.* 84:301-309 (1981).

Prieve, D. C., and E. Ruckenstein, Role of surface chemistry in particle deposition, *J. Colloid Interface Sci.* 60:337-348 (1977).

Pupisky, H., and I. Shainberg, Salt effects on the hydraulic conductivity of a sandy soil, *Soil Sci. Soc. Am. J.* 43:429-433 (1979).

Quirk, J. P., and L. A. Aylmore, Swelling and shrinkage of clay-water systems, *Proc. 7th Int. Cong. Soil Sci.* 378-386 (1960).

Quirk, J. P., and R. K. Schofield, The effect of electrolyte concentration on soil permeabiilty, *J. Soil Sci.* 6:163-178 (1955).

Rabinovich, Y. I., B. V. Derjaguin, and N. V. Churaev, Direct measurements of long range surface force in gas and liquid media, *Adv. Colloid Interface Sci.* 16:63-78 (1982).

Rausell-Colom, J. A., El hinchamiento de la montmorillonita-sodia y del complejo montmorillonita-krillium en electrolitos, Doctoral thesis, University of Madrid (1958).

Rolfe, P. F., and L. A. G. Aylmore, Water and salt flow through compacted clays: I. Permeability of compacted illite and montmorillonite, *Soil Sci. Soc. Am. J.* 41:489-495 (1977).

Rolfe, P. F., and L. A. G. Aylmore, Water and salt flow through compacted clays: II. Electrokinetics and salt sieving, *J. Colloid Interface Sci.* 79:301-307 (1981).

Ruckenstein, E., On the thermodynamics of double-layer forces in open and closed system, *J. Colloid Interface Sci.* 82:490-498 (1981).

Ruckenstein, E., and P. Krape, On the enzymatic superactivity in ionic reverse micelles, *J. Colloid Interface Sci.* 139:408-436 (1990).

Sands, R., and C. P. P. Reid, The osmotic potential of soil water in plant/soil systems, *Aust. J. Soil Res.* 18:13-25 (1980).

Sato, T., T. Watanabe, and R. Otsuka, Effects of layer charge, charge location, and energy change on expansion properties of dioctahedral smectites, *Clays Clay Minerals* 40:103-113 (1992).

Suarez, D. L., J. D. Rhoades, R. Lavado, and C. M. Grieve, Effect of pH on saturated hydraulic conductivity and soil dispersion, *Soil Sci. Soc. Am. J.* 48:50-55 (1984).

Schenkel, J. H., and J. A. Kitchener. A test of the Derjaguin-Verwey-Overbeek theory with a colloid suspension, *Trans. Faraday Soc.* 62:161-173 (1960).

Sengupta, A. K., and K. D. Papadopoulos, Electrical double-layer interaction between two eccentric spherical surfaces, *J. Colloid Interface Sci.* 149:135-152 (1992).

Shainberg, I., and W. D. Kemper, Transport numbers and mobilities of ions in bentonite membranes, *Soil Sci. Soc. Am. Proc.* 36:477-582 (1972).

Shainberg, I., E. Bresler, and Y. Kausner, Studies on Na/Ca montmorillonite system: I, *Soil Sci.* 111:214-219 (1971).

Slade, P. G., J. P. Quirk, and K. Norrish, Crystalline swelling of smectite samples in concentrated NaCl solutions in relation to layer charge, *Clays Clay Minerals* 39:234-238 (1991).

Slade, P. G., and J. P. Quirk, The limited crystalline swelling of smectite in $CaCl_2$, $MgCl_2$, and $LaCl_3$ solutions, *J. Colloid Interface Sci.* 144:18-26 (1991).

Slater, J. G., and J. G. Kirkwood, The van der Waals forces in gases, *Phys. Rev.* 37:682-697 (1931).

Smith, F. G., and W. M. Deen, Electrostatic double-layer interactions for spherical colloids in cylindrical pores, *J. Colloid Interface Sci.* 78:444-465 (1980).

Smith, F. G., and W. M. Deen, Electrostatic effects on the partitioning of spherical colloids between dilute bulk solution and cylindrical pores, *J. Colloid Interface Sci.* 91:571-590 (1983).

Tabor, D., and R. H. S. Winterton, Surface forces: direct measurement of normal and retarded van der Waals forces, *Nature* 219:1120-1121 (1968).

Tabor, D., and R. H. S. Winterton, The direct measurement of normal and retarded van der Waals forces, *Proc. R. Soc. London* A321:435-450 (1969).

Thomas, I. L., and K. H. McCorkle, Theory of oriented flocculation, *J. Colloid Interface Sci.* 36:110-118 (1971).

Usui, S., Interface of electric double layers at constant surface charge, *J. Colloid Interface Sci.* 44:107-113 (1973).

Van Olphen, H., Stabilization of montmorillonite sols by chemical treatments: Part II, *Rec. Trav. Chim. Pays Bas.* 69:1308-1322 (1950).

Van Olphen, H., Rheological phenomena of clay sols in connection with the charge distribution of the micelles, *Discuss. Faraday Soc.* 11:82-84 (1951).

Van Olphen, H., Interlayer forces in bentonite, *Proc. 3rd Natl. Conf. Clays Clay Minerals*, Natl. Acad. Sci., 204-224 (1956).

Van Olphen, H., Forces between suspended bentonite particles, *Proc. 5th Natl. Conf. Clays Clay Minerals*, Natl. Acad. Sci., 418-437 (1954).

Van Olphen, H., *Introduction to Clay Colloid Chemistry*, Interscience Publishers (1963).

Van Riemsdijk, W. H., G. H. Bolt, L. K. Koopal, and J. Blaakmeer, Electrolyte adsorption on heterogeneous surfaces: Adsorption models, *J. Colloid Interface Sci.* 109:219-228 (1986).

Verwey, E. J. W., and J. T. G. Overbeek, *Theory of the Stability of Lyophobic Colloids*, Elsevier, New York (1948).

Viani, B. E., P. F. Low, and C. B. Roth, Direct measurement of the relation between interlayer force and interlayer distance in the swelling of montmorillonite, *J. Colloid Interface Sci.* 96:229-244 (1983).

Warkentin, B. P., and R. K. Schofield, Swelling pressure of dilute Na-montmorillonite pastes, *Proc. 7th Natl. Conf. Clays Clay Minerals*, Natl. Acad. Sci., 343-349 (1958).

Warkentin, B. P., and R. K. Schofield, Swelling pressure of montmorillonite in NaCl solutions, *J. Soil Sci.* 13:98-105 (1962).

Warkentin, B. P., G. H. Bolt, and R. D. Miller, Swelling pressure of montmorillonite, *Soil Sci. Soc. Am. Proc.* 21:495-497 (1957).

Wiese, G. R., and T. W. Healy, Effect of particle size on colloid stability, *Trans. Faraday Soc.* 66:490-499 (1970).

Wilemski, G., Weak repulsive interaction between dissimilar electrical double layer, *J. Colloid Interface Sci.*, 88:111-116 (1982).

Yong, R. N., O. Taylor, and B. P. Warkentin, Swelling pressure of sodium montmorillonite at depressed temperatures, *Proc. 11th Natl. Conf. Clays Clay Minerals*, Natl. Acad. Sci., 268-281 (1963).

4

CAPILLARITY

4.1 AIR-WATER INTERFACES

The capillary forces involved in water retention and transmission in soils will be considered in this chapter.

4.1.1 Laplace Equation

The fundamental equation of capillarity, the Laplace equation is (Adam, 1938)

$$\Delta P = P_w - P_a = \sigma \left(\frac{1}{R_1} + \frac{1}{R_2} \right) \qquad (4.1)$$

Therefore,

$$P_w = P_a + \sigma \left(\frac{1}{R_1} + \frac{1}{R_2} \right) \qquad (4.2)$$

where ΔP is the pressure difference across a curved air-liquid surface, P_a the pressure on the air side and P_w on the liquid side, σ the surface tension of the air-liquid interface, and R_1 and R_2 the main radii of curvature. The sign of R is positive on a convex surface and negative on a concave surface.

The liquid pressure will be higher than the air pressure on a convex surface, and the reverse on a concave surface (Fig. 4.1). When the air pressure is equal to atmospheric pressure, taken as $P_a = 0$, the liquid pressure under the concave surface is called a negative pressure. The magnitude of water pressure is determined by the values of the main radii of curvature.

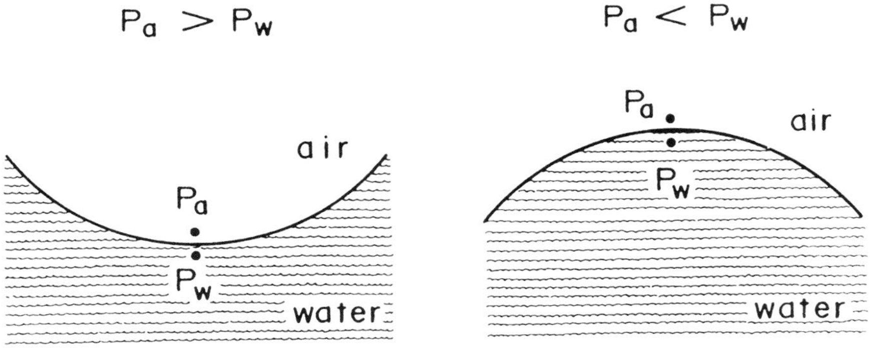

Fig. 4.1. Air-water interfaces for concave and convex surfaces.

4.1.2 Capillary Rise in a Tube

In a cylindrical tube the radii of curvature R_1 and R_2 of the meniscus, or curved air-water interface, are equal. Then Eq. (4.2) can be written

$$P_w = -\sigma\frac{2}{R} \qquad where\ P_a = 0 \tag{4.3}$$

The radius of curvature can be rewritten in terms of the radius of a capillary tube r and the contact angle θ, as shown in Fig. 4.2.

$$R = \frac{r}{\cos\theta} \tag{4.4}$$

Substituting the value of R from Eq. (4.4) into Eq. (4.3) yields

$$P_w = -\sigma\,\frac{2\cos\theta}{r} \tag{4.5}$$

Due to the negative pressure, water rises in the tube until it attains equilibrium (Fig. 4.3) at

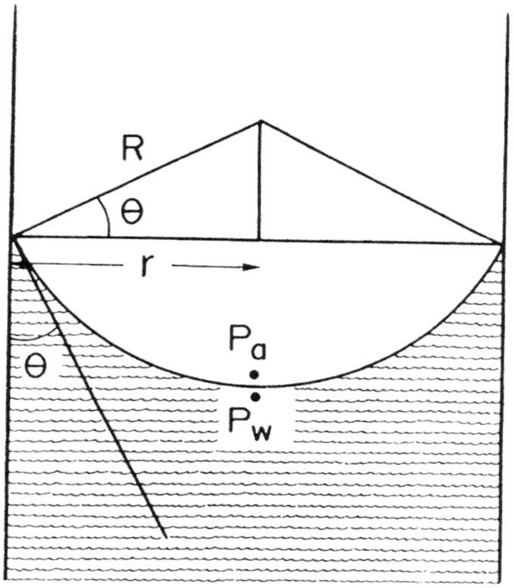

Fig. 4.2. Meniscus in a cylindrical tube.

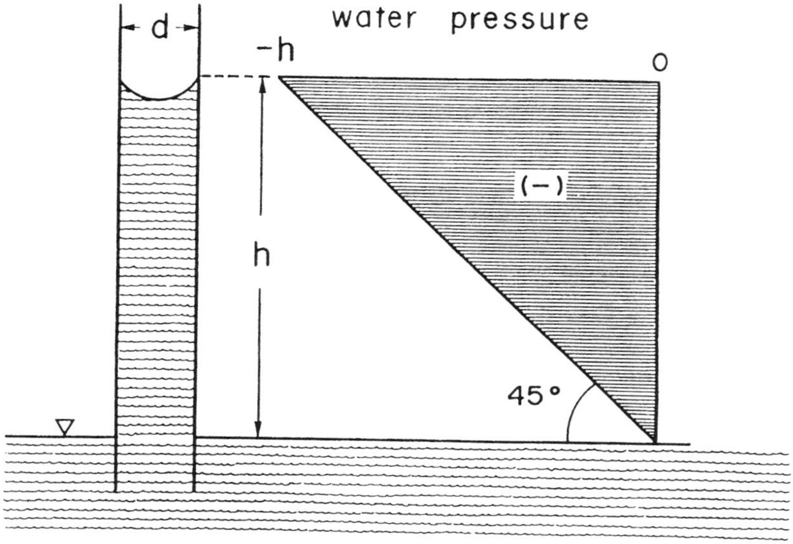

Fig. 4.3. Water pressure in capillary rise.

$$P_w = -\rho g h \tag{4.6}$$

where ρ is the density of water. Substituting Eq. (4.5) into Eq. (4.6) yields

$$\rho g h = \frac{2\sigma \cos\theta}{r} \tag{4.7}$$

Therefore,

$$h = \frac{2\sigma \cos\theta}{\rho g r} \tag{4.8}$$

When a liquid such as water completely wets the surface, the contact angle equals zero. For water, $\sigma = 74$ dyne/cm, $\rho = 1$ g/cm^3, and g = 980 cm/sec^2. Then Eq. (4.8) becomes

$$h \doteq \frac{0.15}{r} = \frac{0.3}{d} \tag{4.9}$$

where d is the diameter of the capillary tube in centimeters. Thus the height of water rise and the water pressure at the meniscus in the capillary tube are determined by its diameter.

4.1.3 Lens Water Between Spheres

The air-water interface for water between spheres has a bi-concave lens shape with a convex-concave lateral surface (Fig. 4.4a). The sign of the radius R_1 is positive and that of R_2 negative. The Laplace equation is

$$P_w = P_a + \sigma\left(\frac{1}{R_1} - \frac{1}{R_2}\right) \tag{4.10}$$

If R_2 is smaller than R_1, the water pressure P_w will be smaller than P_a. Negative capillary pressure of lens water h_c is given as

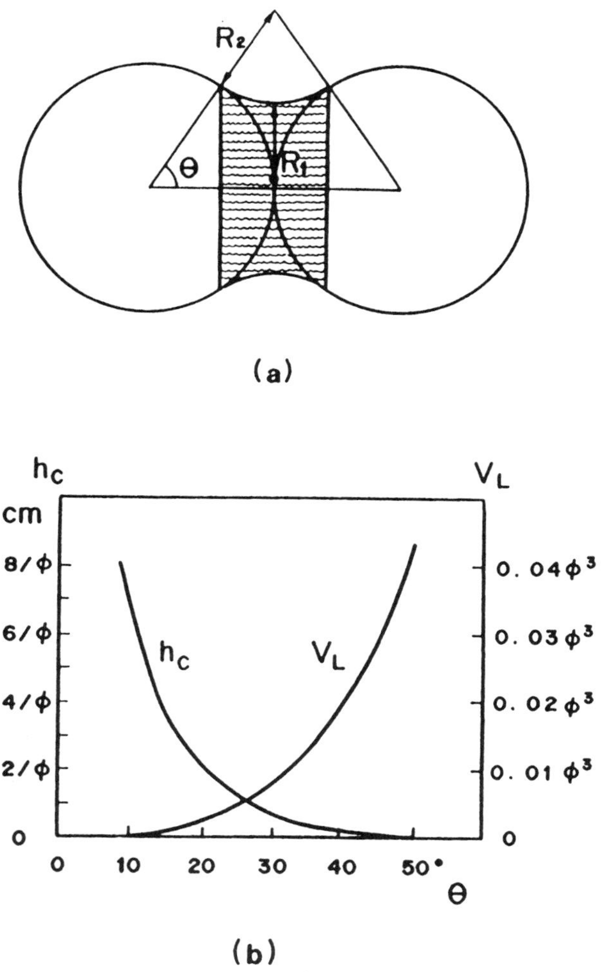

Fig. 4.4. Lens water and its capillary pressure h_c and volume V_L.

$$h_c = -\frac{\sigma}{\rho g}\left(\frac{1}{R_1} - \frac{1}{R_2}\right) = -\ 0.075\left(\frac{1}{R_1} - \frac{1}{R_2}\right) \tag{4.11}$$

The values of R_1 and R_2, and the volume of lens water V_L, can be calculated approximately from the equations (Dallavalle, 1943)

$$R_1 = \frac{\phi}{2} \tan \theta - R_2 \tag{4.12}$$

$$R_2 = \frac{\phi}{2} (\sec \theta - 1) \tag{4.13}$$

$$V_L = \frac{\pi \phi^3}{4} (\sec \theta - 1)^2 \left[1 - \left(\frac{\pi}{2} - \theta\right) \tan \theta\right] \tag{4.14}$$

where ϕ is the diameter of the sphere. The calculated values of h_c decrease rapidly with increase of θ, and become zero at $\theta = 53°$ (Fig. 4.4b).

Fisher (1926) reported that the results obtained from these equations were quite close to measured values. However, Waldron et al. (1961) found that the measured amount of water retained by a packing of glass beads greatly exceeds the amount predicted by the Fisher theory. They ascribed the lack of agreement to the omission of an adsorptive mechanism to account for a layer of adsorbed water on all surfaces of the spheres.

4.1.4 Meniscus Within a Pore Cell

In cubic packing of spheres, a pore cell is surrounded by eight spheres (Fig. 4.5). The vertical and horizontal sections of the cells are shown in Fig. 4.6. The diameter of an inscribed circle, d, varies with the distance z, which is the height from the neck defined in Fig. 4.6:

$$d = d(z) = d_n + \phi - (\phi^2 - 4z^2)^{1/2} \tag{4.15}$$

where d_n is the diameter (cm) of an inscribed circle in the neck and ϕ is the diameter (cm) of the sphere. The size of the neck, d_n, is

$$d_n = 0.414\phi \tag{4.16}$$

Hence Eq. (4.15) is rewritten

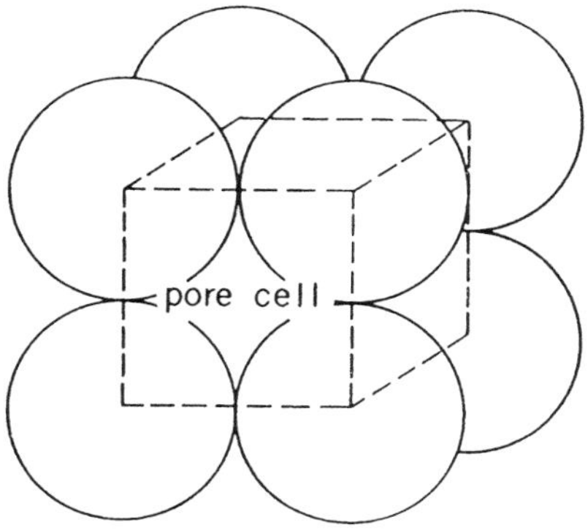

Fig. 4.5. Cubic packing of spheres.

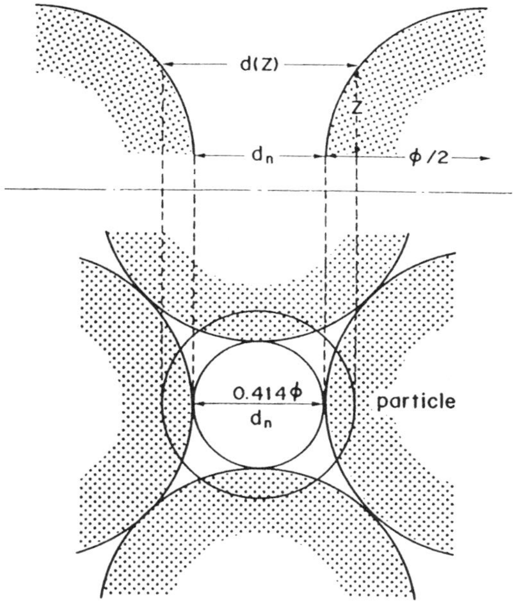

Fig. 4.6. Vertical and horizontal sections of a unit pore cell in cubic packing of spheres.

$$d(z) = 1.414\phi - (\phi^2 - 4z^2)^{1/2} \tag{4.17}$$

In close packing (Fig. 4.7), the neck is formed by three spheres and its diameter is

$$d_n = 0.155\phi \tag{4.18}$$

This value is the smallest diameter of inscribed circles in pore cells.

A meniscus forming at the neck will take the shape shown in Fig. 4.8. This meniscus has usually been considered a spherical surface, and its capillary pressure h_n calculated from Eq. (4.9). However, it is preferable to approximate it as a spheroidal surface, and the pressure equation is (Tabuchi, 1966a)

$$h_n = \frac{0.24}{d_n} \tag{4.19}$$

The calculated values obtained by Eq. (4.19) are nearer to experimental values than those of Eq. (4.9) (Tabuchi, 1966b). The general equation is

$$h_n = \frac{C_n}{d_n} \tag{4.20}$$

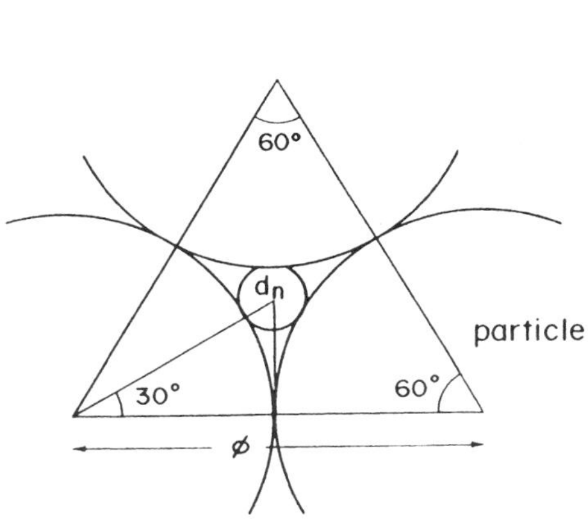

Fig. 4.7. Diameter of a neck d_n in close packing of spheres.

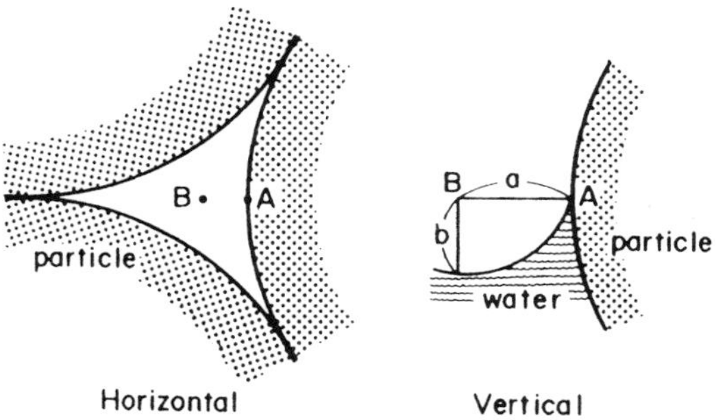

Fig. 4.8. Vertical and horizontal sections of a unit pore cell in close packing of spheres.

where C_n is the capillary coefficient of the meniscus in the neck. In cubic packing, from Eq. (4.16) and Eq. (4.20) we get

$$h_n = \frac{C_n}{0.414\phi} = \frac{2.42\ C_n}{\phi} \tag{4.21}$$

In closest packing (Fig. 4.9) from Eq. (4.18) and Eq. (4.20), we get

$$h_n = \frac{C_n}{0.155\phi} = \frac{6.45\ C_n}{\phi} \tag{4.22}$$

Substituting

$$H_n = h_n\phi \tag{4.23}$$

and using $C_n = 0.24$, Eqs. (4.21) and (4.22) become

$$H_n = 0.58 \text{ and } 1.55 \tag{4.24}$$

or

$$h_n = \frac{0.58}{\phi} \text{ and } \frac{1.55}{\phi} \tag{4.25}$$

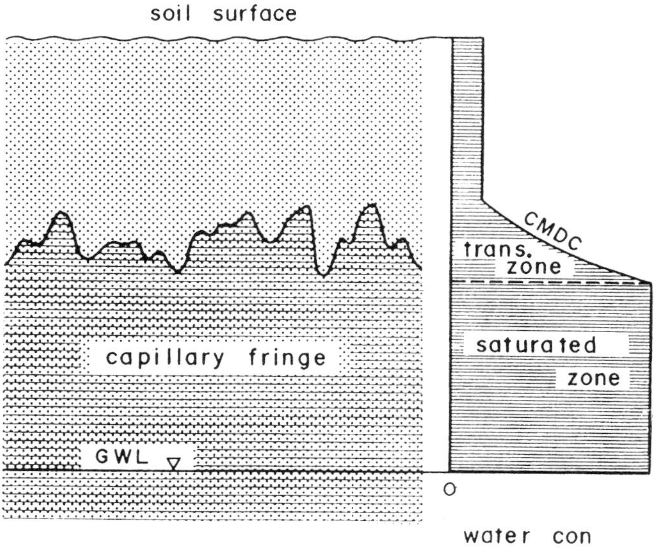

Fig. 4.9. Capillary fringe above ground water level (GWL) and the capillary moisture distribution curve (CMDC).

Hence the largest value of the capillary pressure formed in a neck becomes 1.55/ϕ. When the diameter of the sphere is 1 mm, h_n equals 15.5 cm.

4.2 CAPILLARY WATER IN SOILS

4.2.1 Capillary Rise

The capillary fringe in soil above the groundwater level, GWL, is a well-known phenomenon. The capillary moisture distribution curve (CMDC), is shown is Fig. 4.9. The water in the upper zone is isolated lens water and cannot move easily with changes of GWL. However, the water in the saturated and transition zones is connected with the groundwater and moves with fluctuations of GWL. We call these two zones the "capillary fringe." A model of capillaries with various diameters (Fig. 4.10) can explain the unsaturated zone. The height of capillary rise can be calculated with Eq. (4.9). At any height h_i, larger tubes contain air and smaller tubes contain water.

The capillary rise of water in soil, h_c, is inversely related to particle size, ϕ (Rode, 1969). The height of capillary rise into dry soil is lower than the height

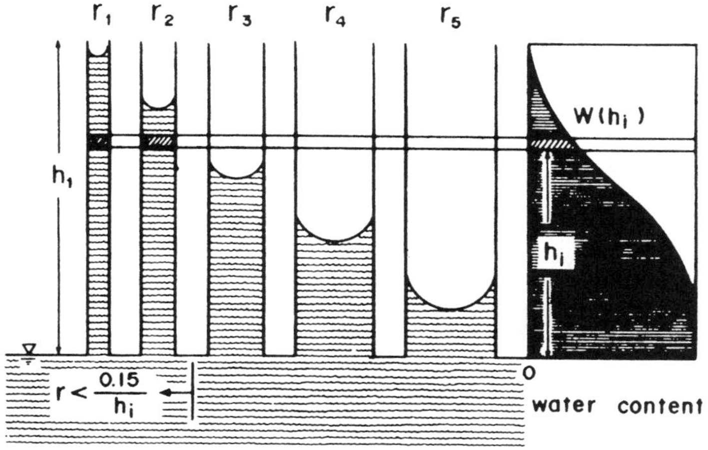

Fig. 4.10. Model of nonuniform capillaries to explain the capillary moisture distribution curve.

measured after drainage from saturated soil. This hysteresis is often explained as the "ink bottle" effect (Fig. 4.11). The passive capillary height h_p after draining is determined by the neck size d_n, and the active capillary height h_a on wetting a dry soil is related to the largest diameter, d_b, of the soil pore. From Eqs. (4.21) and (4.22) for cubic packing,

$$h_p = \frac{2.42\ C_n}{\phi} \tag{4.26}$$

and for closest packing,

$$h_p = \frac{6.45\ C_n}{\phi} \tag{4.27}$$

For C_n, = 0.3,

$$H_p = h_p\phi = 0.73\ or\ 1.94 \tag{4.28}$$

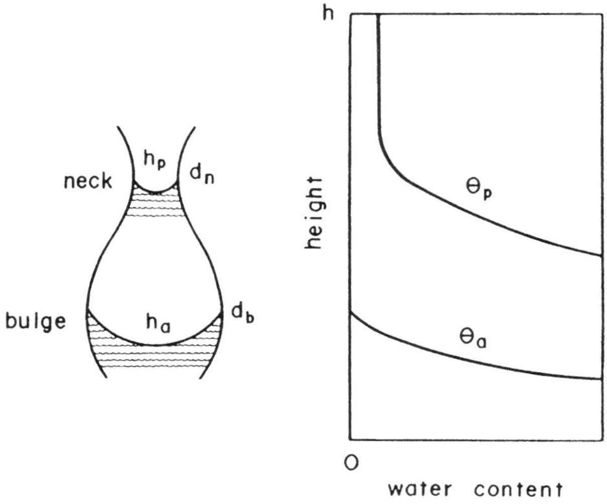

Fig. 4.11. Explanation of hysteresis by the "ink bottle" model.

corresponding to the values obtained by Haines (1927).

Yamanaka (1955) obtained a value of active capillary rise height h_a, using a pore cell model

$$h_a = \frac{0 \sim 0.3}{\phi} \tag{4.29}$$

The value of h_a from Eq. (4.29) is smaller than the values of h_p in Eq. (4.28).

Leverett (1941) examined the CMDC for various liquids and particle packings, and proposed a function j with a parameter for water saturation S_w, as follows:

$$\frac{\rho g h_c}{\sigma}\left(\frac{k}{n}\right)^{1/2} = j\ (S_w) \tag{4.30}$$

This equation can be rewritten

$$h_c = C\left(\frac{n}{K}\right)^{1/2} j\,(S_w) \tag{4.31}$$

where

$$K = k\,\frac{\rho g}{\eta} \qquad C = \sigma\left(\frac{1}{\rho g \eta}\right)^{1/2}$$

Capillary rise, h_c, is defined as a function of porosity, n, and hydraulic conductivity, K, parameters. From the curve of $j(S_w)$ we can obtain the values of capillary rise:

$$h_{c,\max} = 0.7C\left(\frac{n}{K}\right)^{1/2} \tag{4.32}$$

$$h_{c,\min} = 0.42C\left(\frac{n}{K}\right)^{1/2} \tag{4.33}$$

Therefore,

$$\frac{h_{c,\max}}{h_{c,\min}} = 1.66 \tag{4.34}$$

4.2.2 Capillary Moisture Distribution Under Suction

When water is drained by suction from a soil column with a capillary fringe, a capillary moisture distribution curve is formed in the column (Fig. 4.12). After drainage there is a relation W(h) between water content and negative water pressure. If the suction at the bottom of the soil column changes from h_{s1} to h_{s2}, the negative water pressure at height Z also changes from $(h_{s1} + Z)$ to $(h_{s2} + Z)$. Then water content at that height will change. The water volume drained from the soil at height Z, ΔW_z, is the difference between the two values.

$$\Delta W_z = W(h_{s1} + Z) - W(h_{s2} + Z) \tag{4.35}$$

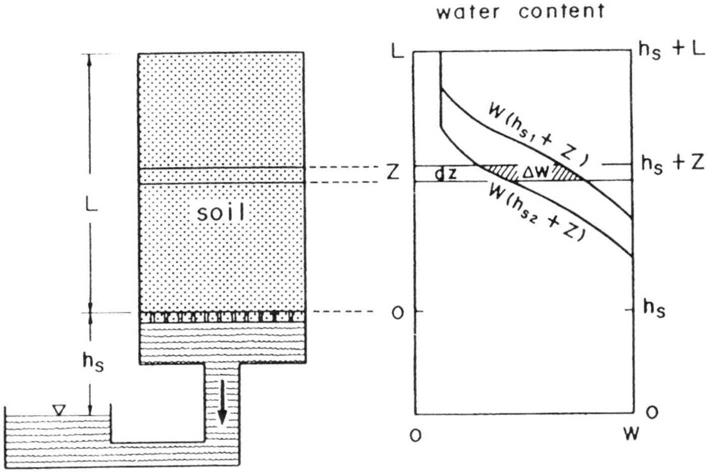

Fig. 4.12. Change of water content in a soil layer on increasing suction.

Drained water volume from the total soil column ΔQ is obtained by integrating Eq. (4.35) over the length of the column.

There is hysteresis in the W(h) curve. If we alternately increase and decrease the suction, a complex suction-moisture content curve may be obtained. Many studies (Poulovassilis, 1962; Poulovassilis and Childs, 1971; Topp, 1969; Nakano, 1976a,b) have analyzed the hysteresis of the suction-moisture curve theoretically using various models. These models consider a small element of soil which has the property to hold water against a suction. However, it is not easy to describe such a property of soil with a simple capillary model. Further, not all small elements always have a certain water-pressure relationship $W = W(h)$, because water content within an intermediate part of a capillary between the bottom and the meniscus at the top is independent of the water pressure.

4.2.3 Suspended Capillary Water in Soil

There are other types of capillary water in soil, such as lens water, capillary water within aggregates, and suspended water in layered soils. Lens water has already been mentioned (Fig. 4.4). Capillary water within aggregates, which is held within their micropores, is illustrated in Fig. 4.13. A meniscus is formed at the outlet of each pore.

Suspended water can exist in the lower part of a fine-grained layer above a coarse-grained layer such as sand (Fig. 4.14). This water has two air-water

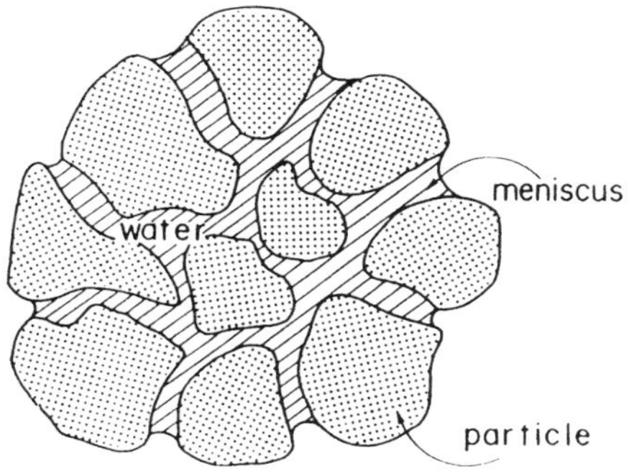

Fig. 4.13. Menisci in an aggregate of soil.

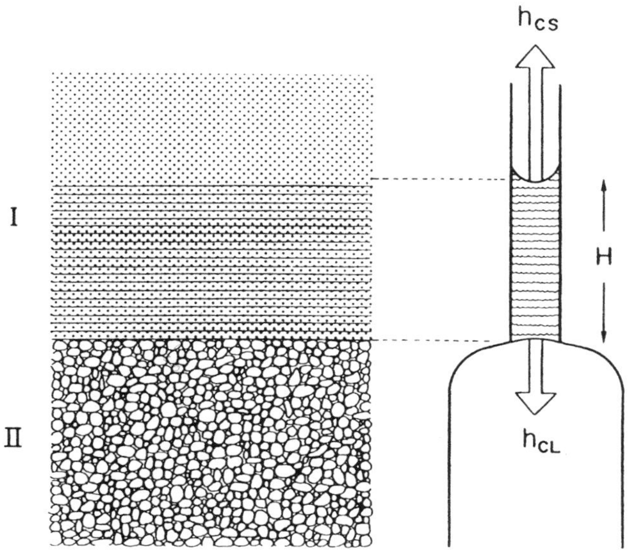

Fig. 4.14. Suspended water in a layered soil.

interfaces; capillary forces act on both surfaces. The capillary force in the upper layer of small particle size, h_{cs}, is larger than that of the lower layer, h_{cL}, so water is held in the soil against the gravitational force.

$$h_{cL} + H < h_{cs} \tag{4.36}$$

where H is the height of suspended water. Therefore,

$$H < h_{cs} - h_{cL} \tag{4.37}$$

When the height of this suspended water is larger than the value of (h_{cs} - h_{cL}), water begins to move downward.

Such suspended water also occurs in a uniform soil due to hysteresis. Because the passive capillary force at the upper face of the wetted zone is larger than the active capillary force at the wetting front, water cannot flow down by gravity.

$$h_{ca} + H < h_{cp} \tag{4.38}$$

This type of suspended water occurs after flooding of a soil surface with a small amount of water. If a large amount of water has infiltrated, the height of the wetted zone would become large and water would begin to flow down.

4.3 CAPILLARY THEORY FOR NONUNIFORM CELL SYSTEMS

4.3.1 Neck Size Distribution Function

Neck size varies with particle packing (Fig. 4.15). The neck size d_{n3}, formed by three spheres in contact, is (see Fig. 4.7)

$$d_{n3} = 0.155\phi \tag{4.18}$$

or

$$D_{n3} = \frac{d_{n3}}{\phi} = 0.155 \tag{4.39}$$

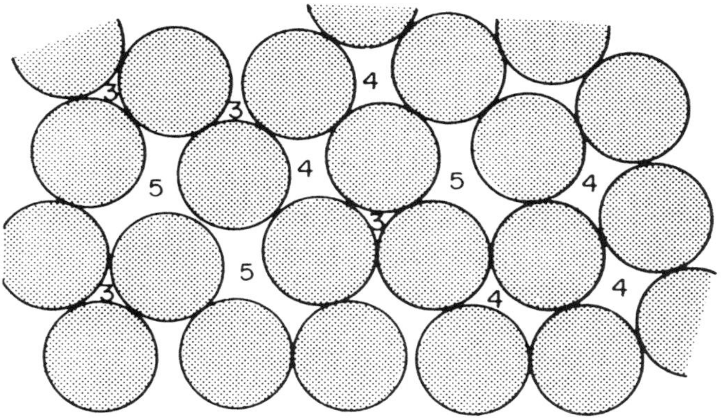

Fig. 4.15. Various types of necks within particle packing.

where ϕ is the diameter of the sphere and D_n is a dimensionless ratio. For four spheres (Fig. 4.16), the neck size D_{n4} is related to the angle θ_1

$$D_{n4} = \frac{d_{n4}}{\phi} = \cos\frac{\theta_1}{2} - 1 \tag{4.40}$$

The value of θ_1 changes from 60° to 90°, so the value of D_{n4} changes from 0.155 to 0.414. Assuming that the probability of any θ_1 is uniform between the values of 60° and 90°, we can calculate the probability of any value of D_{n4}, represented by the function $g_4'(D_n)$. This function can be used to obtain the probability function $g_4(D_n)$, which represents the probability of appearance of a neck size less than the value of D_n

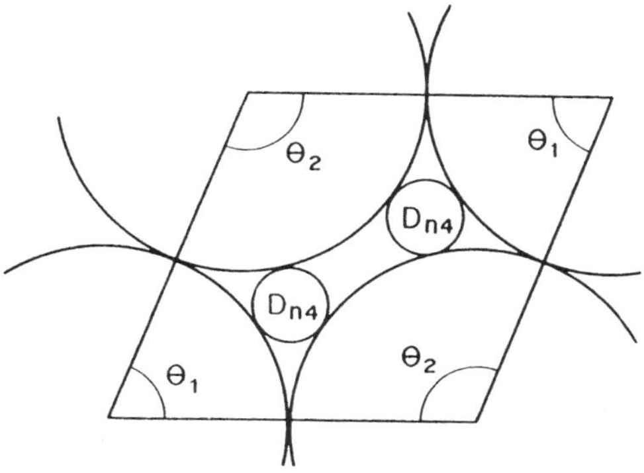

Fig. 4.16. Necks surrounded by four spheres.

$$g_4(D_n) = \int_{D_{n,\min}}^{D_n} g_4'(D_n)\, dD_n \tag{4.41}$$

Three necks are formed by five spheres (Fig. 4.17), and their diameters D'_{n5}, and D''_{n5}, and D'''_{n5} are represented as functions of θ_1 and θ_3. The probability function $g_5(D_n)$ can be obtained by the method used above.

The total function of $g(D_n)$ must be composed of functions for all classes of necks, g_3, g_4, g_5. We have assumed that the probability of necks from six or more spheres is low. Assuming further that the ratio of the number of necks in each class is 1:2:3,

$$g(D_n) \doteqdot \frac{1}{6}(g_3 + 2g_4 + 3g_5) \tag{4.42}$$

This probability function $g(D_n)$ is represented approximately by a fractional function (Tabuchi, 1971a)

$$g(D_n) = -\frac{A_c}{D_n} + 6.45\, A_c \tag{4.43}$$

or

$$g(d_n) = -\frac{A_c}{d_n}\,\phi + 6.45\, A_c \tag{4.44}$$

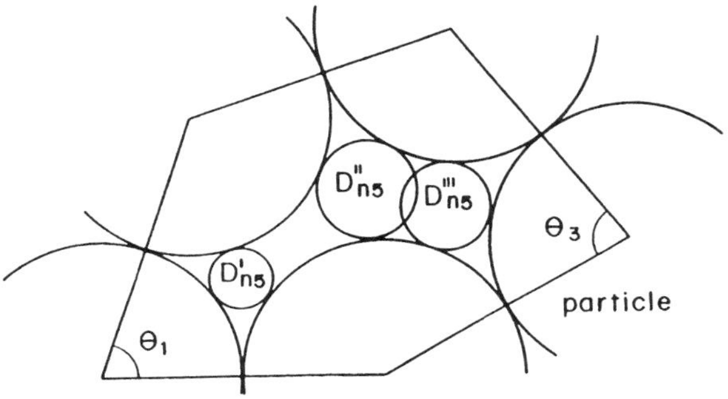

Fig. 4.17. Necks surrounded by five spheres.

Therefore,

$$g'(D_n) = \frac{dg}{dD_n} = \frac{A_c}{D_n^2} \tag{4.45}$$

where A_c is a constant. Equation (4.44) agreed with experimental data from glass beads when the value of A_c is 0.2 (Fig. 4.18). In Fig. 4.18, the function f is defined as

$$f(d_n) = 1.0 - g(d_n) \tag{4.46}$$

Substituting Eq. (4.20) into Eq. (4.44), we obtain

$$g(h_n) = -\frac{A_c}{C_n} h_n \phi + 6.45\, A_c = -\frac{A_c}{C_n} H_n + 6.45\, A_c \tag{4.47}$$

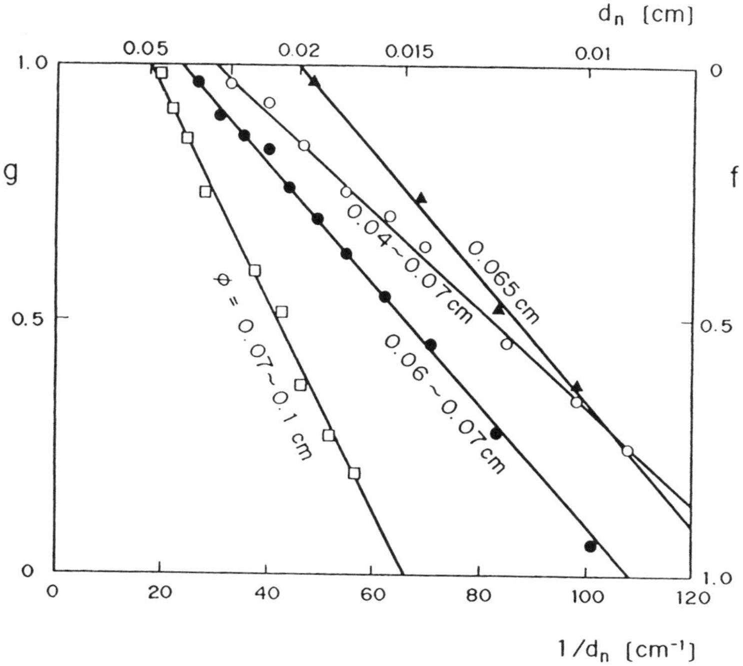

Fig. 4.18. Experimental results of $g(d_n)$ in Eq. (4.44). (From Tabuchi, 1971a)

where $H_n = h_n\phi$. Then the function $f(H_n)$ is

$$f(H_n) = 1 - g(H_n) = \frac{A_c}{C_n} H_n - 6.45 A_c + 1 \qquad (4.48)$$

where f represents the probability of the appearance of a neck in which the capillary pressure is smaller than the value of H_n. This equation represents the probability of necks opening as a function of capillary pressure.

4.3.2 Connecting Systems of Pore Cells

In spherical particle packing, a pore cell is connected with other cells in many ways: six ways in cubic packing and five in rhombohedral cells of closest packing. Since there are many cells, the number of ways air enters into a certain pore cell from the soil surface is infinite. The probability of air entry into a certain pore cell can be calculated from a pore model (Tabuchi, 1966a) with nonuniform necks between two pore cells. At some capillary pressure p, there exist both open necks, where the air-water interface in the neck broken, thus allowing air to pass through, and unopened necks (Fig. 4.19). Air entry through the neck can be described as the probability f(p), which is related to the neck distribution function discussed above.

Second, a system is considered in which unit pore cells are connected regularly and infinitely. Then the phenomenon of air entry into a pore cell (Fig. 4.20) is described as some function with variable f, with the function F(f) determined by the type of pore connecting system.

The fundamental equation of F(f) in cubic packing is

$$F_{0,0,0} = 1.0 - (1.0 - f)\,(1.0 - fG_{1,0,0})^4 \qquad (4.49)$$

$$G_{1,0,0} = 1.0 - (1.0 - f)\,(1.0 - fG_{2,0,0})\,(1.0 - fG_{1,1,0})^2 \qquad (4.50)$$

$$G_{2,0,0} = 1.0 - (1.0 - f)\,(1.0 - fG_{3,0,0})\,(1.0 - fG_{2,1,0})^2 \qquad (4.51)$$

$$\vdots$$

$$G_{\ell,0,0} = 1.0 - (1.0 - f)\,(1.0 - fG_{\ell+1,0,0})\,(1.0 - fG_{\ell,1,0})^2 \qquad (4.52)$$

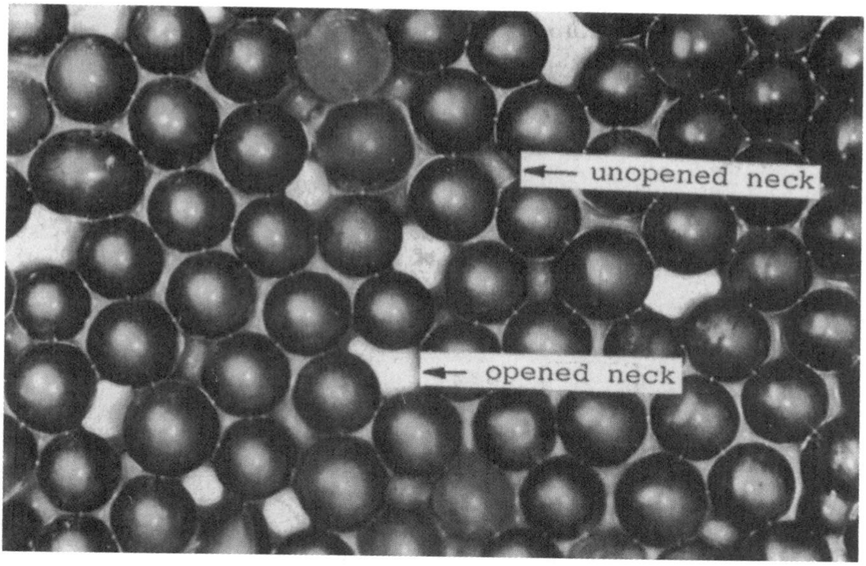

Fig. 4.19. Photograph of meniscus forming at the neck within packed spheres. (From Tabuchi, 1966b)

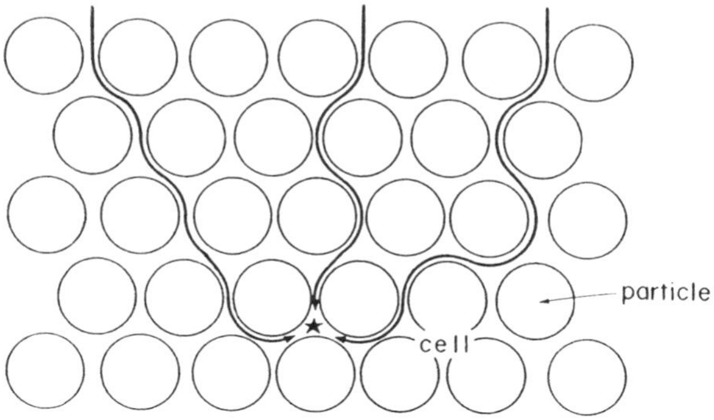

Fig. 4.20. Possible routes of air entry into a cell.

where $F_{0,0,0}$ is the probability of air entry into the cell (0,0,0) of the first layer and $G_{\ell,m,n}$ is that of the cell (ℓ,m,n), as shown in Fig. 4.21.

The probability of air entry into the cell (0,0,n) of the nth layer, $F_{0,0,n}$, is

$$F_{0,0,n} = 1.0 - (1.0 - fF_{0,0,n-1})\,(1.0 - fG_{1,0,n})^4 \tag{4.53}$$

The values of $F_{0,0,n}$ have been calculated by steps, but they become constant when n is in a small range less than 20 (Tabuchi, 1966a). Thus F can be treated as a function of f, which does not depend on the value of n.

The value of F for the cubic system is equal to zero when f is smaller than 0.25 and is 1.0 when f is larger than 0.7 (Fig. 4.22). Consequently, in the cubic system, air entry into the pore cell begins when the value of f reaches 0.25 and becomes complete when f reaches 0.7.

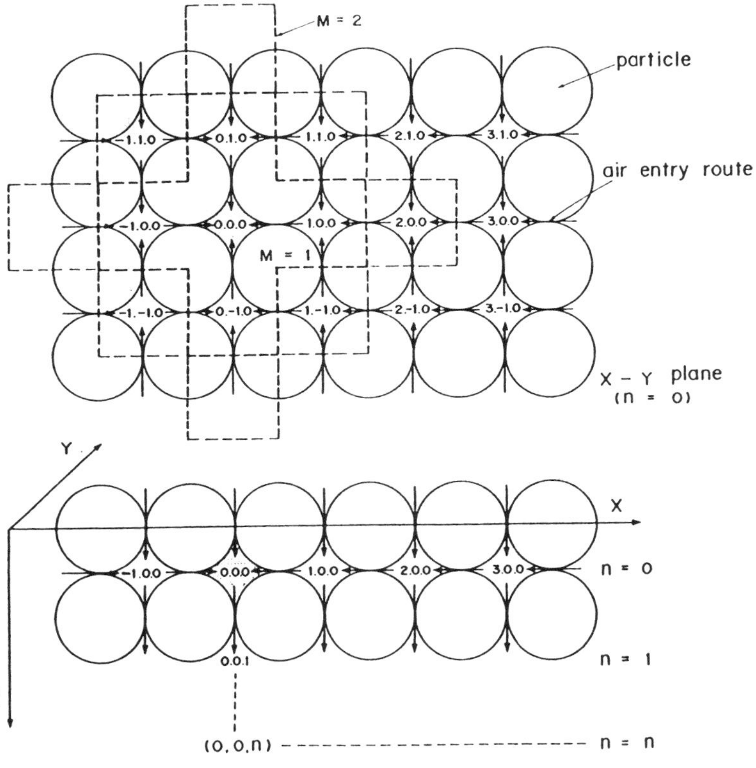

Fig. 4.21. Air entry into a cell (0,0,0). (From Tabuchi, 1966a)

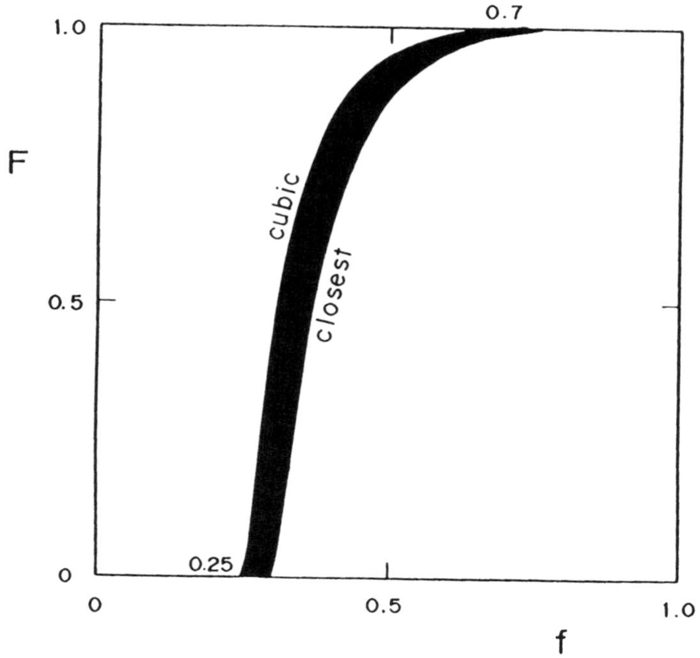

Fig. 4.22. F(f) curves for air entry into cells of cubic and close packing. (From Tabuchi, 1971a)

Using a Weibull distribution function, the approximate equation of F(f) can be expressed as follows (Tabuchi, 1971b):

$$F(f) = 1.0 - \exp\left[-\frac{(f - \gamma)^{\beta}}{\alpha}\right] \tag{4.54}$$

where α, β, and γ are constants of the Weibull distribution function. The use of this approximate equation makes the calculation of F(f) easier. In the cubic system the values of α, β, and γ are

$$\alpha = 0.082 \qquad \beta = 1.0 \qquad \gamma = 0.255 \tag{4.55}$$

In the closest packing system,

$$\alpha = 0.095 \qquad \beta = 1.0 \qquad \gamma = 0.305 \tag{4.56}$$

4.3.3 Drainage Curves

The moisture content as a function of capillary pressure can be obtained from the equations discussed in previous sections. In Eq. (4.48), f(H) represents the probability of a capillary pressure based on a neck size distribution function. Combining f(H) with F(f) from Eq. (4.54), we obtain

$$F(H) = 1.0 - \exp\left[-\frac{(0.83H - 0.29 - \gamma)^{\beta}}{\alpha}\right] \tag{4.57}$$

where $H = h\phi$. In the cubic packing system, substituting the values of α, β, and γ of Eq. (4.55) into Eq. (4.57):

$$F(H) \doteqdot 1.0 - \exp(-10.2H + 6.65) \qquad H \geqq 0.654 \tag{4.58}$$

The value of F represents the probability of air entry into a pore cell corresponding to a given negative water pressure. When the number of pore cells in the packing is extremely large, the value of F can be used as the ratio of the number of cells into which air has entered to the total number of cells. Therefore, the ratio of water saturated cells θ_r can be calculated from the value of F:

$$\theta_r(H) = 1.0 - F(H) \tag{4.59}$$

In the cubic-type system,

$$\theta_r(H) = \exp(-10.2H + 6.65) \qquad H \geqq 0.654 \tag{4.60}$$

or

$$\theta_r(H) = \exp(-10.2h\phi + 6.65) \qquad h\phi \geqq 0.654 \tag{4.61}$$

Water content remaining in the soil at the capillary pressure h is represented by θ_r. The $\theta_r(H)$ curve from Eq. (4.60) is shown in Fig. 4.23. When the capillary pressure attains the air entry pressure H_e, water begins to drain. In the cubic system, the value of H_e is calculated from Eq. (4.60) as

$$H_e = 0.654 \qquad or \qquad h_e = \frac{0.654}{\phi} \tag{4.62}$$

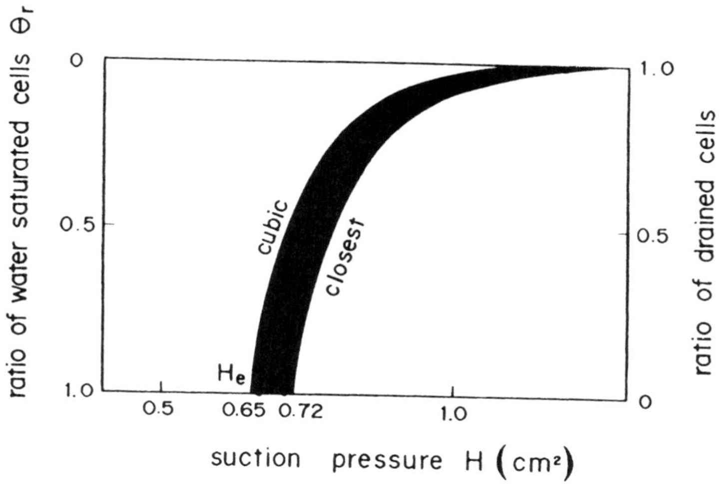

Fig. 4.23. Theoretical curves of the ratio of water saturated cells during drainage by suction.

In drainage designs, these curves are measured experimentally.

4.3.4 Capillary Moisture Distribution Curve After Drainage or After Wetting

We can also sketch the capillary moisture distribution curve (CMDC) after drainage from saturated soil using Eq. (4.60). The value of capillary pressure H in Eq. (4.60) corresponds to the height above GWL. For the cubic packing system,

$$\theta_r(H) = \exp(-10.2H + 6.65) \qquad H \geq 0.654 \tag{4.63}$$

where $H = h\phi$. The curve of Eq. (4.63) is shown in Fig. 4.24. The experimental moisture distribution curve obtained with glass beads agreed better with the theoretical curve of cubic packing than with closest packing (Tabuchi, 1966b).

The height of capillary rise $H_{0.05}$, which corresponds of $\theta_r = 0.05$, and $H_{0.95}$, corresponding to $\theta_r = 0.95$, can be calculated from Eq. (4.63):

$$H_{0.05} = 0.95 \qquad h_{5\%} = \frac{0.95}{\phi} \tag{4.64}$$

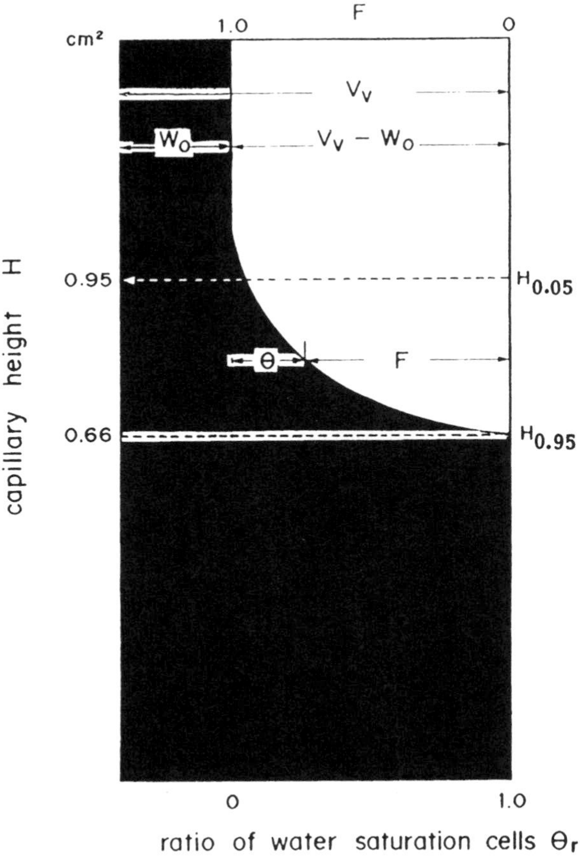

Fig. 4.24. Theoretical capillary moisture distribution curve after drainage in cubic packing. (From Tabuchi, 1971a)

$$H_{0.95} = 0.66 \qquad h_{95\%} = \frac{0.66}{\phi} \tag{4.65}$$

Thus we can obtain the theoretical value of the height of capillary rise, which is inversely related to particle size as indicated in equations obtained by various researchers.

In capillary rise of water into a packing of dry particles we have to consider the entry of water, rather than of air, into the pore cells. However, the mechanism that determines the moisture distribution is the same in both cases and it is possible to use the same analysis. In this case, f(h) corresponds to the passage of water through a bulge of the pore cell, rather than to the entry of air through a neck (Tabuchi, 1971b).

The theoretical equations for the capillary moisture distribution curve obtained using this pore model can add significantly to soil-water research. The relationship $\theta_r = \theta_r(h)$ describes the capillary fringe, where air is contained in pores connecting to the atmosphere, and where water-saturated pore cells are also connected continuously (Fig. 4.25). The number of saturated pore cells can be changed according to water pressure variation to maintain the relation $\theta_r = \theta_r(h)$.

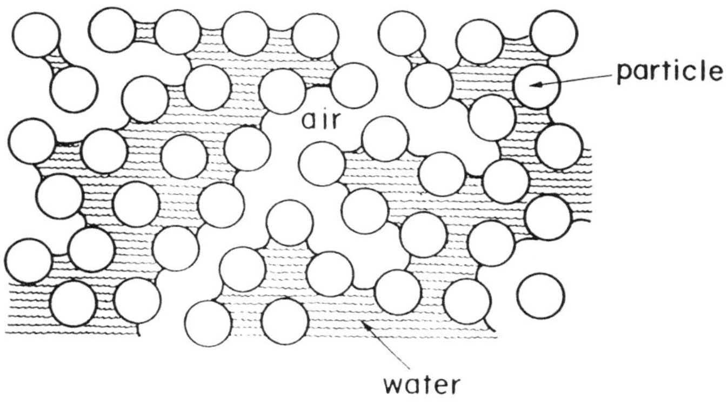

Fig. 4.25. Schematic view of unsaturation in a pore cell model.

ACKNOWLEDGEMENTS

Figures 4.19 and 4.21 are reprinted from Soil Science, Copyright © by the Williams and Wilkins Co., Baltimore. Used with permission.

REFERENCES

Adam, N.K., *The Physics and Chemistry of Surfaces,* Clarendon Press, Oxford (1938).

Dallavalle, J. M. *Micromeritics*, Pitman, London (1943).

Fisher, R. A., On the capillary forces in an ideal soil, *J. Agric. Sci.* 16:492-505 (1926).

Haines, W. B., Studies in the physical properties of soils: IV, *J. Agric. Sci.* 17:264-290 (1927).

Leverett, M. C., Capillary behavior in porous solids, *Trans. AIME* 142:152-169 (1941).

Nakano, M., Pore volume distribution and curve of water content versus suction of porous body. 1. The boundary drying curves, *Soil Sci.* 122:5-14 (1976a).

Nakano, M., Pore volume distribution and curve of water content versus suction of porous body. 2. Two boundary wetting curve, *Soil Sci.* 122:100-106 (1976b).

Poulovassilis, A., Hysteresis of pore water, an application of the concept of independent domains, *Soil Sci.* 93:405-412 (1962).

Poulovassilis, A., and E. C. Childs, The hysteresis of pore water: the non-independence of domains, *Soil Sci.* 112:301-312 (1971).

Rode, A. A., *Theory of Soil Moisture*, Vol. 1, Israel Program for Scientific Translations, Jerusalem (1969).

Tabuchi, T., Theory of suction drain from the saturated ideal soil, *Soil Sci.* 102:161-166 (1966a).

Tabuchi, T., Experiment on suction drain from an ideal soil, *Soil Sci.* 102:329-332 (1966b).

Tabuchi, T., Infiltration and capillarity in the particle packing, *Rec. Land Reclam. Res.* 19:1-121 (1971a). (Japanese)

Tabuchi, T., Theory of suction drain from the saturated idea soil: 2, *Soil Sci.* 112:448-453 (1971b).

Topp, G. C., Soil water hysteresis measured in sandy loam and compared with the hysteretic domain model, *Soil Sci. Soc. Am. Proc.* 33:645-651 (1969).

Waldron, L. J., J. L. McMurdie, and J. A. Vomocil, Water retention by capillary forces in an ideal soil, *Soil Sci. Soc. Am. Proc.* 25:265-267 (1961).

Yamanaka, K., Studies on soil cohesion, *Bull. Nat. Inst. Agric. Sci.* B6:1-139 (1955). (Japanese)

WATER FLOW THROUGH SOIL

5.1 GENERAL CONCEPTS OF WATER FLOW

Water flow through soil in which both water and air exist is called unsaturated flow. In this chapter we discuss unsaturated flow near the region of saturated flow, where water exists in small pores in the soil while air occupies the large pores. In some cases, air exists only as isolated air bubbles. It can be assumed that Darcy's law is applicable in this type of unsaturated flow.

Unsaturated flow can be classified into a type where air exists as air bubbles, and a second type in which air-filled pores are continuous in the soil. Yamazaki (1948) called the former a closed system and the latter, unsaturated flow in an open system.

In the open system, water content θ is related to water pressure p through the moisture characteristic curve. If the water pressure changes, the pressure at the meniscus changes until a new equilibrium is attained. However, in the closed system the relation between θ and p cannot be established because the meniscus cannot move. When the water pressure changes, the air pressure in the bubble also changes before the meniscus moves. Thus we can have various values of water pressure for the same water content in the closed system. Positive water pressure can occur in the closed system. The difference between the open and closed systems, especially close to saturation, should be emphasized since it has often been incorrectly stated that an unsaturated condition always has negative water pressure.

Where a change from a closed system to an open system occurs, the limiting negative pressure, P_L, corresponds to the air entry pressure P_{ent}. If the water pressure decreases below the entry pressure, air enters the soil, and an open system results (Fig. 5.1). A model has not yet been established for the state of water in an unsaturated open system where both continuous air and water exist, and a large number of menisci are formed. Krisher's (1938) model has capillaries connected to each other on all sides. Pore cell models where water and air exist continuously are discussed in Section 4.3.4, Fig. 4.25.

unsaturated		saturated
air continuous	air isolated	none
open	closed	
$\theta(p)$	—	θ_s
$P < P_L$	P_L	$P > P_L$

Fig. 5.1. Pressure conditions in the two types of unsaturated flow systems.

5.1.1 Historical Review

In 1856, H. Darcy formulated Darcy's law while studying water filtration through a sand column. Water discharge Q is proportional to the cross-sectional area A of a sand column, to the hydraulic head difference $\Delta\psi$, and inversely proportional to the length of the soil column L.

$$Q = K \frac{A}{L} \Delta\psi \tag{5.1}$$

Forchheimer (1930) introduced a non-Darcy-type formula in which the relation between flow velocity and gradient is not linear:

$$J = \alpha v + \beta v^2 + \gamma v^3 \tag{5.2}$$

where J is the hydraulic gradient, v the velocity, and α, β, and γ are constants.

The general validity of Darcy's law has been confirmed and applied to many problems of groundwater movement and seepage into soil. The fundamental flow equation was introduced using Darcy's law and many researchers attempted to solve the flow equation mathematically for various boundary conditions (e.g. Muskat, 1937; Polubarinova-Kochina, 1962; Harr, 1962). Many studies (e.g. Pavlovskii, 1956; Yoshida, 1960) have attempted to derive the equation of flow from the Navier-Stokes equation, and hence to give physical meaning to Darcy's law.

Hydraulic conductivity, which plays an important role in Darcy's law, has been examined and the factors affecting hydraulic conductivity have been investigated. Hazen (1892) proposed grain size as the main factor. Slichter (1898) introduced porosity as one of the factors, and examined pore sizes within different particle packing. He used a Hagen-Poiseuille equation for water flow through a capillary tube. Kozeny (1932) established a connection between flow through a capillary tube and flow in soil using a concept of hydraulic radius. He introduced an equation that showed the relation between hydraulic conductivity and porosity. Afterward, he added the surface area of soil particles to the factors determining hydraulic conductivity.

Baver (1938) investigated the relation between hydraulic conductivity and pore sizes in soils; pore size was represented by the value of the equilibrium negative pressure. Purcell (1949) proposed an equation for hydraulic conductivity which contained a pore distribution function.

The research on saturated flow initiated by Darcy's law was extended to unsaturated flow. Wyckoff and Botset (1936) showed a relation between hydraulic conductivity and water saturation in their experiments with mixtures of gas and liquid. Hydraulic conductivity decreases with decrease of the water saturation ratio. Based on this fact, much research has been reported concerning unsaturated hydraulic conductivity. Aberiyanov (1949) introduced an equation for unsaturated flow based on a model where water flowed through a partial section of a capillary while air existed in the central part. Krisher (1938), in the field of chemical engineering, proposed a model of nonuniform capillaries connected to each other. This model could explain the observation that the water content of a particle packing differed with height above groundwater level, and that water could move in the packing not only in a vertical but also in a horizontal direction. Childs and Collis-George (1950) explained unsaturated flow with a model of capillaries of nonuniform diameters. They introduced an equation for unsaturated hydraulic conductivity using the probability sequences of pores of different sizes. The effect of air bubbles on hydraulic conductivity has been examined experimentally and theoretically by many researchers (e.g. Orlob and Radhakrishna, 1956; Yawata, 1960; Kuroda, 1965). In Japan, research was reported on unsaturated flow under layered conditions and on capillary flow (Akiba, 1938; Yamazaki, 1958; Tabuchi, 1961).

The Darcy type of equation was used by Green and Ampt (1911) and Budagovskii (1955) to describe infiltration of water into soil. Tabuchi (1961) proposed a theoretical analysis of infiltration into a dry, layered sand column.

In the last 20 years a number of books on soil water have been written by soil physicists in many countries (Baver et al., 1972; Marshall, 1959; Childs, 1969; Hillel, 1980a,b; Kirkham and Powers, 1972; Yong and Warkentin, 1975; Rode, 1969; Nerpin and Chudnovskii, 1970; Kachinskii, 1970; Dallavalle, 1943; Scheidegger, 1957; Collins, 1961; De Wiest, 1969; Luikov, 1966; Yamazaki, 1969; Yawata, 1975; Campbell, 1985; Jury and Roth, 1990; Jury et al., 1991; Miyazaki, 1993). Valuable data have accumulated, and theoretical analyses have been developed. However, there remain unsolved problems in soil water movement; for example, the soil water-plant root system relations, water and associated heat transfer, and physicochemical properties of percolating water.

5.1.2 Darcy's Law

Darcy's law is one of the well-known laws used in soil physics:

$$Q = KA \frac{h_0 - h_b + L}{L} \tag{5.3}$$

where Q is the discharge (cm^3/sec), K the hydraulic conductivity (cm/sec), L the length of a soil column (cm), A the cross-sectional area of soil (cm^2), and h_0 and h_b are water pressures (cm) at the ends of a soil column, as shown in Fig. 5.2. Equation (5.3) is often rewritten as

$$Q = KA \frac{\Delta\psi}{L} = KAJ \tag{5.4}$$

where $\Delta\psi$ is the total head loss (cm) and J is the hydraulic gradient. The differential form of Eq. (5.4) is

$$Q = -KA \frac{d\psi}{dx} \tag{5.5}$$

where x is in the flow direction. The discharge per unit area, q (cm/sec), is the rate of seepage or percolation, sometimes called the apparent velocity

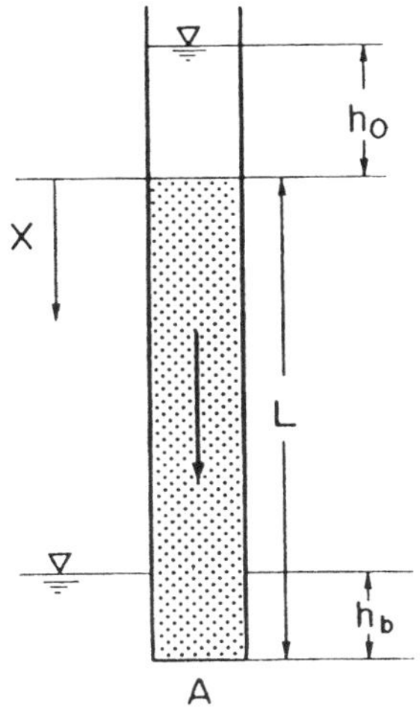

Fig. 5.2. Flow of water through a soil column to illustrate Darcy's law.

$$q = \frac{Q}{A} = KJ = -K\frac{d\psi}{dx} \tag{5.6}$$

Velocity of flow through a soil, v, is different from the value of q since the actual cross-sectional area of water flow, A_f, is less than that of the soil, A (Fig. 5.3). If the soil is saturated, the cross-sectional area of water flow is

$$A_f = nA \tag{5.7}$$

where n is porosity (1.0 > n > 0). Since

$$vA_f = qA \tag{5.8}$$

then

$$vn = q \qquad v > q \tag{5.9}$$

The water velocity is always larger than the percolation rate. Total head loss $\Delta\psi$ is the sum of water pressure Δh and elevation Δz

$$\Delta\psi = \Delta h + \Delta z \tag{5.10}$$

Darcy's law is widely applied not only to water but also to other liquids, and can be rewritten

$$Q = k\frac{\rho g}{\eta}A\frac{\Delta\psi}{L} \tag{5.11}$$

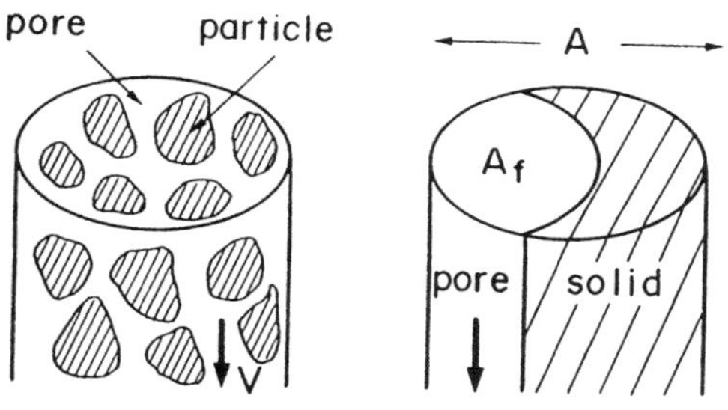

Fig. 5.3. Cross-sectional areas for flow velocity.

where ρ is the density of the liquid, g the acceleration constant due to gravity, η the viscosity of the liquid, and k the permeability coefficient. The coefficient k is a constant defined by the arrangement of particles, is not dependent on the type of liquid, and has dimensions of (cm^2).

5.1.3 Equation of Continuity

The fundamental equation determining water flow through soil is the Laplace equation. Consider water flow within a microelement in the soil, as shown in Fig. 5.4. The sides of the element are Δx, Δy, and Δz; q_x, q_y, and q_z are discharges per unit area in each direction. Discharge entering the element in the x direction is q_x multiplied by area Δy Δz. Discharge emerging from the element in the x direction is

$$\left(q_x + \frac{\partial q_x}{\partial x}\Delta x\right)\Delta y\,\Delta z \tag{5.12}$$

The differences in discharge flowing through the two faces in each direction are as follows:

$$x\ direction:\quad q_x\,\Delta y\,\Delta z - \left(q_x + \frac{\partial q_x}{\partial x}\Delta x\right)\Delta y\,\Delta z$$

$$= -\frac{\partial q_x}{\partial x}\Delta x\,\Delta y\,\Delta z$$

$$y\ direction:\quad -\frac{\partial q_y}{\partial y}\Delta x\,\Delta y\,\Delta z$$

$$z\ direction:\quad -\frac{\partial q_z}{\partial z}\Delta x\,\Delta y\,\Delta z \tag{5.13}$$

These differences equal the change of water content within the element:

$$\frac{\partial W}{\partial t}\Delta x\,\Delta y\,\Delta z = \left(-\frac{\partial q_x}{\partial x} - \frac{\partial q_y}{\partial y} - \frac{\partial q_z}{\partial z}\right)\Delta x\,\Delta y\,\Delta z \tag{5.14}$$

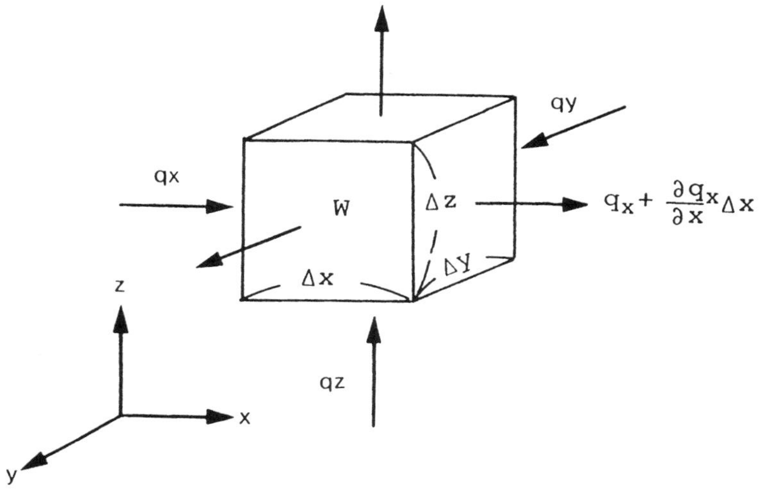

Fig. 5.4. Unit element of soil.

where W is the water content within the element and t is time. This equation is rewritten as

$$\frac{\partial W}{\partial t} + \frac{\partial q_x}{\partial x} + \frac{\partial q_y}{\partial y} + \frac{\partial q_z}{\partial z} = 0 \tag{5.15}$$

Substituting Darcy's law [Eq. (5.6)] into Eq. (5.15), we obtain the fundamental flow equation:

$$\frac{\partial W}{\partial t} + \frac{\partial}{\partial x}\left(-K_x \frac{\partial \psi}{\partial x}\right) + \frac{\partial}{\partial y}\left(-K_y \frac{\partial \psi}{\partial y}\right) + \frac{\partial}{\partial z}\left(-K_z \frac{\partial \psi}{\partial z}\right) = 0 \tag{5.16}$$

where K_x, K_y, and K_z are hydraulic conductivities for the three directions x, y, z. If the soil is homogeneous and isotropic, we have

$$K_x = K_y = K_z = K = constant \tag{5.17}$$

and

$$\frac{\partial W}{\partial t} - K\left(\frac{\partial^2 \psi}{\partial x^2} + \frac{\partial^2 \psi}{\partial y^2} + \frac{\partial^2 \psi}{\partial z^2}\right) = 0 \tag{5.18}$$

If the flow is in a steady state in which soil water content is constant, Eqs. (5.15) and (5.18) become

$$\frac{\partial q_x}{\partial x} + \frac{\partial q_y}{\partial y} + \frac{\partial q_z}{\partial z} = 0 \tag{5.19}$$

$$\frac{\partial^2 \psi}{\partial x^2} + \frac{\partial^2 \psi}{\partial y^2} + \frac{\partial^2 \psi}{\partial z^2} = 0 \tag{5.20}$$

This is the Laplace differential equation. The solution of this equation, $\psi(x,y,z)$, represents the total head or the flow potential at (x,y,z), from which we can sketch the flow net. The flow net, which consists of equipotential lines and flow lines, gives us potential distribution, discharge, and flow direction. Many problems have been solved under various boundary conditions (Muskat, 1937; Polubarinova-Kochina, 1962).

One-dimensional steady flow is the simplest case:

$$\frac{dq_x}{dx} = 0 \tag{5.21}$$

Therefore,

$$q_x = const \tag{5.22}$$

Equation (5.22) shows that the discharge is constant at each section vertical to the flow direction. Substituting Eq. (5.6) into Eq. (5.22), we obtain

$$KJ = -K\frac{d\psi}{dx} = const \tag{5.23}$$

If K is constant,

$$J = -\frac{d\psi}{dx} = const \tag{5.24}$$

5.2 PRESSURE DISTRIBUTION DURING WATER FLOW

Pressure distribution in steady flow can be obtained by solving Eq. (5.19) or (5.20) under given boundary conditions.

5.2.1 Steady Vertical Downward Flow in a Uniform Soil

Taking the z axis as the direction of downward flow (positive up), the fundamental flow equation (5.19) is

$$\frac{dq_z}{dz} = 0 \tag{5.25}$$

Therefore,

$$q_z = const \tag{5.26}$$

Since the value of K_z is constant, we arrive at the following solution:

$$\frac{d\psi}{dz} = const \tag{5.27}$$

Integrating Eq. (5.27), we obtain

$$\psi = Az + B \tag{5.28}$$

where A and B are constants.

In this case, total head ψ is a linear function of z. Using boundary conditions $\psi = \psi_a$ at $z = z_a$ and $\psi = \psi_b$ at $z = z_b$, we obtain the values for A and B.

$$A = \frac{\psi_a - \psi_b}{z_a - z_b} \tag{5.29}$$

$$B = \frac{\psi_b z_a - \psi_a z_b}{z_a - z_b} \tag{5.30}$$

The distribution of $\psi(z)$ is shown in Fig. 5.5. ψ at z is defined by

$$\psi = h + z \tag{5.31}$$

where h is pressure head expressed as the height of water and z is the elevation. Substituting Eq. (5.31) into Eq. (5.28), we obtain

$$h = \frac{h_a - h_b}{z_a - z_b} z + \frac{z_a h_b - z_b h_a}{z_a - z_b} \tag{5.32}$$

The values of h are shown in Fig. 5.5.

Under conditions where $z_b = 0$, $h_b = 0$, and $\psi_b = 0$, we obtain the values of ψ and h from Eqs. (5.28) to (5.32):

$$\psi = \frac{\psi_a}{z_a} z \tag{5.33}$$

$$h = \frac{h_a}{z_a} z \tag{5.34}$$

The lines of ψ and h are shown in Fig. 5.6a.

Water pressure changes according to the boundary conditions. Negative values of h, calculated from Eq. (5.32), are shown in Fig. 5.6b and c. Even if a negative water pressure appears, the soil remains in a saturated condition if air cannot enter the soil through the permeameter wall.

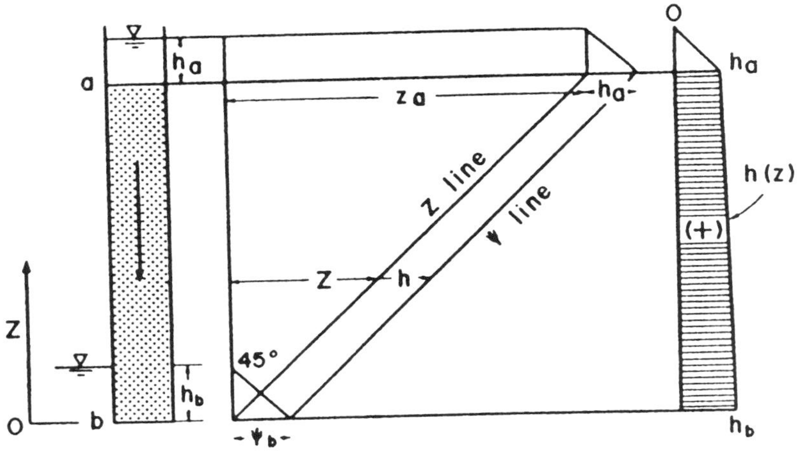

Fig. 5.5. Distribution of total head and pressure in a uniform soil column.

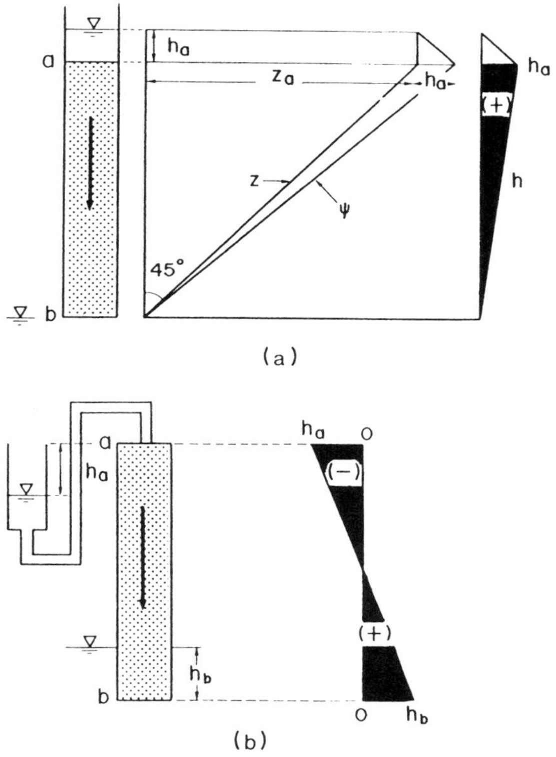

Fig. 5.6. Examples of pressure distribution during downward flow.

5.2.2 Horizontal and Upward Flow in a Uniform Soil

Let us take the x axis as the direction of horizontal flow. The Laplace equation becomes

$$\frac{\partial^2 \psi}{\partial x^2} = 0 \tag{5.35}$$

Integrating Eq. (5.35), we obtain

$$\psi = Ax + B \tag{5.36}$$

where A and B are constants.

In this horizontal case, z is constant. If we take the value of z as zero, total head equals pressure head.

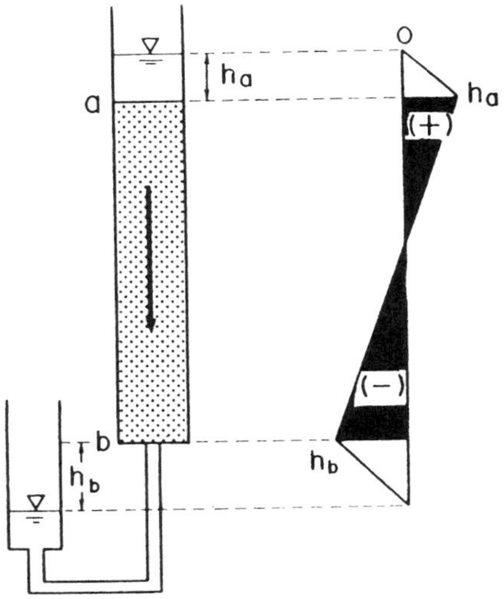

(c)

$$\psi = h \tag{5.37}$$

Using the boundary conditions $h = h_a$ at $x = 0$, and $h = h_b$ at $x = L$, we obtain

$$\psi = h = \frac{h_b - h_a}{L} x + h_a \tag{5.38}$$

5.2.3 Flow in a Layered Soil

The flow equation for steady vertical downward flow through a layered soil is

$$K(z) \frac{d\psi}{dz} = const \tag{5.39}$$

where K is not constant but a function of z. When we integrate Eq. (5.39), using Darcy's law, we get

$$\psi = q \int \frac{dz}{K(z)} + C \tag{5.40}$$

Assume a two-layered soil, with constant hydraulic conductivities K_1 and K_2. The relations between ψ and z in each soil layer are linear (Fig. 5.7), as in the uniform soil. Since the discharge is constant through the whole profile, the discharge through the first layer q_1 is equal to that through the second layer q_2. Using Darcy's law, we have

$$K_1 \frac{\psi_a - \psi_b}{L_1} = K_2 \frac{\psi_b - \psi_c}{L_2} \tag{5.41}$$

where L_1 and L_2 are the lengths of each layer and ψ_a, ψ_b, and ψ_c are the total heads at points a, b, c, respectively (shown in Fig. 5.7). Solving Eq. (5.41), we have

$$\psi_b = \frac{K_1 \psi_a L_2 + K_2 \psi_c L_1}{K_2 L_1 + K_1 L_2} \tag{5.42}$$

If we take z = 0 at the point c, ψ_b is defined as

$$\psi_b = h_b + L_2 \tag{5.43}$$

Then pressure head h_b is calculated from Eqs. (5.42) and (5.43), that is,

$$h_b = \frac{L_2(K_1 L_1 + K_1 h_a - K_2 L_1) + K_2 h_c L_1}{K_2 L_1 + K_1 L_2} \tag{5.44}$$

When $h_c = 0$, we get

$$h_b = \frac{L_2(K_1 L_1 + K_1 h_a - K_2 L_1)}{K_2 L_1 + K_1 L_2} \tag{5.45}$$

If $K_1L_1 + K_1h_a - K_2L_1 < 0$, h_b becomes negative. Transforming the inequality above, we obtain

$$K_2 > K_1 \left(\frac{h_a}{L_1} + 1 \right) \tag{5.46}$$

This condition indicates that negative water pressure occurs in the soil.

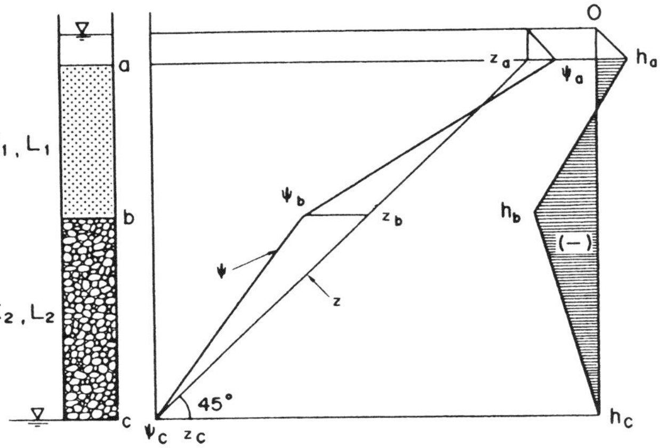

Fig. 5.7. Distribution of total head and pressure for flow through layered soil.

5.3 HYDRAULIC CONDUCTIVITY

Hydraulic conductivity, defined by Darcy's law, is one of the most important measurements in the physics of water flow through soil. Hydraulic conductivity varies with factors such as soil porosity, pore size, and water temperature. Many researchers have attempted to clarify the relationships of hydraulic conductivity to these factors.

5.3.1 Temperature Effect

Hydraulic conductivity is defined by Eq. (5.11):

$$K = k \frac{\rho g}{\eta} \qquad cm/sec \tag{5.47}$$

Water density ρ and viscosity η vary with temperature T. If the hydraulic conductivity at T°C is K_T and at 20°C is K_{20}, then:

$$\frac{K_T}{K_{20}} = \frac{\nu_{20}}{\nu_T} \tag{5.48}$$

where

$$\nu_{20} = \left(\frac{\eta}{\rho}\right)_{20} \qquad and \qquad \nu_T = \left(\frac{\eta}{\rho}\right)_T \tag{5.49}$$

The values of ν_T/ν_{20}, shown in Fig. 5.8, decrease with increasing temperature, so hydraulic conductivity increases with temperature. K at 35°C is twice as large as K at 7°C.

5.3.2 Particle Size

Empirical evidence and intuitive reasoning indicate that conductivity increases with increasing particle size. It was generally found that hydraulic conductivity is proportional to the second power of particle size

$$K \propto \phi^2 \tag{5.50}$$

where ϕ is particle diameter. If ϕ changes 10 times, hydraulic conductivity will change 100 times. Hazen (1892) replaced ϕ with the effective diameter ϕ_e, because the diameter of soil particles is not uniform. He used the largest diameter of the lower 10% of the particles as ϕ_e.

Small particles mixed into a soil will decrease its hydraulic conductivity. Kachi and Suezawa (1937) proposed an experimental equation for hydraulic conductivity of such a mixture

$$\frac{K}{K_0} = C \exp(as) \tag{5.51}$$

where K_0 is hydraulic conductivity of the original soil, s the volume of small particles, and C and a are constants.

These relationships have not been found to be sufficiently reliable for general use, so hydraulic conductivity is usually measured experimentally for soils.

5.3.3 Pore Properties

Slichter (1898) introduced an equation based on a particle packing model that included both particle size and porosity:

$$K \propto \phi^2 \, n^{3.3} \tag{5.52}$$

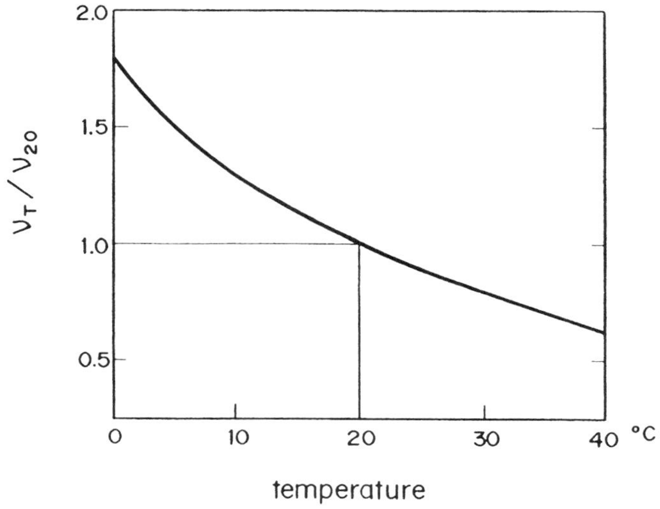

Fig. 5.8. Values of v_T/v_{20} for water, where v is the ratio of viscosity to density.

where ϕ is the particle size and n is the porosity. Kozeny (1932) introduced a similar equation,

$$K = \frac{\rho g}{c\eta} \phi^2 \frac{n^3}{(1 - n)^2} \tag{5.53}$$

where c is a constant and η is the viscosity. Later he substituted surface area for particle size,

$$K = \frac{\rho g}{c\eta} \frac{1}{U^2} \frac{n^3}{(1 - n)^2} \tag{5.54}$$

where U is the surface area of the soil per unit volume.

The expected relation between hydraulic conductivity and porosity is not always seen, especially in volcanic ash soil. Tada (1965) compacted Kantō loam at various values of water content W and measured the apparent density and saturated hydraulic conductivity. Hydraulic conductivity decreased gradually with increasing water content to a minimum value at W = 145% (Fig. 5.9). However, the porosity increased. Tada concluded that this unexpected observation was due to the change of pore size. Compacting at high water content decreased large pores, so the compacted soil showed a low value of K

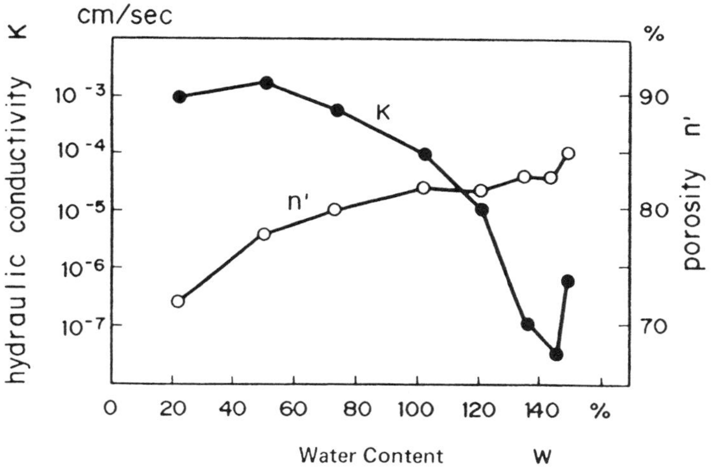

Fig. 5.9. Change of hydraulic conductivity and porosity of volcanic ash soil on compaction at different water contents. (From Tada, 1965)

despite its high porosity. Soma et al. (1983) have also reported that compacted volcanic ash soil shows a minimum value of K, similar to that shown in Fig. 5.9.

Pore size as well as total porosity is an important factor determining hydraulic conductivity. Nelson and Baver (1940) related hydraulic conductivity to volume of pores drained at -40 cm water pressure. However, it is difficult to measure actual pore sizes in soils. The results shown in Table 5.1 and Fig. 5.10 were obtained from microscopic observation on undisturbed soil samples by Tabuchi (1963). These results will be used to discuss the influence of pore size on water flow.

The macropores larger that 0.1 mm are 20% of the total porosity in Kantō volcanic ash soil, and large pores over 1 mm also exist. The hydraulic conductivity of this soil is high. Macropores form less than 10% of the porosity of the diluvial soil. The hydraulic conductivity K_{ws}, air entry pressure h_e, and volume of water drained at air entry, W_e, were measured. From h_e, the size of the pore d_e corresponding to the air pressure was calculated

$$d_e = \frac{3}{h_e} \quad mm \tag{5.55}$$

Table 5.1. Some characteristics of soil pores and hydraulic conductivity. (From Tabuchi (1963); Tabuchi et al. (1963))

	Depth (cm)	Porosity (%)	Macropore porosity (%) >1.0 mm	Macropore porosity (%) 1-0.5 mm	Macropore porosity (%) 0.5-0.25 mm	Macropore porosity (%) 0.25-0.1 mm	Entry pores d_e (mm)	Entry pores W_e (%)	Hydraulic conductivity K_{we} (cm/sec)	Hydraulic conductivity K_{ws} (cm/sec)	Structure
	15 A	80	0	6	12	3	0.5	5	0.05	0.03	Granular
Volcanic ash	100 B	78	0	7	12	4	0.5	2	0.006	0.04	Massive
soil,	150 A_b	81	4	2	10	3	0.6	4	0.05	0.08	Massive
Kantō loam	180 B_b	78	7	8	5	4	0.4	1	0.002	0.01	Massive
(north)	240 D	87	18	7	0	0	1.5	14	0.04	0.15	Pumice
	0 A	75	0	0	11	3	0.4	5	0.02	0.02	Aggregate
Volcanic ash	30	76	0	5	14	3	0.3	3	0.03	0.02	Aggregate
soil,	43	80	3	11	12	2	0.2	5	0.002	0.01	Nutty
Kantō loam	80 B	83	4	6	13	4	0.5	4	0.008	0.03	Massive
(south)	140 A_b	80	0	2	18	2	0.8	3	0.02	0.11	Massive
Diluvial	5 A	56	2	2	2	3	0.5	10	0.01	0.03	Massive
red-yellow	30 B	51	0	1	3	2	0.2	1	<0.0001	0.0009	Massive
soil	70 C	50	0	3	2	1	0.1	2	<0.0001	0.002	Massive
(Shizuoka)	90	--	0	3	5	1	--	--	--	--	Massive

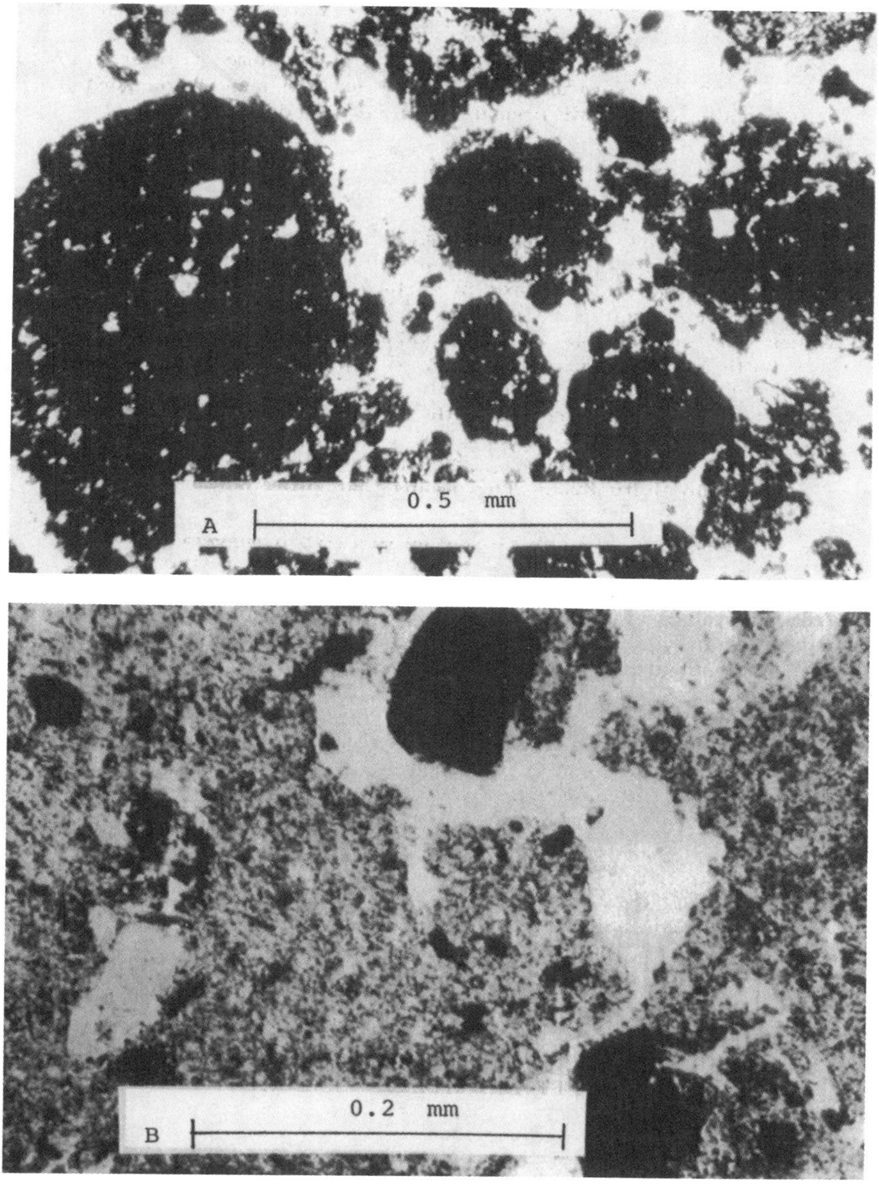

Fig. 5.10. Photomicrographs of undisturbed samples of Kantō loam volcanic ash soil: (A) surface soil; (B) subsoil. (From Tabuchi, 1963)

This calculated value d_e represents the size of the largest pores in the soil. The value of d_e agreed well with the observed value of the maximum diameter of pores in surface samples of the volcanic ash soil, but for subsoils and for the nonvolcanic ash soil, calculated values were smaller than observed values. This may be due to an "ink bottle" type of pore structure, with the value of d_e representing the diameter of the necks.

After air entered the soil at the entry pressure, the conductivity K_{we} of entry pores was measured with air (Table 5.1). In the surface sample of Kantō loam K_{we} is nearly equal to the saturated hydraulic conductivity K_{ws}. This means that entry pores play the dominant role in water movement. But in the subsoil, the values of K_{we} are 20 to 30% of K_{ws}. Other pores play an important role in permeability. Various sizes of tubular pores occur in the subsoil.

The hydraulic conductivity of the entry pores can be calculated from the values of d_e and W_e, assuming capillary behavior. The calculated values are much larger than the measured values, probably due to the tortuosity and pore size variation related to soil structure.

Since the pore sizes of soils are not uniform, we can use the pore size distribution as a physical quantity representing the pores (Fig. 5.11). Several papers have attempted to analyze hydraulic conductivity from the pore size distribution (e.g. Purcell, 1949). The number of capillaries n_i corresponding to each diameter d_i is calculated using the suction-drainage curve, W_d (h):

$$n_i = \frac{4\,\Delta W_d}{\pi d_i^2} \tag{5.56}$$

where d_i is $3/h_i$ (mm) and the value of ΔW_d is the water volume (%) drained at the negative pressure h_i. Hydraulic conductivity due to all capillaries is obtained by applying the Hagen-Poiseuille equation to each group of capillaries as follows:

$$K = \frac{\rho g \pi}{128\eta}\,\frac{1}{10^4}\sum_{i=1}^{\infty}\left[d_i^4 \times n_i(d_i)\right] \tag{5.57}$$

If we replace n_i by the distribution function f(d), we have

$$K = \frac{\rho g \pi}{128\eta}\,\frac{N}{10^4}\sum_{i=1}^{\infty}\left[d_i^4 \times f(d_i)\right] \tag{5.58}$$

where

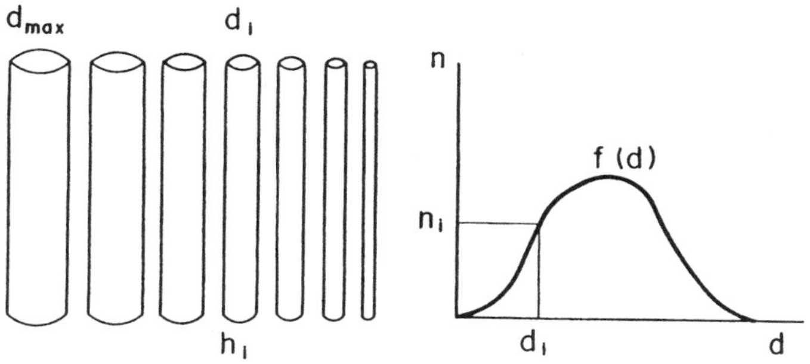

Fig. 5.11. Model of nonuniform capillaries.

$$f(d_i) = \frac{n_i}{N} \tag{5.59}$$

and N is the total number of capillaries.

5.3.4 X-Ray Stereoradiograph Method

The true state of coarse pores can be seen in radiographs using a new soft X-ray method (Tokunaga et al., 1984, 1985, 1988). In the paddy field soil, root pores formed by rice are well-developed (Fig. 5.12). Vertical pores with maximum 0.7-0.8 mm diameter are present at 1 to 1.5 per cm^2. Horizontal pores have nearly uniform diameters of 60-90 μm, with pore numbers of 2-5 per cm^2. The calculated values of permeability coefficients by the Poiseuille formula for the vertical pore distribution closely approach the measured values. Entry pores of volcanic ash subsoil were examined by this method (Sasaki, 1991). Pores had a tubular shape with 0.3 to 3 mm diameter and hydraulic conductivity K_{we} less than 30% of K_{ws}.

5.3.5 Other Approaches to Permeability

Recently, attempts have been made to analyze permeability using stochastic models. The Monte Carlo method has been applied to percolation problems on regular lattice networks by many researchers (e.g. Frisch et al., 1962; Dean, 1963). Particles are distributed randomly over the lattice with a probability of occupation P_r at each site, and cluster sizes of particles linked together are calculated by computer for various values of P_r. The critical probability P_c at which the lattice becomes porous is obtained.

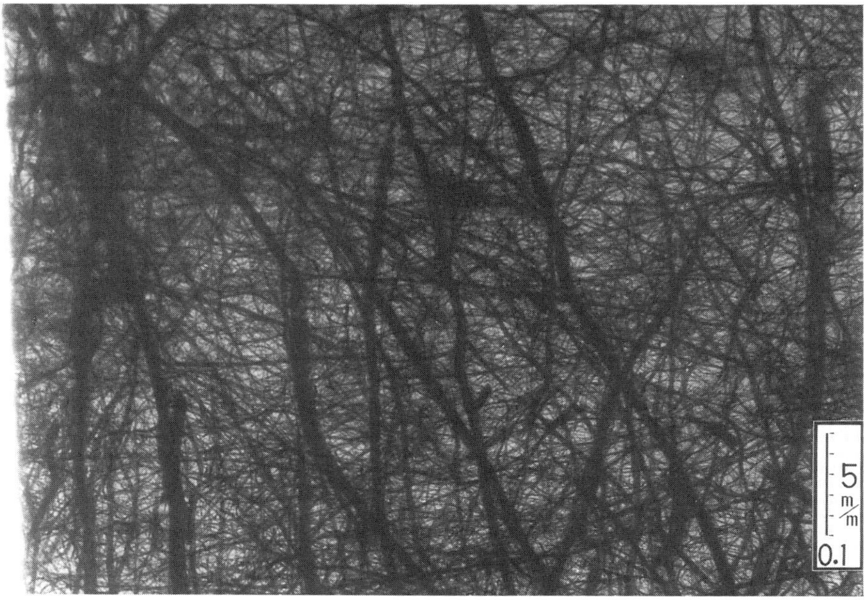

Fig. 5.12. Radiograph of macropores in surface soil (15 cm depth) in clayey paddy field. (From Tokunaga et al., 1985)

Onizuka (1975) has calculated the conductivity of a block consisting of many microelements. Each microelement is occupied by a particle or a pore, determined randomly by computer according to the probability that equals the porosity. The continuity of pores is calculated. The bulk electrical conductivity of the block is calculated by replacing pore elements with resistors. The results are:

$$K = constant \bullet (P_r - P_c)^a$$
$$a = 1.725 \pm 0.005 \qquad (5.60)$$
$$P_c = 0.318 \pm 0.002$$

where P_r is the probability of the pore phase and equals porosity in this case. Equation (5.60) indicates one of the relations between porosity and saturated hydraulic conductivity. Takeuchi (1971) has proposed a method of calculating hydraulic conductivity based on unit capillaries, where the size, length, and direction are obtained by the simulation of random numbers.

5.3.6 Hydraulic Gradient Effects

Several researchers have reported that hydraulic conductivity changes with the hydraulic gradient J. Darcy's law predicts that hydraulic conductivity remains constant when the hydraulic gradient changes. It has been pointed out that Darcy's law is not applicable in a high-velocity region because of turbulent flow. The change in hydraulic conductivity at low values of hydraulic gradient (Kimura, 1972) may be due to the change of effective pores and the relation to the critical hydraulic gradient. It is well known that water does not move below a certain critical threshold value of hydraulic gradient, especially through a clayey soil (Swartzendruber, 1962). This is discussed more completely in Chapter 6.

5.3.7 Average Hydraulic Conductivity of Layered Soil

The average hydraulic conductivity of a layered soil is

$$K_{ave} = \frac{\Sigma\ L_i}{\Sigma\ (L_i/K_i)} = \frac{\Sigma\ L_i}{\int dz/K(z)} \qquad (5.61)$$

where L_i is the thickness of each layer and K_i the hydraulic conductivity.

Fujikawa (1968) considered the hydraulic conductivity of a mixture of clayey and sandy soil. Let the depth from the top surface of the mixed soil be z and the volume of a thin layer dz of clayey soil be $M'(z)\ dz$ (Fig. 5.13). Then the volume of the mixed clayey soil in the unit pore is

$$S = \frac{M'(z)\ dz}{n\ dz} = \frac{M'(z)}{n} \qquad (5.62)$$

where n is porosity and $M'(z) = dM/dz$. The hydraulic conductivity of the mixed soil is given in Eq. (5.51). Substituting Eq. (5.62) into Eq. (5.51), we have

$$K = CK_0 \exp\left[\frac{aM'(z)}{n}\right] \qquad (5.63)$$

The average hydraulic conductivity is obtained by substituting Eq. (5.63) into Eq. (5.61). That is,

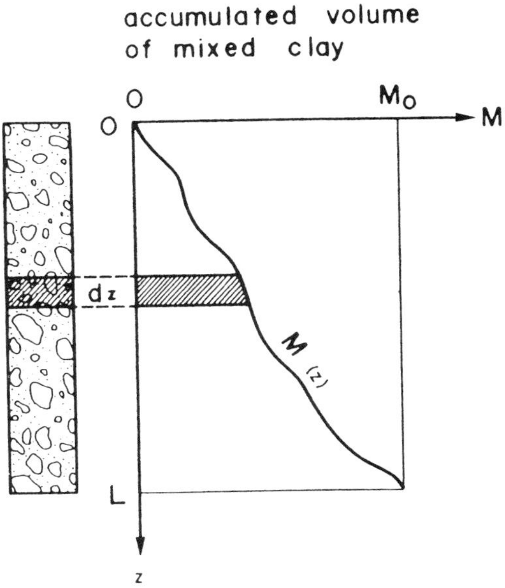

Fig. 5.13. Distribution of clay particles through a soil column.

$$K_{ave} = \frac{L}{\int_0^L K^{-1}\, dz} = \frac{L}{\int_0^L C^{-1} K_0^{-1} \exp\left[\dfrac{-aM'(z)}{n}\right] dz} \tag{5.64}$$

Assuming that

$$C^{-1} K_0^{-1} \exp\left[-an^{-1}M'(z)\right] = F \tag{5.65}$$

we obtain

$$K_{ave} = \frac{L}{\int_0^L F(z)\, dz} \tag{5.66}$$

5.3.8 Unsaturated Hydraulic Conductivity

Hydraulic conductivity of unsaturated soils has been measured by many researchers. Wyckoff and Botset (1936) used a mixed fluid of gas and water. Other investigators have measured unsaturated hydraulic conductivity by various methods.

The values of hydraulic conductivity K follow the general relation (Budagovskii, 1955)

$$\frac{K_u}{K_s} = m^a \tag{5.67}$$

$$m = \frac{W - W_H}{n' - W_H} \tag{5.68}$$

where K_u is unsaturated hydraulic conductivity, K_s the saturated hydraulic conductivity, a is a constant, m is the effective saturation ratio, W the water content (%), W_H the content of immobile water (%), and n' the porosity (%). The value of K decreases rapidly with decreasing water content, as shown in Fig. 5.14. Values of the constant a found by various researchers are:

Slichter (1898):	a = 3.3
Kozeny (1932):	a = 4
Leibenzon (1947):	a = 3.7
Budagovskii (1955):	a = 4

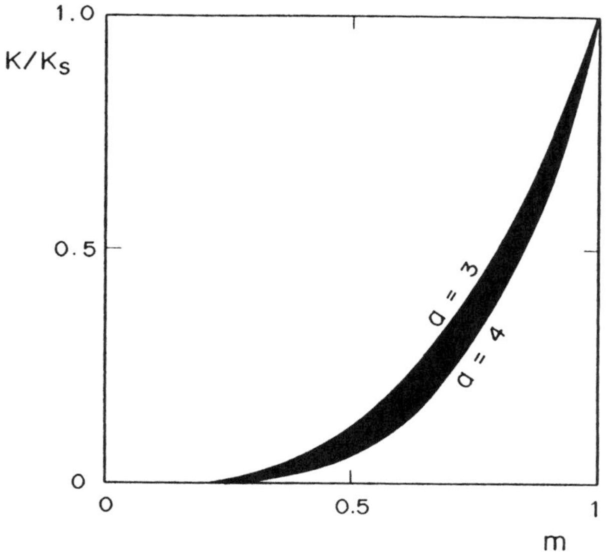

Fig. 5.14. Decrease of unsaturated hydraulic conductivity with saturation ratio.

5.4 SOME SPECIAL PROBLEMS

5.4.1 Entrapped Air

Flowing water usually contains dissolved air, which may be released within pores in the soil. Air bubbles will decrease hydraulic conductivity, but when flowing water dissolves air bubbles within the pores, the hydraulic conductivity increases. Releasing and dissolution of air are important factors in water flow in soil. Yawata (1960) found experimentally that:

1. The accumulation of air bubbles under low water pressure is greater than under high water pressure.
2. Air bubbles decrease seepage discharge.
3. Hydraulic conductivity will decrease in a zone of air accumulation.
4. The zone of air accumulation develops from the top surface of the soil to the lower part during downward percolation.

These facts are reflected in the water pressure distributions shown in Fig. 5.15. At $t = t_0$, air bubbles are not present in the saturated soil, and K is constant throughout the layer. After some hours, at $t = t_1$, air bubbles appear and accumulate to a depth b from the soil surface. The decreasing value of K in this zone changes the total head and water pressure. Head loss within this zone becomes larger than in the lower part of the saturated zone, and water pressure at b decreases. Sometimes, p_b becomes negative. Then the zone of air accumulation will develop further down and total head will change. With the continued release of air bubbles, the air volume increases and bubbles become connected to each other.

The release of air bubbles is easier at low pressure than at high pressure. In a layered soil, air bubbles will be released where water pressure is negative. A negative water pressure would therefore be useful in preventing seepage loss. To increase permeability it would be necessary to increase water pressure or to release the air in a vacuum before water enters the soil. Deaired water could dissolve air within soil pores and the hydraulic conductivity would increase.

Kuroda (1965) has described the release of air bubbles related to water pressure p and temperature T based on Henry's law. Air volume released from percolating water is

$$\Delta V_a = qAX(p,T)\ \Delta t \tag{5.69}$$

where q is the percolation rate, A the cross-sectional area, and X(p,T) is air volume at p and T released from a unit volume of water ($p = p_0$, $T = T_0$). Therefore,

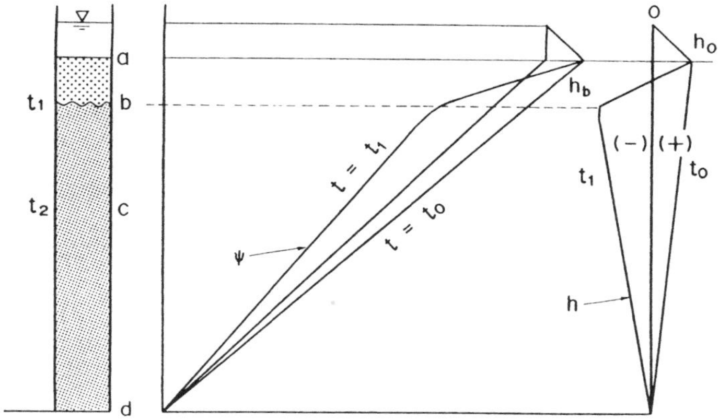

Fig. 5.15. Change of total head and water pressure due to entrapped air.

$$V_a = A \int_0^t Xq\ dt \tag{5.70}$$

If we assume that released air bubbles are contained uniformly through the soil layer, the effective saturation ratio m [Eq. (5.68)] after air is released is

$$m = \frac{n' - 100V_a / AL - W_H}{n' - W_H} \tag{5.71}$$

where L is soil length. Since unsaturated hydraulic conductivity is a function of m [Eq. (5.67)], we can arrive at

$$K_u = K_s \left[1 - \frac{100\ V_a}{(n' - W_H)AL} \right]^a \tag{5.72}$$

Based on this equation, we can obtain the equation that represents the decrease of discharge due to entrapped air as follows. The value of q is obtained by Darcy's law,

$$q = K_u\ J \tag{5.73}$$

where J is the hydraulic gradient. Substituting Eqs. (5.72) into (5.73), we get

$$(1 - C_1 V_a)^{-a} \Delta V_a = Q_0 X(p,T) \Delta t \tag{5.74}$$

where

$$C_1 = \frac{100}{(n' - W_H)AL} \qquad Q_0 = K_s JA$$

If the values of p and T are constant, we can integrate Eq. (5.74) to obtain

$$K_u = K_s(1 + Bt)^{-z} \tag{5.75}$$

where

$$B = Q_0 X(p,T) (a - 1)C_1 \qquad z = \frac{a}{a - 1}$$

Then

$$Q = Q_0(1 + Bt)^{-z} \tag{5.76}$$

Equation (5.76) shows the relation between the value of Q and the time t, indicating the decrease of Q due to entrapped air.

However, water pressure p is not constant throughout a soil layer under the usual conditions of downward percolation, and the volume of air bubbles released is not uniform. So the analysis of flow is more complicated. Furthermore, we have to consider that the zone of air accumulation can contain continuous air-filled spaces as well as isolated air bubbles. Nakamura (1969) has reported that after sufficient air has been released, the air spaces become continuous, and water flows under a unit hydraulic gradient. The relation $\theta(p)$ between the water content and water pressure must also be taken into account in the analysis, a factor that complicates the calculation still further.

5.4.2 Capillary Syphon Flow

When soil layers are as shown in Fig. 5.16, water will flow from a to e. Water rises from a to b and flows downward to e. This type of flow is likened to the action of a syphon and is called capillary syphon flow (Akiba, 1938). Total head ψ decreases with distance from a to e. If hydraulic conductivity is constant throughout the soil, the total head line is as shown in Fig. 5.16, and we can

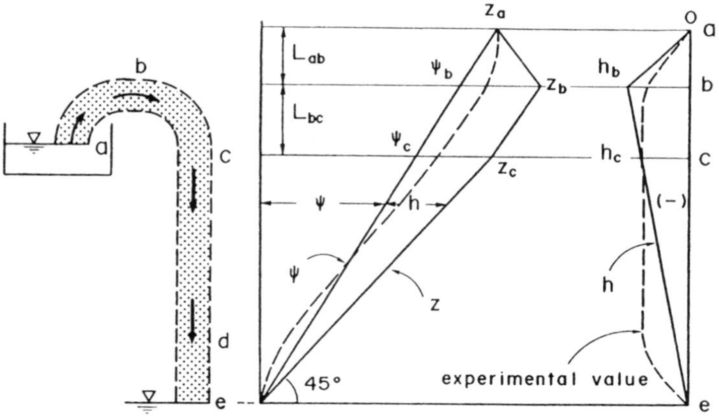

Fig. 5.16. Distribution of total head and water pressure in capillary syphon flow. (From Yamazaki, 1943)

obtain the distribution of water pressure from $\psi = z + h$. The water pressure becomes negative through all layers, as shown in Fig. 5.16.

If the permeameter wall is made of pervious material through which air can pass, the water content in the soil attains equilibrium with h. If the length L_{ab} is sufficient, this flow becomes unsaturated flow in an open system. Consequently, we can apply the relation $\theta = \theta(p)$ to flow analysis.

$$p_b < p_a, p_e \tag{5.77}$$

Therefore,

$$\theta_b < \theta_a, \theta_e \tag{5.78}$$

Hydraulic conductivity K at b is smaller than K at a and e.

$$K_b < K_a, K_e \tag{5.79}$$

Accordingly, the total head is not a straight line, but a curve. Head loss at point b is much larger than that near a and e. Yamazaki (1943) has shown this in experiments with sand. Measured values are shown as dashed lines in Fig. 5.16. Negative water pressure in the zone b-d is almost constant:

$$J_{bd} \doteqdot 1 \tag{5.80}$$

Therefore,

$$q \doteqdot K_{bd} \tag{5.81}$$

Accordingly, the discharge is determined by the hydraulic conductivity in the zone having a constant negative water pressure.

Capillary syphon flow occurs under natural conditions. Saturated flow takes place near the surface of groundwater and subsurface drains, but in other places water flow is unsaturated. A plant to purify polluted water was designed using this mechanism (Niimi and Arimizu, 1977). Polluted water will be purified under aerobic conditions during unsaturated flow.

5.4.3 Capillary Fringe and Water Flow

Flow with a free water surface may occur from seepage through earth dams or dikes or percolation from canals or paddy fields. The water flows not only through the saturated zone but also through the capillary zone with a negative water pressure.

The region of water flow is divided into a saturated and an unsaturated zone. In the former, hydraulic conductivity is uniform, but in the latter, hydraulic conductivity varies. Yoshida (1966, 1968) has investigated the characteristics of the steady flow, including the unsaturated capillary flow. From his experiment it was verified that the total potential ψ changes continuously from a saturated zone to an unsaturated zone, and that the streamlines are perpendicular to the equipotential lines. The fundamental equation of the flow, including unsaturated flow, was proposed as

$$q = -K(\theta) \; grad \; \psi \tag{5.82}$$

where K is a function of water saturation. Furthermore, the characteristic curve between water content and negative pressure in a zone of capillary flow was nearly equal to that in static capillary rise. Yoshida has also attempted to solve this water problem by means of a relaxation method.

5.4.4 Unsaturated Downward Flow in Layered Soil

Unsaturated flow in a layered soil where less pervious soil overlies more pervious soil was examined by Yamazaki (1948). Water pressure usually becomes negative. If the permeameter wall is made of a pervious material, air can enter into the soil when the water pressure is less than air entry pressure p_{ent}.

As a result, water flows in an unsaturated condition, where the hydraulic gradient is equal to unity and hydraulic conductivity depends on water pressure. These results are summarized by the following equations when $p < p_{ent}$,

$$K_u = K(\theta) \quad and \quad \theta = \theta(p) \tag{5.83}$$

$$J = 1 \tag{5.84}$$

Therefore,

$$q = K(p) \tag{5.85}$$

Total head and water pressure distribution for this type of flow are shown in Fig. 5.17.

Unsaturated conditions in soil can result from air bubble accumulation, from capillary rise, from air entry, and from partial flow after infiltration into layered soil (see Sec. 5.5.3). Each results in a different water flow. The important relation between water content θ and water pressure p is not the same in each case. In an open system there is a defined relation between θ and p which does not exist when bubbles are present.

Takagi (1960) analyzed unsaturated downward flow using $\theta(p)$ and $K(p)$ relations in the region of negative water pressure. Zaslavsky (1964) has analyzed flow using the critical value of the negative pressure in order to distinguish the two systems of flow. He has called it the critical capillary head.

Here we shall analyze downward flow for three cases. P_{L1} and P_{L2} are the limiting negative pressures of each layer. Because the first layer is less pervious than the second layer,

$$0 > P_{L2} > P_{L1} \tag{5.86}$$

Case I: $0 > P_b > P_{L2} > P_{L1}$

Because water pressure is larger than P_L, water flows in a closed system. If we neglect air bubbles, the water content of the first layer, θ_1, and that of the second layer, θ_2, equal the saturated values, and hydraulic conductivity K_1 and K_2 are the saturated values. Discharge Q is calculated according to Darcy's law for saturated flow:

$$Q = K_{ave} A\left(\frac{h_0}{L} + 1\right) \tag{5.87}$$

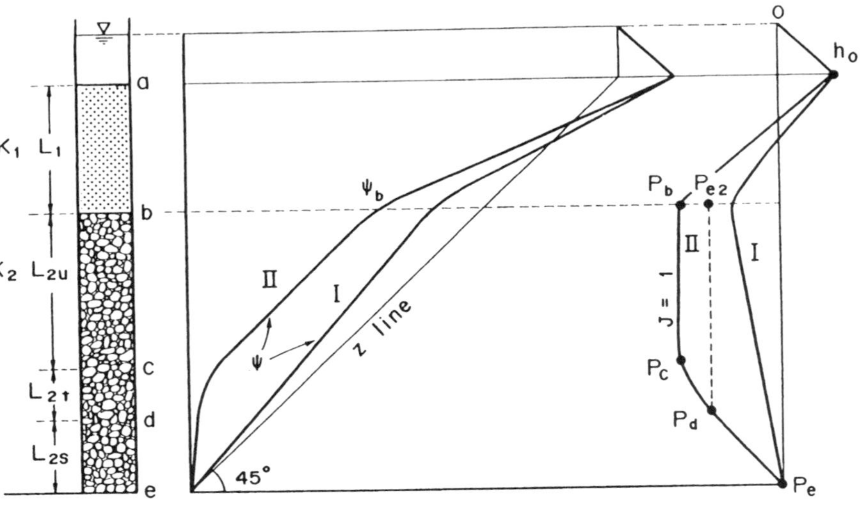

Fig. 5.17. Distributions of total head and pressure for unsaturated downward flow in a layered soil. (From Yamazaki, 1948)

where

$$L = L_1 + L_2 \qquad K_{ave} = \frac{L}{L_1/K_{1s} + L_2/K_{2s}}$$

Water pressure distribution is shown in Fig. 5.18 (Case I).

Case II: $0 > P_{L2} > P_b > P_{L1}$

Water flows in an open system through the second layer because $P_{L2} > P_b$. However, in the lower zone (d-e) of the second layer it flows in a closed system because the water pressure is larger than P_{L2} due to the boundary condition $p_e = 0$. Accordingly, the flow through layers a-b, b-d, and d-e is:

$$Q = K_{1s}A\left(\frac{\rho g h_0 - P_b}{\rho g L_1} + 1\right) \tag{5.88}$$

$$Q = K_2(p)A\left(\frac{dp}{\rho g\, dz} + 1\right) \tag{5.89}$$

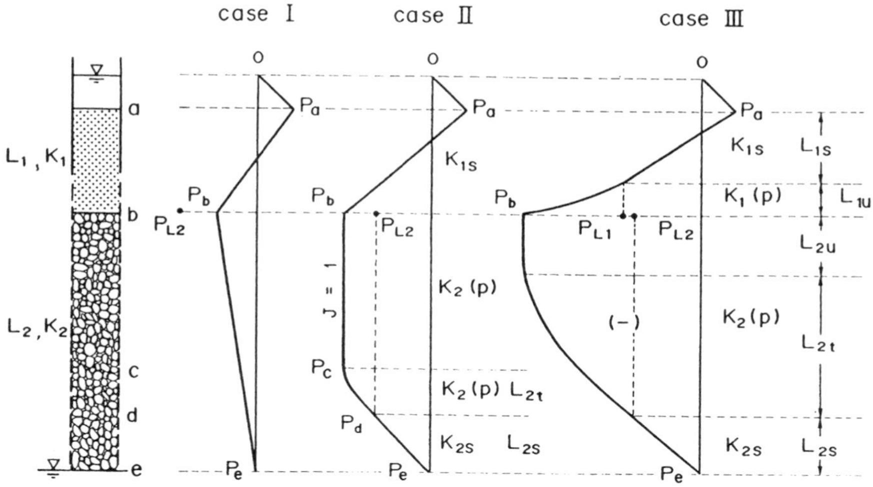

Fig. 5.18. Pressure distribution for unsaturated downward flow through a two-layered soil.

$$Q = K_{2s} A \left(\frac{P_d - P_e}{\rho g L_{2s}} + 1 \right) \tag{5.90}$$

$$= K_{2s} A \left(\frac{P_{L2}}{\rho g L_{2s}} + 1 \right) \tag{5.91}$$

The values of K_{1s}, K_{2s}, A, h_0, P_{L2}, L_1, and L_2 are known; the unknown values are Q, P_b, and L_{2s}. From Eq. (5.89), we obtain

$$\rho g \int dz = \int \frac{dp}{Q/AK_2(p) - 1} \tag{5.92}$$

If the function K(p) is known, this equation can be integrated and the unknown values calculated from Eqs. (5.88), (5.91), and (5.92). The water pressure distribution p(z) will also be obtained.

In an open system, for p(z) from Eq. (5.89) we obtain

$$\frac{dp}{dz} = \left[\frac{Q}{K_2(p)A} - 1 \right] \rho g \tag{5.93}$$

When p increases, the value of K(p) increases and the value of the right side decreases. Accordingly, the value of dp/dz decreases with an increase of p. When p decreases, dp/dz increases. When the value of dp/dz is positive, the curve of p(z) becomes line a in Fig. 5.19, and when the value of dp/dz is negative, it is line b. A curve of p(z) such as line c in Fig. 5.19 cannot exist. If it does exist, the discharge at A and B becomes unequal, which contradicts the condition of a steady state. In downward flow where the water pressure at the bottom of the soil layer equals zero, curve a is not acceptable; only curve b in Fig. 5.19 satisfies the condition. Consequently, water pressure decreases in an upward direction. The gradient dp/dz always increases for the upward direction and gradually approaches zero—in other words, constant water pressure. Accordingly, when the unsaturated zone is long enough, the zone with constant water pressure (J = 1) will appear in the upper part.

Case III: $0 > P_{L2} > P_{L1} > P_b$

In this case, the lower part of the first layer also becomes an open system. We obtain the equation

$$Q = K_{1,uns} A \left(\frac{p_{L1} - P_b}{\rho g L_{1u}} + 1 \right) \tag{5.94}$$

$$K_{1,uns} = K_1(p) \tag{5.95}$$

The equations of other zones are those mentioned above.

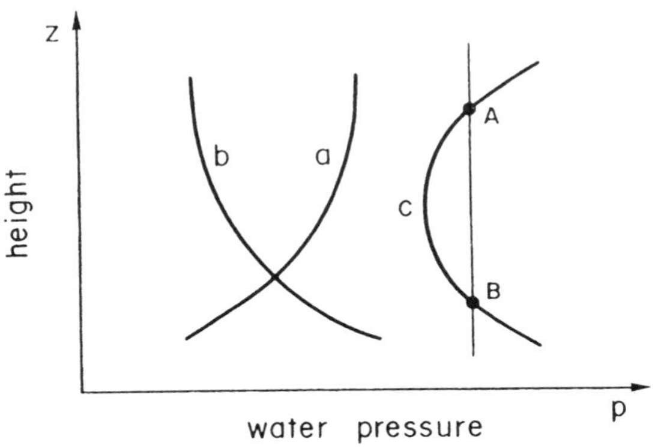

Fig. 5.19. Three curves of possible changes of pressure distribution in Fig. 5.18 for Case II.

In laboratory experiments with layered soils, water flow in an open system was observed where water pressure was smaller than P_L (Ishihata and Tokunaga, 1985; Sasaki, 1992). Air entered into the layers horizontally through the permeameter wall, and dissolved oxygen in seepage water increased rapidly. The amount in percolating water depends on the water flow system.

This type of water flow in an open system was also observed under flooded conditions in upland paddy fields (Tokunaga and Sasaki, 1990). Water pressures were -15 cm ~ -25 cm H_2O in the subsoil, which was smaller than entry pressure, and air pressures in the subsoil were equal to atmosphere pressure. This means that air in the subsoil is connected with the atmosphere even under flooded conditions in an open system.

5.5 INFILTRATION

5.5.1 General Concepts

Infiltration refers to water entry into the soil from rainfall or irrigation, and is the first stage of water movement in soil. It is necessary in any irrigation plan or for any runoff problem to know the infiltration rate and the soil water content after infiltration.

Research on infiltration can be considered in three groups. The first relates to water entry rate into soil as measured in the field. The "intake rate" is represented in an empirical equation. Horton's experimental equation is

$$q = q_c + (q_0 - q_c) \exp(-Bt) \tag{5.96}$$

where q_0 is the initial intake rate, q_c the intake rate at steady state, t the time, and B is a constant. The second group relates to experimental measurements of infiltration rate, wetting front, water content, and water pressure during infiltration. The third group contains theoretical studies.

The Darcy equation has been applied to infiltration (Green and Ampt, 1911; Onstad et al., 1973). Budagovskii (1955) examined infiltration theory with an infiltration equation of the Green and Ampt type, based on results from a model of water movement through a capillary tube. Mathematical considerations of infiltration have been reported by Philip (1956, 1969), Klute (1952), and others based on diffusion equations. Infiltration is related to water potential gradients due to differences of water content. Infiltration based on the Green and Ampt model is discussed in this chapter, and infiltration based on the diffusion equations is considered in Chapter 6.

5.5.2 Downward Infiltration into a Uniform Soil

The zone of wet soil during infiltration has been divided into three parts: the transmission zone, the wetting zone, and the wetting front (Bodman and Coleman, 1944; Hansen, 1955). Water content is approximately as shown in Fig. 5.20, which indicates that the water content is almost uniform in the transmission zone, but decreases rapidly near the wetting front. Therefore, the following equation can be used:

$$V_W \doteqdot \frac{A(W_i - W_0)Y}{100} \tag{5.97}$$

where A is the cross-sectional area of the soil (cm^2), W_i is the water content during infiltration (%), W_0 the initial water content before infiltration (%), Y the distance of the wetting front from a surface (cm), and V_W the volume of water entering the soil (cm^3). From this equation we can obtain

$$\frac{1}{A}\frac{dV_W}{dt} = \frac{W_i - W_0}{100}\frac{dY}{dt} \tag{5.98}$$

$$q_e = \frac{(W_i - W_0)v_f}{100} \tag{5.99}$$

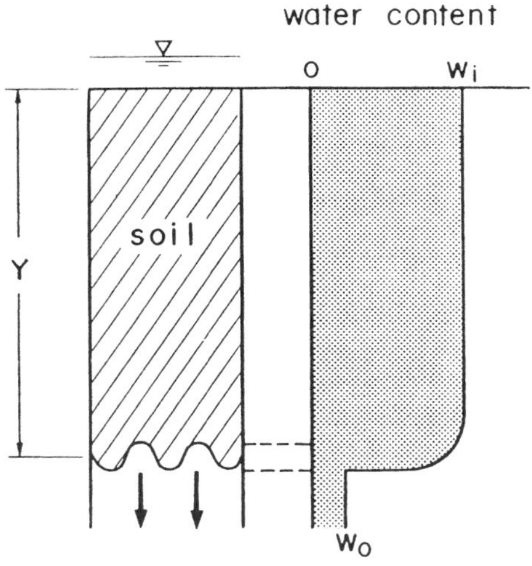

Fig. 5.20. Distribution of water content in a soil column during infiltration.

where q_e is the water entry rate ($cm^3/cm^2 \cdot sec$) and v_f is the velocity of the wetting front (cm/sec).

Infiltration can be represented by a capillary model as shown in Fig. 5.21. The infiltration equation is obtained from the Hagen-Poiseuille equation, adding the value of the capillary force into the head:

$$Q = \frac{\pi r^4}{8\eta} \frac{\rho g(h_0 + Y + h_c)}{Y} \tag{5.100}$$

where h_c is a capillary force due to the meniscus at the wetting front (cm). Equation (5.100) is rewritten

$$Q = AK' \frac{H_c + Y}{Y} \tag{5.101}$$

where $A = \pi r^2$, $K' = \rho g r^2/8\eta$, $H_c = h_0 + h_c$.

$$v_f = \frac{dY}{dt} = \frac{Q}{A} \tag{5.102}$$

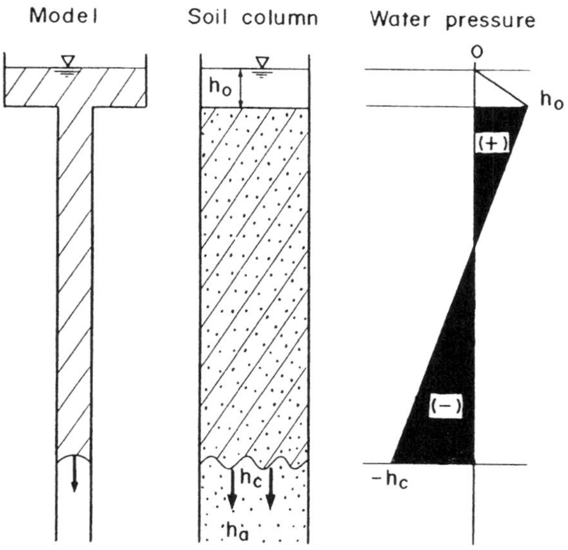

Fig. 5.21. Pressure distribution during infiltration from a capillary model.

Therefore,

$$\frac{dY}{dt} = K' \frac{H_c + Y}{Y} \tag{5.103}$$

Integrating Eq. (5.103), we obtain

$$t = \frac{1}{K'} \left[Y - H_c \log_e \left(1 + \frac{Y}{H_c} \right) \right] \tag{5.104}$$

which gives the relation between t and Y or v_f and Y.

The infiltration equation can be obtained from the analysis of forces acting on the water (Budagovskii, 1955).

$$F_g = \rho g(Y + h_0) \tag{5.105}$$

$$F_r = \alpha v Y \tag{5.106}$$

$$F_c = \rho g h_c \tag{5.107}$$

$$F_i = \frac{\rho Y dV}{dt} \tag{5.108}$$

$$F_a = \rho g h_a \tag{5.109}$$

where F_g is the gravitational force, F_r the resisting force, F_c the capillary force, F_i the inertia force, F_a the air pressure, h_0 the water depth, α is a constant, v the water velocity, h_c the advancing capillary force, and h_a the air pressure ahead of the wetting front. Since the combined force is zero, we have

$$F_g - F_r + F_c - F_i - F_a = 0 \tag{5.110}$$

Assuming that F_i approaches zero in soil water movement, Eq. (5.110) is rewritten

$$F_g - F_r + F_c - F_a = 0 \tag{5.111}$$

Substituting the values of forces into this equation, we obtain

$$v = \frac{\rho g}{\alpha} \frac{Y + h_0 + h_c - h_a}{Y} \tag{5.112}$$

Then multiplying the value of $(W_i - W_H)/100$ yields

$$v \frac{W_i - W_H}{100} = \frac{\rho g}{\alpha} \frac{W_i - W_H}{100} \frac{Y + h_0 + h_c - h_a}{Y} \tag{5.113}$$

Therefore,

$$q = K \frac{Y + h_0 + h_c - h_a}{Y} \tag{5.114}$$

where

$$q = \frac{W_i - W_H}{100} \left(\frac{dY}{dt} \right) \qquad K = \frac{\rho g}{\alpha} \frac{W_i - W_H}{100} \tag{5.115}$$

W_H is the nonmobile water content (%), K the hydraulic conductivity during infiltration (cm/sec), and q the infiltration rate ($cm^3/cm^2 \cdot sec$). In dry sand, h_a is negligible. Then

$$q = K \frac{Y + h_0 + h_c}{Y} \tag{5.116}$$

This addition of the capillary force h_c to the driving force in Darcy's equation produces a Green-Ampt type of infiltration equation. The values of K and h_c are important factors in infiltration. The Green-Ampt equation (5.116) is similar to Eq. (5.101), which was introduced using a capillary model and the Hagen-Poiseuille equation. Accordingly, the use of this equation for infiltration assumes that water flow through the soil is a set of capillary flows, each of which has a meniscus at its front. This is different from the water flow of the thin film type or diffusion type.

Because the hydraulic conductivity in the transmission zone is uniform due to uniform water content, the hydraulic gradient also becomes uniform, as shown in a linear water pressure distribution (Fig. 5.21). Since water pressure at the wetting front is defined by the meniscus at the front, the water pressure distribution is as shown in Fig. 5.21. The capillary force in this case is not a static value but a dynamic and average value formed from many meniscus fronts.

When water is supplied at a rate less than the intake rate, the infiltration is not as described above. Budagovskii (1955) has reported that the downward velocity of the wetting front becomes constant and the water content in the transmission zone is constant. Then the infiltration rate also becomes constant.

$$q = K(1 + \beta) \tag{5.117}$$

The value of K varies with the intensity of rainfall I_r

$$K = K(I_r) \tag{5.118}$$

K increases and approaches a value K_H with an increase of I_r, where K_H is hydraulic conductivity of soil with water ponded on the surface. Swartzendruber (1974) has examined the relationship of intensity I_r to the infiltration rate.

5.5.3 Downward Infiltration into a Layered Soil

Natural soil may have a layered structure, with differences in porosity, hydraulic conductivity, and capillary force. Many researchers have investigated infiltration into layered soils, both experimentally and theoretically (e.g. Coleman and Bodman, 1945; Budagovskii, 1955; Tabuchi, 1961; Miller and Gardner, 1962; Hill and Parlange, 1972).

The saturated hydraulic conductivity of the first layer, K_{1s}, may be larger or smaller than that of the second layer, K_{2s}.

1. $K_{1s} > K_{2s}$: In this case the water content within the lower layer is not affected by the upper layer. Infiltration is the same as in a uniform soil. Infiltration rate q decreases gradually with the increase of Y and approaches the value of K_{2s}.

2. $K_{1s} < K_{2s}$: In this case, experimental data show that q becomes constant after the entry of the wetting front into the lower layer. The water content within the lower layer W_2 also becomes constant, and consequently K_2 will be constant.

Tabuchi (1961) has attempted to clarify why the infiltration rate becomes constant and why certain conditions are necessary for this phenomenon to occur. A special flow condition, termed "partial flow," was observed in experiments with dry sand in which water flows only in a partial section of the permeameter with an air-water boundary on each side (Fig. 5.22). The apparent water content is lower than that of saturated sand, but the flow section is almost saturated.

The analysis of this partial flow is as follows. Assume a small vertical flow pipe having a unit cross-sectional area within the partial flow as shown in Fig. 5.22. The infiltration equation is

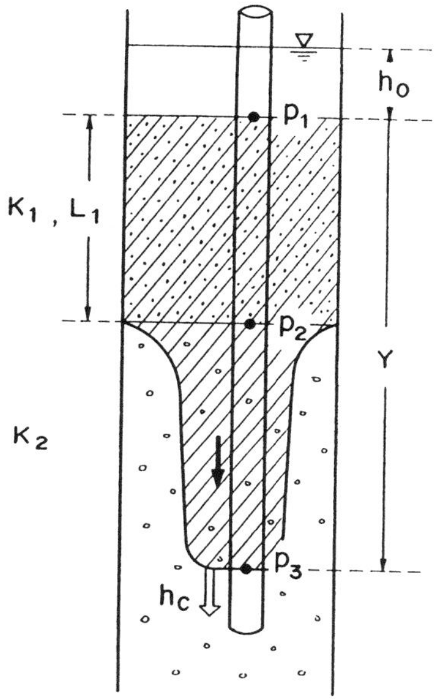

Fig. 5.22. Occurrence of partial flow during infiltration through a layered sand column. (From Tabuchi, 1961)

$$q = K_{ave}\left(1 + \frac{h_0 + h_{c2}}{Y}\right) \tag{5.119}$$

where h_{c2} is the advancing capillary force of the lower layer. With two layers, we obtain the value of K from Eq. (5.61):

$$K_{ave} = \frac{Y}{L_1/K_1 + (Y - L_1)/K_2} \tag{5.120}$$

where K_1 and K_2 are hydraulic conductivity of the infiltration stage of the upper layer and lower layer, respectively. From Eq. (5.119) we have

$$v = \frac{K_1K_2}{n_2}\frac{Y + H_{c2}}{L_1K_2 + K_1(Y - L_1)} \tag{5.121}$$

where

$$H_{c2} = \frac{p_1 - p_3}{\rho g} = h_0 + h_{c2} \tag{5.122}$$

and n_2 is the porosity of the lower layer and v the velocity of the wetting front. Equation (5.121) can be rewritten as follows:

$$\frac{dv}{dY} = \frac{K_1 K_2}{n_2} \frac{L_1 K_2 - L_1 K_1 - K_1 H_{c2}}{(L_1 K_2 - L_1 K_1 + Y K_1)^2} \tag{5.123}$$

The necessary condition for $dv/dY < 0$ is

$$L_1 K_2 - L_1 K_1 - K_1 H_{c2} < 0 \tag{5.124}$$

This can be rewritten as

$$q_L > K_2 \tag{5.125}$$

where

$$q_L = K_1 \left(\frac{H_{c2}}{L_1} + 1 \right) \tag{5.126}$$

We can obtain the following results from Eq. (5.123):

$$\begin{aligned} &\textit{I.} \quad \textit{When } q_L > K_2, \ dv/dY < 0 \\ &\textit{II.} \quad \textit{When } q_L = K_2, \ dv/dY = 0 \\ &\textit{III.} \quad \textit{When } q_L < K_2, \ dv/dY > 0 \end{aligned} \tag{5.127}$$

When $q_L > K_2$ (Case I), the farther the wetting front advances, the slower the downward velocity of the wetting front. When $q_L < K_2$ (Case III), the farther the wetting front advances, the more the downward velocity of the wetting front increases. The velocity of the forward part of the wetting front is always greater than that of the trailing part. Then it appears that the roughness of the wetting front increases with the increase of Y in Case III.

The value of water pressure p_2 in the flow pipe can be given as

$$p_2 = p_3 + (q_L - K_2)L_1 \frac{\rho g(Y - L_1)}{L_1K_2 + K_1(Y - L_1)} \tag{5.128}$$

From this equation we can obtain the following results:

$$\begin{aligned} &I. \quad q_L > K_2, \quad dp_2/dY > 0 \\ &II. \quad q_L = K_2, \quad dp_2/dY = 0 \\ &III. \quad q_L < K_2, \quad dp_2/dY < 0 \end{aligned} \tag{5.129}$$

Accordingly, in Case III ($q_L < K_2$) the water pressure of the trailing part is greater than that of the leading part; water flows to the leading part, the roughness of the wetting front will increase, and partial flow will occur. The cross-sectional area of partial flow gradually decreases, and the distribution of water pressure approaches the distribution in the second case where the water pressure within the lower layer is constant and hydraulic gradient J_2 equals 1. Finally, the partial flow attains equilibrium in which water flows only in a vertical direction.

In the stable state,

$$J_2 = 1 \tag{5.130}$$

Therefore,

$$p_2 = p_3 = -\rho g h_{c2} \tag{5.131}$$

Accordingly,

$$J_1 = \frac{p_1 - p_2}{\rho g L_1} + 1 = \frac{h_0 + h_{c2}}{L_1} + 1 \tag{5.132}$$

Discharge flowing through the upper layer is

$$Q_1 = K_1 J_1 A = K_1 \left(\frac{h_0 + h_{c2}}{L_1} + 1 \right) A \tag{5.133}$$

Therefore,

$$Q_1 = q_L A \tag{5.134}$$

Discharge flowing through the lower layer is

$$\begin{aligned} Q_1 &= K_2 J_2 A_p \\ &= K_2 A_p \qquad (since\ J_2 = 1) \end{aligned} \tag{5.135}$$

Since the discharge of the lower layer equals the discharge flowing through the upper layer in this case, we can obtain the cross-sectional area A_p of partial flow as follows:

$$q_L A = K_2 A_p$$

Therefore,

$$A_p = \frac{q_L A}{K_2} \tag{5.136}$$

The infiltration rate q becomes constant after partial flow attains equilibrium in Case III ($q_L < K_2$). But in Case I ($q_L > K_2$), partial flow does not appear and the decrease in q continues. These results are shown in Fig. 5.23. In Eq. (5.119), the advancing capillary force, h_{c2}, is used. However, in the stable state, the static capillary force should be used because the air-water interface on the partial flow side is static. Then the infiltration rate increases slightly after the wetting front enters the second layer, as shown in Fig. 5.23.

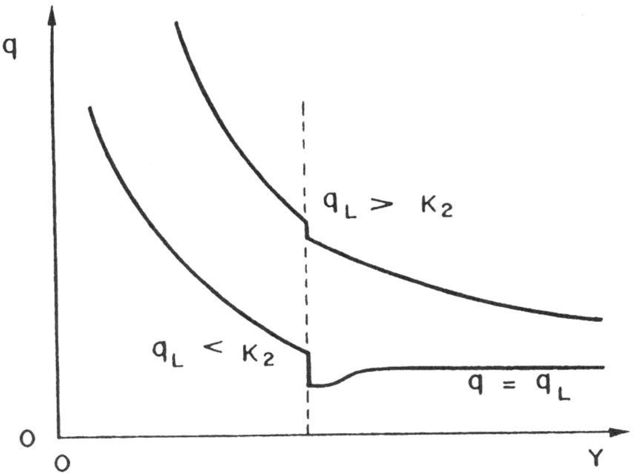

Fig. 5.23. Infiltration rate changes during partial flow. (From Tabuchi, 1961)

The comparison of q_L and K_2 is very important to distinguish the infiltration type in a two-layer soil. If $q_L < K_2$, partial flow occurs (Case III). The value of q_L is generally less than K_2 because the value of K_1 is considerably less than the value of K_2 in a fine-over-coarse-layered soil. However, we can make a condition where the value of q_L is larger than K_2 by taking a larger value of surface water depth h_0. From Eq. (5.125) we get

$$h_0 > \frac{(K_2 - K_1)L_1}{K_1} - h_{c2} \tag{5.137}$$

In this case, partial flow does not occur (Case I).

Experimental data were obtained with dry sands and glass beads. Infiltration rate q and water volume added to the column, V_w, were measured. In Case III ($q_L < K_2$) with small surface water depth, q became constant after the front entered into the lower layer, as shown in Fig. 5.24a. The values of q agreed with theoretical values. V_w increased regularly, but less in the lower layer than in the upper layer. This is due to partial flow in a cross-sectional area which is smaller than that of the permeameter. Experimental results for Case I ($q_L > K_2$) were obtained for hydraulic conductivity of 1.4×10^{-2} cm/sec for the upper layer, 4×10^{-2} cm/sec for the lower layer, and a surface water depth of 66 cm. Partial flow did not occur and q decreased continuously even after the wetting front entered the lower layer (Fig. 5.24b).

After the wetting front reached the bottom of the second layer, partial flow continued in the upper part of the second layer. Near the bottom of the column, partial flow diminished and the soil became saturated because water pressure at the bottom was equal to atmospheric pressure in this experiment. Hence water pressure increased slightly and the water pressure distribution through the column became similar to that for unsaturated flow in an open system shown in Fig. 5.18 (Case II). However, this was partial flow and its flow state was quite different from that of an open system, despite the similarity of the water pressure distribution curves.

The length of the saturated zone formed in the lower part of the second layer, L_{2s}, is obtained as follows. The discharge through this zone is

$$Q = K_2 \left(\frac{-h_{c2} - h_b}{L_{2s}} + 1 \right) A \tag{5.138}$$

where h_b is the water pressure at the bottom and usually zero. From Eqs. (5.138) and (5.134), we get

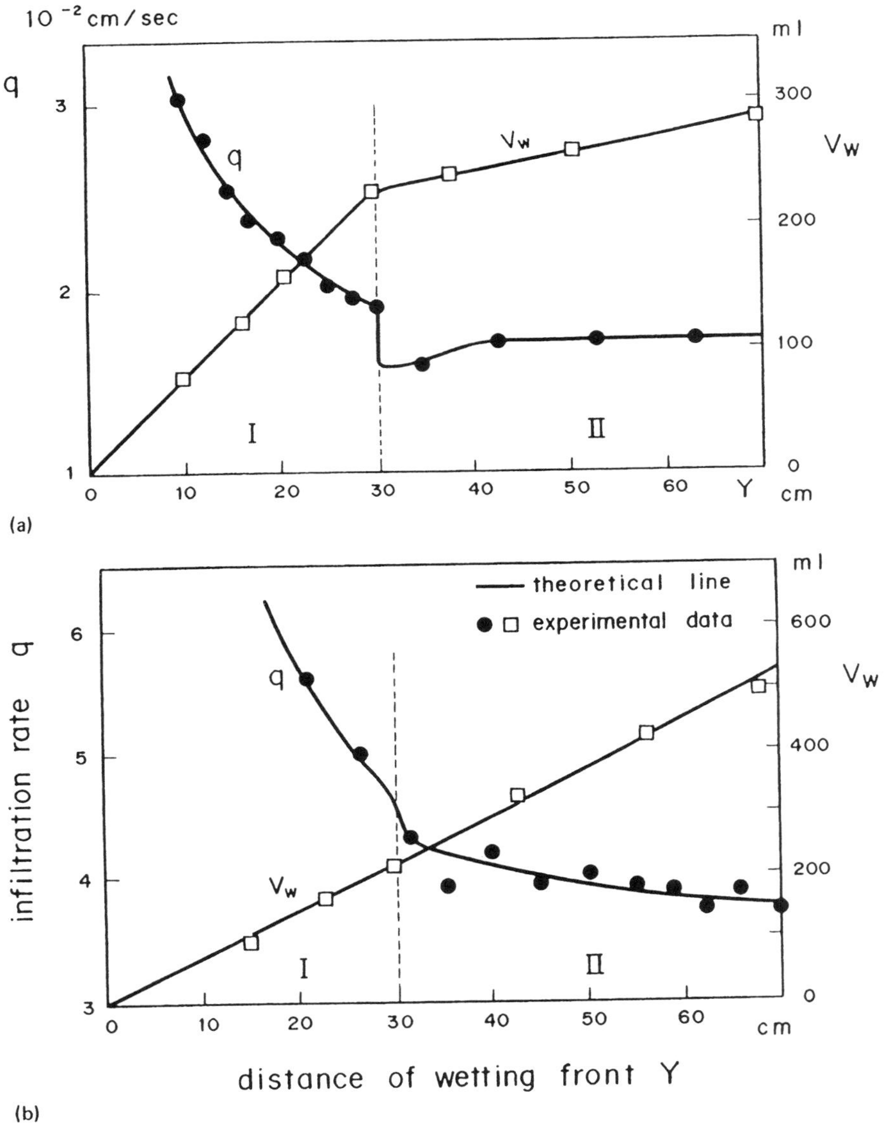

Fig. 5.24. Comparisons of theoretical curves and experimental values for infiltration rate, q, into layered soil: (a) partial flow; (b) no partial flow. (From Tabuchi, 1961)

$$L_{2s} = \frac{K_2(h_{c2} + h_b)}{K_2 - q_L} \tag{5.139}$$

Tamai et al. (1987) discussed fingering of liquid movement in homogeneous porous media after the ponded liquid vanished, and Miyazaki (1988) found this partial-type flow in water infiltration into layered soil on slopes. Hillel and Baker (1988) discussed the conditions for occurrence of fingering, and Baker and Hillel (1990) tested fingering during infiltration into layered soil consisting of different particle sizes. Predictions regarding the onset of fingering and the wetted fractional volume of fingers were both validated.

Let us consider a three-layered, dry soil as shown in Fig. 5.25. The second layer is more permeable than the first and third layers. Where $q_L < K_2$, partial flow occurs in the second layer. However, it diminishes in the lower part of the second layer because water pressure increases due to the presence of the third layer underneath. When the wetting front passes into the third layer, the infiltration equation may be obtained for each layer and we obtain the value of L'_{2s}.

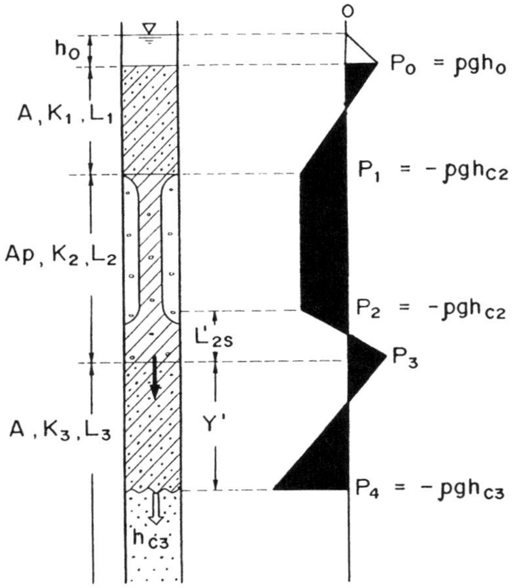

Fig. 5.25. Pressure distribution during infiltration into a three-layered soil column.

$$L'_{2s} = \frac{K_2[K_3(h_{c2} - h_{c3}) + Y'(q_L - K_3)]}{K_3(K_2 - q_L)} \tag{5.140}$$

If the value of L'_{2s} is smaller than the length of the second layer L_2, partial flow remains in the second layer. The value of the water pressure p_1 is kept constant and the infiltration rate is also constant.

After the wetting front arrives at the bottom of the third layer, the value of L_{2s} is expressed as

$$L_{2s} = \frac{K_2[K_3(h_{c2} + h_b) + L_3(q_L - K_3)]}{K_3(K_2 - q_L)} \tag{5.141}$$

where L_3 is the length of the third layer and h_b is the water pressure at the bottom (Tabuchi, 1971). Thus we can generally describe the necessary conditions for partial flow in multilayered soil as

$$L_{2i} > \frac{K_{2i}}{K_{2i+1}} \left[\frac{K_{2i+1}(h_{c2i} - h_{c2i+2}) + L_{2i+1}(q_L - K_{2i+1})}{(K_{2i} - q_L)} \right] \quad i = 1,2,3,... \tag{5.142}$$

where K_{2i} is hydraulic conductivity of a more permeable layer (even-numbered layers), L_{2i} is its length, K_{2i+1} is the hydraulic conductivity of a less permeable layer, and L_{2i+1} is its length. Experimental data are insufficient and a theoretical analysis has not yet been made for infiltration into a wetted layered soil.

5.5.4 Horizontal Water Flow

The infiltration equation for horizontal flow is obtained from Eq. (5.116), omitting the gravitational force.

$$q = K \frac{H_c}{Y} \tag{5.143}$$

where $H_c = h_0 + h_c$. If the height of soil, D, is large, the wetting front forms an inclined line as shown in Fig. 5.26. This line is approximately expressed by

$$Y(x) = \frac{Y_0}{H_c} (H_c - x) \tag{5.144}$$

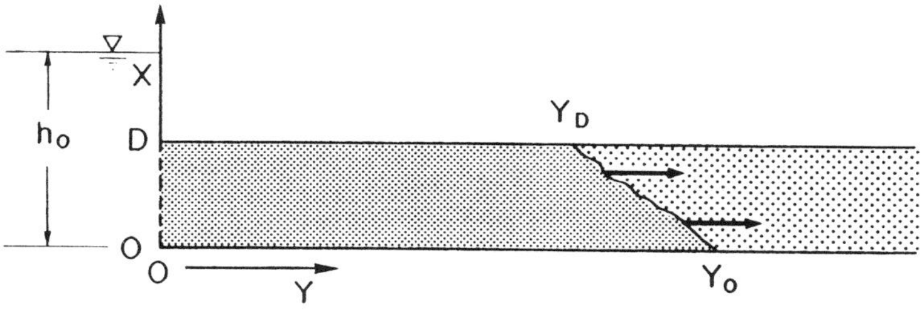

Fig. 5.26. Wetting front in horizontal infiltration into a sandy soil column. (From Tabuchi, 1971)

If it is assumed that the vertical component of flow does not exist, the infiltration rate at the height, x, is

$$q_x = K \frac{H_c - x}{Y(x)} \tag{5.145}$$

Total discharge for horizontal infiltration is calculated from Eqs. (5.144) and (5.145).

$$Q = \int_0^D q_x \, dx = \int_0^D K \frac{H_c}{Y_0} \, dx = \frac{KDH_c}{Y_0} \tag{5.146}$$

Therefore,

$$QY_0 = KDH_c = const \tag{5.147}$$

5.5.5 Upward Flow

Since the gravitational force acts opposite to the direction of water flow (Fig. 5.27), the infiltration equation for upward flow is

$$q = K \frac{H_c - Y}{Y} \tag{5.148}$$

where $H_c = h_0 + h_c$. From Eq. (5.148) we obtain

$$qY = K(H_c - Y) \tag{5.149}$$

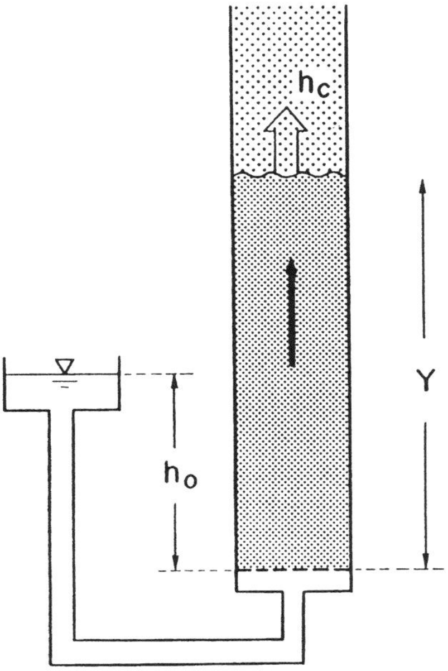

Fig. 5.27. Upward infiltration.

The values of qY have a linear dependence on Y, and the theoretical line agrees with the experimental data of infiltration into dry sand (Fig. 5.28). The gradient of this line indicates the value of K. When Y attains H_c, q becomes zero, so capillary rise should stop at that height.

$$Y \leqq H_c = h_0 + h_c \tag{5.150}$$

However, capillary rise continues beyond this height. The reason may be the difference between the advancing and the static capillary force. The advancing capillary force is a dynamic and average value of the meniscus in each pore at the front. When the wetting front rises, large pore cells will remain unsaturated and small pore cells will be saturated. Accordingly, a meniscus is formed only in small cells. The increase of the capillary force may be caused by this selection of pore cells.

Additionally, it is reported (Bikerman, 1958) that the contact angle depends on the rate of movement of the three-phase boundary along the solid. Muto (1976) has observed the change of a meniscus in a capillary and found that the contact angle was larger than zero while the meniscus was moving and approached zero when water rise had stopped. He concluded that the capillary force had gradually increased.

Thus the value of H_c in the infiltration equation tends to increase during capillary rise.

$$H_c = h_0 + h_{c,dymanic} \dashrightarrow h_0 + h_{c,static} \quad (5.151)$$

The data (Fig. 5.28) show this tendency, in which H_c increases when Y approaches H_c.

$$h_{c,static} > h_{c,dynamic} \quad (5.152)$$

5.6 RECENT DEVELOPMENTS IN FINGERING RESEARCH

A gravity driven partial flow, described in Section 3.5.3 was found by Tabuchi (1961). Since then, many researchers have studied this phenomenon. The concept of fingering or unstable wetting front used in these papers has the same physical meaning as partial flow. These studies will be reviewed under characteristics, criteria, and mechanisms.

5.6.1 Characteristic Features of Fingering

Collins (1961) discussed displacement for two liquids in porous media and analyzed "fingerlike flow" by a method similar to that used by Tabuchi (1961). Miller and Gardner (1962) described the phenomenon of an initially dry sand layer wetting in only a few places. Hill and Parlange (1972) also pointed out fingerlike flow in their studies on infiltration into layered sands. They found that the finger consisted of a saturated inner core with an unsaturated outer layer.

Three stages have been recognized in the evolution of the unstable flow field in fingering (Glass and Steenhuis, 1984; Glass et al., 1987; Glass et al., 1989c). The initial stage is dominated by rapid downward movement of finger "core" areas. In the second stage the core areas continue to transmit most of flow, but there is also slow lateral movement of moisture from finger core areas. The lateral movement is slow, having a time scale on the order of days; the downward finger growth in the first stage is much faster, on the order of minutes. The final stage is a steady-state flow field in which core and fringe areas coexist for long periods of steady infiltration.

Glass and Steenhuis (1984) and Glass et al. (1987) demonstrated the persistence of finger areas from one infiltration cycle to the next. After an interruption in water supply, drainage to field capacity, and rewetting, fingers form in the same locations as in the first cycle and have the same core areas, which continue to conduct almost all the water. Fringe-area contribution is

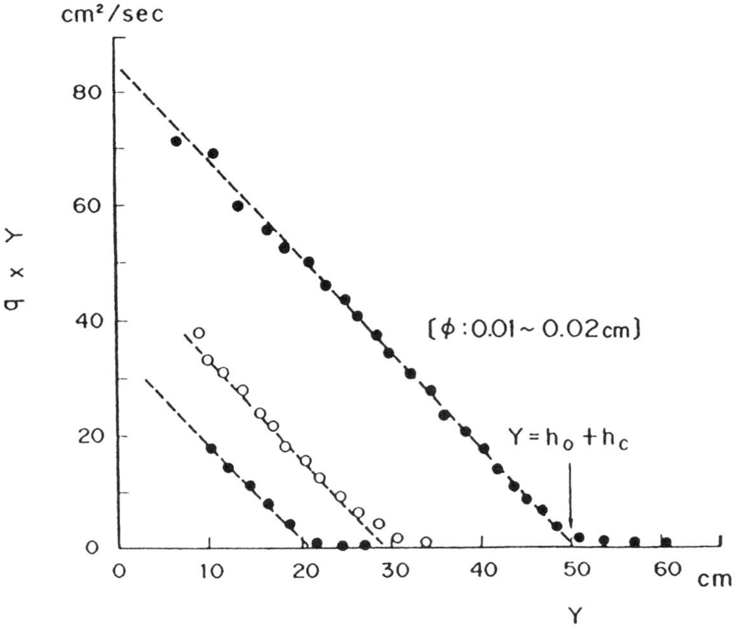

Fig. 5.28. Relation between qY and Y in upward infiltration. (From Tabuchi, 1971)

higher than in the first cycle, and the steady flow field is achieved much more rapidly.

Hillel and Baker (1988) and Baker and Hillel (1990) studied water penetration into fine over coarse-layered air-dry sands in transparent laboratory chambers. The hypothesis they tested was that the wetting front, with a high suction, will not penetrate into a coarse-textured layer until the suction at the interlayer plane falls to the effective water entry suction of the lower layer—a suction value that is predictable from the hydraulic properties of the lower layer.

In their first paper, they supposed that suction at water entry might be measured by a separate static determination, namely by the limit of capillary rise of water in the sand. They considered that an assembly of pores has a characteristic water entry suction, determined by the narrowest pores that form a continuous network in the matrix. However, dynamic measurements of flux and suction at the interface plane during infiltration revealed that the effective water entry suction is characteristic of the predominant pore size in the sublayer and can be predicted from its median particle size. Table 5.2 shows the observed entry suctions and the limits of capillary rise for the sands they used. Table 5.3 indicates the downward advancing capillary forces and the limits of capillary rise obtained experimentally for glass beads by Tabuchi (1961). Because of gravity,

Table 5.2. Values of water entry suction and maximum capillary rise for different sand sizes. (From Baker and Hillel, 1989)

Sand size (μm)	Water entry suction (cm)	Maximum capillary rise (cm)
1000-2000	1.9	8
710-1000	5.6	10
500-710	6.7	14
355-500	11.4	22
250-355	15.5	26
106-250	23.8	47

Table 5.3. Values of advancing capillary force and maximum capillary rise for glass beads. (From Tabuchi, 1961)

Diameter of glass beads (μm)	Advancing capillary force (cm)	Maximum capillary rise (cm)
400-700	1	2.5
300-400	3	4.5
200-300	5	8
100-200	8	22

the radius of curvature of the water-air interface moving downward in a capillary is considered to be larger than that under a static condition in capillary rise. Consequently, it may be more reasonable to use the suction corresponding to the downward advancing capillary force instead of the limit of capillary rise as the entry suction for a sand. Though Baker and Hillel (1990) failed in relating water entry suction directly to the limit of capillary rise for the sand, the concept of water entry suction must play an important role in the flow field produced by fingering and also in the mechanism of finger persistence discussed later. In addition, their hypothesis regarding the wetted fraction of the lower layer was confirmed.

Diment et al. (1982) and Diment and Watson (1983, 1985) published a series of papers on stability analysis of water movement in unsaturated porous materials. In the first paper, a linear perturbation equation for vertical water movement with nonsharp fronts was derived from principles used in hydrodynamic stability analysis. The second paper discussed the application of this analysis to the stability of redistribution following infiltration, infiltration into a scale-heterogeneous medium, and infiltration into a fine-over-coarse stratified profile. Although instability was expected for these systems and a noticeable trend toward instability was present, the numerical results did not predict the occurrence of such a condition. They concluded that this must have been caused by the initial water content of 0.05 $cm^3\ cm^{-3}$ used in the simulation. In this they recognized the significant effect of initial water content on occurrence of fingering. In the third paper, they studied the effect of initial water content and grain size on wetting front instability. They measured water redistribution following infiltration for initially dry sands of three sizes: coarse, with sieve sizes between no. 14 and no. 48; medium, between no. 28 and no. 100; and fine, between no. 120 and no. 240. The saturated hydraulic conductivities of the three sands were approximately 16, 2.4, and 0.48 cm/min. Instability was found for the coarse sand. A marked shift toward stability was observed for the medium sand, and no instability exhibiting closely spaced fingers was found for the fine sand. They also investigated wetting front patterns during continuous infiltration into a stratified profile consisting of the fine sand over coarse sand under three initial water contents: 0, 0.01, and 0.02 $cm^3\ cm^{-3}$. The results (Fig. 5.29) clearly show an unstable pattern where the initial water content is zero. The initial water content of 0.01 $cm^3\ cm^{-3}$ shows the gradual development of an unstable front, but far less pronounced than under the initial dry condition. No unstable pattern was observed at 0.02 $cm^3\ cm^{-3}$.

Fingering has also been observed in the field. The phenomenon of an initially dry layer wetting in only a few places was found in a layered volcanic ash soil where the infiltration rate became constant after the wetting front had reached the second layer (Tabuchi et al., 1970). Measurements of the infiltration of water containing dyes as tracers, followed by careful excavation, have shown fingering to occur (Starr et al., 1978; Starr et al., 1986; Glass et al., 1988; Hendrikx et al., 1988). Van Ommen et al. (1988) observed fingers with a technique using iodide as a tracer. Partial-type flow was also found in water infiltration into a layered soil on a slope by Miyazaki (1988).

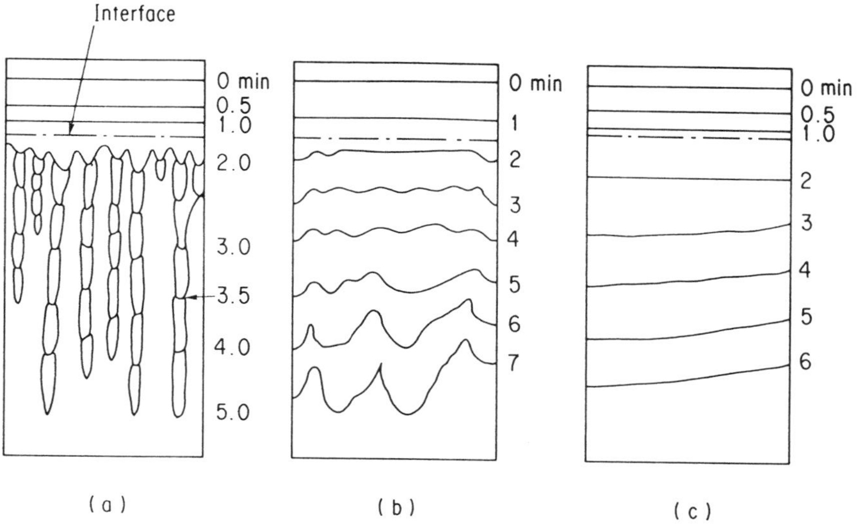

Fig. 5.29. Wetting front patterns during continuous infiltration into a stratified profile consisting of 10 cm of fine sand over 70 cm of coarse sand at different initial water contents, (a) 0 cm^3/cm^3, (b) 0.01 cm^3/cm^3, and (c) 0.02 cm^3/cm^3. (From Diment and Watson, 1985)

5.6.2 Criteria for Wetting Front Instability and Finger Width

The criterion for wetting front instability proposed by Tabuchi (1961) was that $q_2 < K_2$ where q_2 is the water entry rate and K_2 is the hydraulic conductivity during infiltration into the lower layer (Section 5.6.3). Raats (1973) derived a criterion for instability based on the simple hydraulic model used by Green and Ampt (1911). The criterion is: the wetting front is unstable whenever the pressure head at the soil surface is smaller than at the wetting front. He considered the implications of the criterion for some special cases. Assuming that the contact angle of a liquid with solid grains is generally smaller on the receding side than the advancing side, Tamai et al. (1987) derived a criterion for wetting front instability produced when the ponded liquid vanishes. The criterion is similar to that of Tabuchi (1961).

The general theory of hydrodynamic stability analysis which addresses the question of whether a given fluid flow is stable relative to imposed disturbances has also been used to analyze wetting front instability phenomena. Saffman and Taylor (1958) performed a stability analysis on a steady, constant-velocity, one-dimensional, vertical downward displacement of one fluid by another. The criterion used by Tamai et al. (1987) is easily derived by replacing one liquid with air in their result (Glass et al., 1989c).

Philip (1975a,b) derived a criterion which is more general than that of Raats, using a hydrodynamic stability analysis of a delta-function model of infiltration (Green and Ampt model). The criterion is that the water pressure gradient immediately behind the front should oppose the flow. He also developed descriptions for various specific circumstances in which the flow may become unstable: penetration to a non-wetting stratum; compression of air ahead of the front; redistribution following infiltration; and increase of conductivity with depth. Based on this theory, White et al. (1976, 1977) studied wetting front stability from changes of pressure gradient. Hillel and Baker (1988), and Baker and Hillel (1990), paying attention to a physical cause rather than a formalistic criterion for the instability, proposed the criterion $K_u(\psi_e) > Q_t$ where K_u is the hydraulic conductivity of the sublayer, ψ_e is the water entry suction of the sublayer, and Q_t is the flow rate through the upper layer.

The criteria developed from these studies use a pressure condition or a flow rate condition. Though the criteria belonging to the latter have an identical form in which $q < K_2$ where q is the flow rate through the top layer and K_2 is the hydraulic conductivity of the second layer under a given condition, it should be noted that the hydraulic conductivities used in the respective criteria are different. Tamai et al. (1987) and Glass et al. (1989c) use the saturated hydraulic conductivity, the hydraulic conductivity of the infiltration stage. Tabuchi (1961) and Hillel and Baker (1988) use the hydraulic conductivity corresponding to the water entry suction. These reflect differences in the concept of the physical mechanism of fingering.

Hydrodynamic stability analyses have also been used to predict width of fingers. Parlange and Hill (1976), assuming that the speed of the perturbed front is affected by its curvature, determined the dependence of finger size on soil properties by an immediate extension of Saffman and Taylor's (1958) stability analysis. The result indicates that the finger size should widen when the soil texture is finger and the initial moisture content is larger. Glass et al. (1989b) confirmed experimentally that the finger width closely approximated the stability analysis by Parlange and Hill (1976). Glass et al. (1989a) from classical dimensional analysis, derived expressions for finger width (diameter) and finger velocity that are functions of porous media properties and initial boundary conditions. The scaling theory of Miller and Miller (1956) was used to generalize the analysis. The results explain why fingering has been noted only in coarse sand, where the minimum finger size is most likely to be less than the size of the experimental chambers. For fine sands, the minimum finger size is probably larger than the size of most experimental chambers (on the order of 1 m or more). Later, Glass et al. (1991) proposed an analysis to investigate finger diameter in three dimensional, axisymmetric disturbances.

5.6.3 Mechanism of Finger Persistence

All the criteria for wetting front instability described in the preceding section consider only downward flow. Does fingering occur in every soil for which these criteria are satisfied, that is, is such a criterion a necessary and sufficient condition for occurrence of fingering?

Once formed, fingers are stable for long periods of steady infiltration [e.g. Tabuchi (1961); Glass and Steenhuis (1984)]. In uniform infiltration, water moves along a water pressure gradient, and stops when the pressure at all points in the soil becomes identical. On the other hand, in fingering, even with the discontinuity of water content between the inside and the outside of a finger in the horizontal direction, water does not move from the inside toward the outside. Thus, finger persistence is an interesting phenomenon. Only a few papers have attempted to clarify the mechanism of finger persistence.

Clearly, the stable persistence of a finger is assured by a stationary air-water interface at the side of the finger. Two conditions are required. The suction at any point at a same level in the finger must have a constant value, i.e. no lateral suction gradient. Glass et al. (1989c) explained, on the basis of hysteresis in the moisture characteristic relationship, why a nearly-saturated core area and a less-saturated fringe area surrounding the core in a finger coexist. However, this explanation does not refer to the second condition for finger persistence.

This second condition is that the suction at the interface must be equal to the lateral water entry suction. Three concepts may be considered related to water entry suction into a porous medium, corresponding to mutual relations between the respective flow directions and the direction of gravity. The first concept is that the upward water entry suction has a flow direction opposite to that of gravity. This may be regarded as approximately equal to the suction corresponding to the limit of capillary rise. The second is the lateral water entry suction, which is fundamentally free from gravity. The third is the downward water entry suction, which may almost correspond to the advancing capillary force. Judging from the mutual relations of the respective flow directions to the direction of gravity, it may be reasonable that inequal relationships of the upward water entry suction > the lateral water entry suction > the downward water entry suction are valid.

The lateral water entry suction must have a unique value for a porous medium with a uniform initial water condition. From these two conditions it can be concluded that in a stable, persisting finger the suction at any point except at the vicinity of the wetting front has a constant value, i.e. the lateral water entry suction. In other words, finger flow has unit gradient. Tabuchi (1961) in Section 3.5.3 conjectured that a slight increase of q after the wetting front enters the second layer (Fig. 5.23) was produced by the change of the driving force from the advancing capillary force h_{c2} to the static capillary force h_s. Since the air-water interfaces on the sides of partial flows are static, he considered the suction at the sides to be equal to the static capillary rise force. This idea is regarded to be reasonable though the lateral water entry suction may not correspond to the static capillary rise force.

Baker and Hillel (1989) measured suction at the interlayer plane during infiltration into air-dry layered soils as a function of time. The results (Fig. 5.30) show that in each case where fingering does not occur the suction at the interlayer plane decreases with time after the breakthrough. However, the suction increases with time in each case where fingers occur. This increase of suction seems to confirm the conjecture by Tabuchi.

What about the conditions where fingering did not occur? White et al. (1976) measured stability of infiltration flows into coarse or fine sands, where the flows were perturbed by suddenly changing the pressure gradient behind the wetting front from negative to a positive value. Fingering occurred in the coarse sand, but not in the fine sand. The moisture characteristic curve (Fig. 5.31) for the coarse sand suggests clearly that the wetting front would be sharp, and the moisture profile would differ little from a step function. The delta-function model of Philip (1975) can be adopted. Judging from the moisture characteristic, the fine sand does not satisfy the conditions of the model.

As described in Section 5.6.1, the experiments by Diment and Watson (1985) showed quite clearly that even low initial water contents had a strongly inhibiting effect on the development of instability patterns. They said that the presence of even very thin water films considerably inhibits the incipient development of the fingers.

Based on the experimental results described, we will attempt a brief discussion of the mechanism of finger persistence. Suppose water infiltration into a fine-over-coarse layered sand system satisfying criteria for wetting front instability, and consider the process in which a stable finger is formed after the wetting front reaches the interface between the layers. Water begins to enter into the sublayer and form a finger when the suction at the interface reduces to that corresponding to the advancing capillary force (advancing capillary suction). The suction in the finger at that time is approximately equal to the advancing capillary suction.

If the lateral water entry suction would be less than the advancing capillary suction, the finger will become stable. However, the results by Tabuchi (1961) and Baker and Hillel (1989) suggest that the advancing capillary suction is less than the water entry suction. In this case, lateral water movement occurs. After the suction inside the finger reaches the lateral water entry suction, the finger will persist in stable form. Then, the advancing capillary suction may increase toward the lateral water entry suction. The mechanism which brings about the increase in suction has not yet been made clear. However, if the suction does not increase, lateral water movement continues for a long period, and the finger must finally disappear.

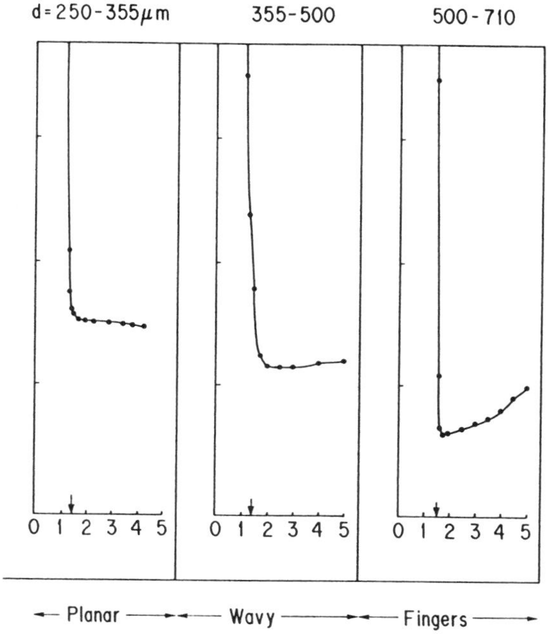

Fig. 5.30. Suction as a function of time, measured at the interlayer plane during infiltration into air-dry layered soils: 45 to 106 μm top layer over sublayer of different particle diameter, d (μm). Data points are direct instrument readings; solid lines are from chart record of continuous scans; dashed lines are interpolations between scans. Arrows indicate breakthrough. (From Baker and Hillel, 1990)

It should be noted that this process can occur only under a condition where the suction can be transmitted without changing the water content in the finger. That is, the suction at the air-water interface on the finger side must influence the advancing capillary suction, or the suction at the wetting front must govern the suction in the part behind the wetting front. This condition is guaranteed only when the finger is nearly saturated. If the wetting front is not sharp, and differs considerably from a step-function, the suction at the wetting front cannot affect suction behind the wetting front, and the influence of the suction at the wetting front is realized only through changes of water content in that part. Consequently, since the suction in the main part behind the wetting front is much lower than at the wetting front, large lateral water movement occurs, and the finger finally disappears.

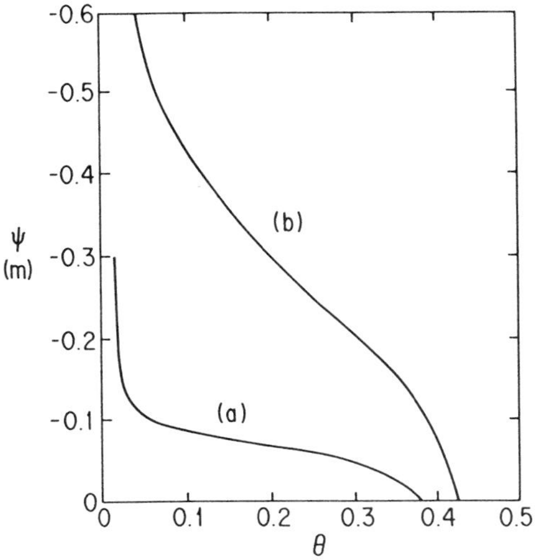

Fig. 5.31. Moisture characteristics (wetting curves) for (a) coarse sand, (b) fine sand. ψ is moisture potential, θ is volumetric moisture content. (From White et al., 1976)

The following description by Raats (1973) suggests such a mechanism. In fine-textured soil with a wide distribution of pore sizes, the water content changes gradually with the pressure head so that the wetting front is not sharp and the water content behind the wetting front is not restricted to a unique value of θ_s. As a result, infiltration of nonponding rainfall may actually be stable in such soil. Also, small perturbations of a wetting front will to some extent be damped by lateral movement due to a gradient of the pressure head.

The mechanism of the effect of initial water content on fingering has also not been clarified. The change of contact angle of water on soil particles with an increase of water content might play an important role in the mechanism through changing the advancing capillary suction and the lateral water entry suction. In addition, since fingering is brought about by local irregularities, analyses from a microscopic viewpoint may be required in addition to those from macroscopic points of view. Further studies are necessary.

REFERENCES

Aberiyanov, S. F., Relation between hydraulic conductivity and air content of soil, *Rep. Acad. Nauk. SSSR* 2:141-144 (1949). (Russian)

Akiba, M., Capillary syphoning and capillary water movement induced by ground water movement, *J. Agric. Eng. Soc. Jpn.* 10:29-39 (1938). (Japanese)

Baker, R. S. and D. Hillel, Laboratory tests of a theory of fingering during infiltration into layered soil, *Soil Sci. Soc. Am. J.* 54:20-30 (1990).

Baver, L. D., Soil permeability in relation to non-capillary porosity, *Soil Sci. Soc. Am. Proc.* 3:52-56 (1938).

Baver, L. D., W. H. Gardner, and W. R. Gardner. *Soil Physics*, Wiley, New York (1972).

Bikerman, J. J., *Surface Chemistry*, Academic Press, New York, p. 349 (1958).

Bodman, G. B., and E. A. Coleman, Moisture and energy conditions during downward entry of water into soils, *Soil Sci. Soc. Am. Proc.* 8:116-122 (1944).

Budagovskii, A. I., Infiltration into the soil, *Izv. Acad. Nauk. SSSR* 136 pp. (1955). (Russian)

Campbell, G. S., *Soil Physics with Basic*, Elsevier, Amsterdam (1985).

Childs, E.C., *The Physical Basis of Soil Water Phenomena*, Wiley, New York (1969).

Childs, E. C., and N. Collis-George, The permeability of porous materials, *Proc. R. Soc. London* 201A:392-405 (1950).

Collins, R. E., *Flow of Fluids through Porous Materials*, Reinhold, New York (1961).

Coleman, E. A., and G. B. Bodman, Moisture and energy conditions during downward entry of water into moist and layered soils, *Soil Sci. Soc. Am. Proc.* 9:3-11 (1945).

Dallavalle, J. M., *Micromeritics*, Pitman, New York (1943).

Darcy, H., *Les Fontaines Publique de la Ville de Dijon*, Victor Dalmont, Paris (1856).

Dean, P., A new Monte Carlo method for percolation problems on a lattice, *Proc. Cambridge Phil. Soc.* 59:397-410 (1963).

De Wiest, J. M. ed., *Flow through Porous Media*, Academic Press, New York (1969).

Diment, G. A., and K. K. Watson, Stability analysis of water movement in unsaturated porous materials, 2. Numerical studies, *Water Resour. Res.* 19:1002-1010 (1983).

Diment, G. A., and K. K. Watson, Stability analysis of water movement in unsaturated porous materials, 3. Experimental studies, *Water Resour. Res.* 21:979-984 (1985).

Diment, G. A., K. K. Watson, and P. J. Blennerhassett, Stability analysis of water movement in unsaturated porous materials, 1. Theoretical considerations, *Water Resour. Res.* 18:1248-1254 (1982).

Forchheimer, P., *Hydraulik*, B. G. Teubner, Leipzig (1930).

Frisch, H. L., J. M. Hammersley, and D. J. A. Welsh, Monte Carlo estimates of percolation probabilities for various lattices, *Phys. Rev.* 126:949-951 (1962).

Fujikawa, M., On the mixing for seepage control, *Rec. Land Reclam. Res.* 7:19-28 (1968). (Japanese)

Glass, R. J., and T. S. Steenhuis, Factors influencing infiltration flow instability and movement of toxics in layered sandy soils, *Tech. Pap. 84-2508*, Am. Soc. Agric. Eng., St. Joseph, MI (1984).

Glass, R. J., T. S. Steenhuis, and J.-Y. Parlange, Wetting front instability as a rapid and far-reaching hydrologic process in the vadose zone, *J. Contam. Hydrol.* 3:207-226 (1988).

Glass, R. J., J.-Y. Parlange, and T. S. Steenhuis, Wetting front instability, 1. Theoretical discussion and dimensional analysis, *Water Resour. Res.* 25:1187-1194 (1989a).

Glass, R. J., T. S. Steenhuis, and J.-Y. Parlange, Wetting front instability, 2. Experimental determination of relationships between system parameters and two-dimensional unstable flow field behavior in initially dry porous media, *Water Resour. Res.* 25:1195-1207 (1989b).

Glass, R. J., T. S. Steenhuis, and J.-Y. Parlange, Mechanism for finger persistence in homogeneous unsaturated porous media: Theory and verification, *Soil Sci.* 148:60-70 (1989c).

Glass, R. J., J.-Y. Parlange, and T. S. Steenhuis, Immiscible displacement in porous media: Stability analysis of three-dimensional, axisymmetric disturbances with application to gravity-driven wetting front instability, *Water Resour. Res.* 27:1947-1956 (1991).

Green, W. H., and G. A. Ampt, Studies on soil physics: 1, *J. Agric. Sci.* 4:1-24 (1911).

Hansen, V. E., Infiltration and soil water movement during irrigation, *Soil Sci.* 79:93-105 (1955).

Harr, M. E., *Ground Water and Seepage*, McGraw-Hill, New York (1962).

Hazen, A., Some physical properties of sands with special reference to their use in filtration, Ann. Rpt. Mass. State Bd. of Health, p. 541 (1892).

Henrickx, J. M. H., L. W. Dekker, E. J. van Zuilen, and O. H. Boersma, Water and solute movement through a water repellent sand soil with grasscover, in *Proceedings of the International Conference and Workshop on Validation of Flow and Transport Models for the Unsaturated Zone*, Ruidoso, New Mexico, May 23-26, edited by P. J. Wierenga, pp. 131-146, New Mexico State Press, Las Cruces (1988).

Hill, D. E., and J. Y. Parlange, Wetting front instability in layered soils, *Soil Sci. Soc. Am. Proc.* 36:697-702 (1972).

Hillel, D., *Fundamentals of Soil Physics*, Academic Press, New York (1980a).

Hillel, D., *Applications of Soil Physics*, Academic Press, New York (1980b).

Hillel, D., and R. S. Baker, A descriptive theory of fingering during infiltration into layered soils, *Soil Sci.* 146:51-56 (1988).

Jury, A. W., W. R. Gardner, and W. H. Gardner, *Soil Physics*, Fifth Edition, John Wiley & Sons (1991).

Jury, A. W., and K. Roth, *Transfer Functions and Solute Movement through Soil*, Birkhaüser (1990).

Kachi, K., and H. Suezawa, On the control of percolation of sandy soil, *J. Agric. Eng. Soc. Jpn.* 9:381-393 (1937). (Japanese)

Kachinskii, N. A., *Soil Physics*, Vol. 2, Izd. Vuisshaya Shukoura, Moskva (1970). (Russian)

Kimura, S., On the soil physical approach to ground water movement, *Soil Phys. Cond. Plant Growth Jpn.* 27:15-19 (1972).

Kirkham, D., and W. L. Powers, *Advanced Soil Physics*, Wiley, New York (1972).

Klute, A., A numerical method for solving the flow equation for water in unsaturated materials, *Soil Sci.* 73:105-116 (1952).

Kozeny, J., Die Durchlässigkeit des Bodens, *Kulturtechniker* 35:301-307 (1932).

Krisher, O., Grundgesetze der Feuchtigkeitsbewegung in Trockenguttern, *VDI Z.* 82:373-378 (1938).

Kuroda, M., Unsaturated permeation due to dissolved gas and permeability, *Trans. Agric. Eng. Soc. Jpn.* 13:1-6 (1965). (Japanese with English summary)

Leibenzon, L. S., *Motion of Natural Liquids and Gases through Porous Media*, Gostekhizdat, Moscow, 244 pp. (1947).

Luikov, A. V., *Heat and Mass Transfer in Capillary-Porous Bodies* (translated from Russian, 1961), Pergamon Press, Oxford (1966).

Marshall, T. J., *Relations between Water and Soil*, Commonwealth Bureau of Soils, Harpenden, U.K. (1959).

Miller, D. E., and W. H. Gardner, Water infiltration into stratified soil, *Soil Sci. Soc. Am. Proc.* 26:115-119 (1962).

Miller, E. E., and R. D. Miller, Physical theory for capillary flow phenomena, *J. Appl. Phys.* 4:324-332 (1956).

Miyazaki, T., Water infiltration into layered soil slopes, *Trans. Jpn. Soc. Irrig. Drain. Reclam. Eng.* 133:1-9 (1988).

Miyazaki, T., *Water Flow in Soils*, Marcel Dekker, New York (1993).

Muskat, M., *The Flow of Homogeneous Fluids through Porous Media*, McGraw-Hill, New York (1937).

Muto, K., Personal letter (1976).

Nakamura, T., Fundamental researches on percolation of capillary soil water, *Memoirs Coll. Agric. Ehime Univ.* 14:1-106 (1969). (Japanese with English summary)

Nelson, W. L., and L. D. Baver, Movement of water through soils in relation to the nature of the pores. *Soil Sci. Soc. Am. Proc.* 5:69-76 (1940).

Nerpin, S. V., and A. F. Chudnovskii, *Physics of the Soil* (translated from Russian, 1967), Israel Program for Scientific Translations, Jerusalem (1970).

Niimi, T., and T. Arimizu, *Soil Purification Methods of Waste Water*, Capillary Siphoning Research Society, Tokyo (1977). (Japanese)

Onizuka, K., Computer experiment on a 3D site percolation model of porous materials, its connectivity and conductivity, *J. Phys. Soc. Jpn.* 39:527-535 (1975).

Onstad, C. A., T. C. Olson, and L. R. Stone, An infiltration model tested with monolith moisture measurements, *Soil Sci.* 116:13-17 (1973).

Orlob, G. T., and G. N. Radhakrishna, The effects of entrapped gases on the hydraulic characteristics of porous media, *Trans. Am. Geophys. Union* 39:648-659 (1956).

Parlange, J.-Y., and D. E. Hill, Theoretical analysis of wetting front instability in soils, *Soil Sci.* 122:236-239 (1976).

Pavlovskii, N. N., Reports II, *Izv. Acad. Nauk SSSR* (1956). (Russian)

Philip, J. R., The theory of infiltration; 1-7, *Soil Sci.* vols:83-85 (1956-1958).

Philip, J. R., Theory of infiltration, *Hydroscience* 5:215-296 (1969).

Philip, J. R., Stability analysis of infiltration, *Soil Sci. Soc. Am. Proc.* 39:1042-1049 (1975a).

Philip, J. R., The growth of disturbances in unstable infiltration flows, *Soil Sci. Soc. Am. Proc.* 39:1049-1053 (1975b).

Polubarinova-Kochina, P. Y., *Theory of Ground Water Movement* (translated from Russian, 1952), Princeton University Press, Princeton, NJ (1962).

Purcell, W. R., Capillary pressures, *Trans. AIME* 186:39-46 (1949).

Raats, P. A. C., Unstable wetting fronts in uniform and nonuniform soils, *Soil Sci. Soc. Am. Proc.* 37:681-685 (1973).

Rode, A. A., *Theory of Soil Moisture*, Vol. 1 (translated from Russian, 1965), Israel Program for Scientific Translations, Jerusalem (1969).

Saffman, P. G., and G. Taylor, The penetration of a fluid into a porous medium or Hele-Shaw cell containing a more viscous liquid, *Proc. Roy. Soc. London Ser. A*, 245:312-331 (1958).

Sasaki, C., Studies on entry pores in volcanic ash subsoil (I), *Trans. Jpn. Soc. Irrig. Drain. Reclam. Eng.* 151:65-73 (1991). (Japanese with English summary)

Sasaki, C., On the dissolved oxygen content in seepage water of open and closed system percolation in a stratified soil column, *Trans. Jpn. Soc. Irrig. Drain. Reclam. Eng.* 159:65-71 (1992). (Japanese with English summary)

Scheidegger, A. E., *The Physics of Flow through Porous Media*, University of Toronto Press, Toronto (1957).

Slichter, C. S., Theoretical investigation of the motion of ground waters, *U.S. Geol. Surv., 19th Ann. Rep.* (1898).

Soma, K., T. Maeda, and K. Yamada, The relationship between the effect of soil compaction and hydraulic conductivity for Kuroboku soil, *Trans. Jpn. Soc. Irrig. Drain. Reclam. Eng.* 103:62-67 (1983). (Japanese with English summary)

Starr, J. L., H. C. DeRoo, C. R. Frink, and J.-Y. Parlange, Leaching characteristics of a layered field soil, *Soil Sci. Soc. Am. J.* 42:376-391 (1978).

Starr, J. L., J.-Y. Parlange, and C. R. Frink, Water and chloride movement through a layered field soil, *Soil Sci. Soc. Am. J.* 50:1384-1390 (1986).

Swartzendruber, D., Modification of Darcy's law for the flow of water in soils, *Soil Sci.* 93:22-29 (1962).

Swartzendruber, D., Infiltration of constant-flux rainfall into soil as analyzed by the approach of Green and Ampt, *Soil Sci.* 117:272-281 (1974).

Tabuchi, K., Studies on soil pores through microscopical observation of thin sections, *Trans. Agric. Eng. Soc. Jpn.* 7:21-31 (1963). (Japanese with English summary)

Tabuchi, T., Infiltration and ensuing percolation in columns of layered glass particles packed in laboratory, *Trans. Agric. Eng. Soc. Jpn.* 2:27-35 (1961). (Japanese with English summary)

Tabuchi, T., Infiltration and capillarity in the particle packing, *Rec. Land Reclam. Res.* 19:1-121 (1971). (Japanese)

Tabuchi, T., K. Tabuchi, and N. Nagata, On the relation between the pore characteristics and permeability in the Kanto-Loam volcanic ash soil, *Trans. Agric. Eng. Soc. Jpn.* 7:53-60 (1963). (Japanese with English summary)

Tabuchi, T., M. Nakano, T. Yawata, S. Sasaki, T. Maeda, S. Yazawa, and N. Marutani, Studies on the soil moisture movement in layered volcanic ash soil: 1, *Trans. Jpn. Soc. Irrig. Drain. Reclam. Eng.* 31:1-9 (1970). (Japanese with English summary)

Tada, A., On the compaction curve of the Kanto-Loam and its permeability: 1, *Trans. Agric. Eng. Soc. Jpn.* 14:36-40 (1965). (Japanese with English summary)

Takagi, S., Analysis of the vertical downward flow of water through a two-layered soil, *Soil Sci.* 90:98-103 (1960).

Takeuchi, H., Simulation of flow through porous media by means of stochastic model, *Proc. Jpn. Soc. Civ. Eng.* 187:79-93 (1971). (Japanese)

Tamai, N., T. Asaeda, and C. G. Jeevaraj, Fingering in two-dimensional, homogeneous, unsaturated porous media, *Soil Sci.* 144:107-112 (1987).

Tokunaga, K., H. Naruoka, and T. Fukaya, Study on the soil and its void by X-ray radiograph (I), *Trans. Jpn. Soc. Irrig. Drain. Reclam. Eng.* 114:61-68 (1984). (Japanese with English summary)

Tokunaga, K., T. Sato, H. Kikuchi, and K. Kon, On the real state of coarse pores in clayey paddy field's soil and their permeability, *Soil Phys. Cond. and Plant Growth* 51:49-62 (1985). (Japanese with English summary)

Tokunaga, K., X-ray stereoradiographs using new contrast media on soil macropores, *Soil Sci.* 146:199-207 (1988).

Tokunaga, K., and C. Sasaki, Downward flow in open system in the paddy field on volcanic ash hill, *Jpn. Soc. Irrig. Drain. Reclam. Eng.* 58:29-34 (1990). (Japanese)

van Omen, H. C., L. W. Dekker, R. Dejksam, J. Hulshof, and W. H. van der Molen, A new technique for evaluating the presence of preferential flow paths in non-structured soils, *Soil Sci. Soc. Am. J.* 40:824-829 (1988).

White, I., P. M. Colombera, and J. R. Philip, Experimental study of wetting front instability induced by sudden change of pressure gradient, *Soil Sci. Soc. Am. J.* 40:824-829 (1976).

White, I., P. M. Colombera, and J. R. Philip, Experimental studies of wetting front instability induced by gradual change of pressure gradient and by heterogeneous porous media, *Soil Sci. Soc. Am. J.* 41:483-489 (1977).

Wyckoff, R. D., and H. G. Botset, The flow of gas liquid mixtures through unconsolidated sands, *Physics* 7:325-345 (1936).

Yamazaki, F., On the negative pressure of capillary downward flow, *J. Agric. Eng. Soc. Jpn.* 15:26-40 (1943). (Japanese)

Yamazaki, F., Researches on the vertically downward flow of water through layered soil, *Bull. Tokyo Coll. Agric. For.* 1:1-19 (1948) (Japanese) [2nd ed., *Data Rec. Invest.* 6:1-30 (1958)]. (Japanese)

Yamazaki, F. ed., *Soil Physics*, Yokendo, Tokyo (1969). (Japanese)

Yawata, T., Decrease of hydraulic conductivity by air binding, *Rec. Land Reclam. Res.* 10:1-32 (1960). (Japanese)

Yawata, T., *Physics of Soil*, Tokyo University Press, Tokyo (1975). (Japanese)

Yong, R. N., and B. P. Warkentin, *Soil Properties and Behaviour*, Elsevier, Amsterdam (1975).

Yoshida, S., Fundamental equations of the flow of homogeneous fluids through porous media, *Trans. Agric. Eng. Soc. Jpn.* 1:19-26 (1960). (Japanese with English summary)

Yoshida, S., The law of similarity of the flow through porous media with a free surface, *Trans. Agric. Eng. Soc. Jpn.* 15:12-15 (1966). (Japanese with English summary)

Yoshida, S., The fundamental studies of the flow through porous media, *Bull. Yamagata Univ. (Agric. Sci.)* 5:257-329 (1968).

Zaslavsky, D., Theory of unsaturated flow into a nonuniform soil profile, *Soil Sci.*, 97:400-411 (1964).

6

UNSATURATED WATER MOVEMENT

6.1 LIMITATIONS ON THE VALIDITY OF DARCY'S LAW

Many soil physicists and hydrologists have assumed that Darcy's law and the diffusion equation are applicable to unsaturated flow at any water content. The validity of Darcy's law at low water contents can be questioned because capillary models cannot be reasonably assumed and the water is under adsorptive force fields. The diffusion equation for water flow in an unsaturated soil is valid only when the unsaturated flow obeys Darcy's law. Therefore, validity of the diffusion equation should be confirmed by investigating the validity of Darcy's law for unsaturated flow.

The application of Darcy's law to water flow in unsaturated soil is based on an analogy with Fourier's law and Ohm's law. Analogy is an important means of developing science; it is especially effective at the first stage of study of a subject. However, to advance the study, the validity of the analogy has to be tested experimentally.

6.1.1 Darcy's Law and the Capillary Model

The application of Darcy's law to unsaturated flow by Richards (1931) was based on a capillary model and Hagen-Poiseuille flow in capillary tubes. Krischer (1938) proposed a capillary model of a porous medium (Fig. 6.1) that is different from the model proposed by Childs and Collis-George (1950); the difference exists whether or not capillaries are interconnected. When water is added to a tube, it can move from one tube to another in order to attain equilibrium. Using this model and applying the Hagen-Poiseuille law to water flow in each tube, Krischer (1938) and Otani and Maeda (1964) introduced an equation for water flow from groundwater to the soil surface where water is evaporated at a constant rate. The velocity of water flow, G, at distance x from the groundwater surface is given by

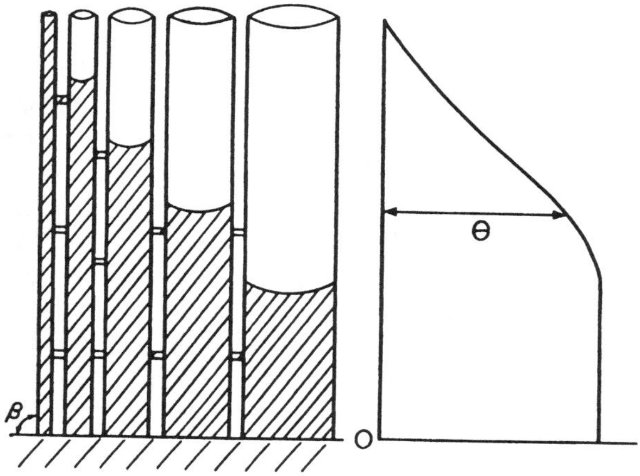

Fig. 6.1. Schematic representation of a capillary model with interconnected capillary tubes.

$$G = \frac{\rho \pi g_e}{8\eta}\left(-\frac{\rho g \sin\beta}{g_e} + \frac{dP_x}{dx}\right)\int_{r_{\min}}^{r(x)} r^4 \frac{dn}{dr}\,dr \tag{6.1}$$

where ρ and η are the density and viscosity of water, respectively; P_x the pressure at x; r_x the radius of the tube corresponding to P_x; $r_{\min}$ represents the minimum value among the radii of the tubes; and dn is the number of tubes between r and r + dr. Using $r = 2\sigma/P$ and $d\theta = \pi r^2 dn$ and assuming that $\beta = \pi/2$, Eq. (6.1) becomes

$$G = \frac{\rho \pi g_e}{8\eta}\int_{P_{\max}}^{P_x}\left(\frac{2\sigma}{P}\right)^2 \frac{d\theta}{dP}\,dP\,\frac{dP_x}{d\theta_x}\,\frac{d\theta_x}{dx} \tag{6.2}$$

where θ is the volumetric water content and θ_x indicates the water content at x.

However, it is questionable whether it is correct to describe water flow at all water contents using capillary models. Two types of water flow may be dominant at low water contents: water flow in film water, and water flow through filled pores under force fields. These water flows do not obey Darcy's law, as will be described later. This leads us to expect that the capillary models may approximate water flow under high water contents, and Darcy's law and the diffusion equation are valid for unsaturated flow under that condition. When water content is low, Darcy's law may not be applicable to unsaturated flow.

6.1.2 Experimental Results

Many researchers have used the diffusion equation to predict change of water content with time in phenomena such as infiltration, drainage, and evaporation. Some of the results were successfully predicted, but there were also predicted values that did not agree well with experimental results. Of the many measurements that have been made, we will consider only experiments in which the relation between hydraulic gradient or moisture gradient and amount of flow was measured directly. If this relation is a straight line through the origin, water flow obeys Darcy's law. Also, the relationship between water content and the diffusivities obtained under different hydraulic gradients or moisture gradients must be measured directly. Where diffusivities are a unique function of water content, Darcy's law will be considered valid for that unsaturated flow.

Table 6.1 shows results from studies that meet these requirements. Figures 6.2, 6.3, and 6.4 represent results by Hasegawa and Maeda (1977), Swartzendruber (1963), and Rawlins and Gardner (1963). Rawlins and Gardner pointed to swelling as a possible cause for the nonuniqueness of D(θ), but the changes in hydraulic gradients with time could result from other causes. Nimmo

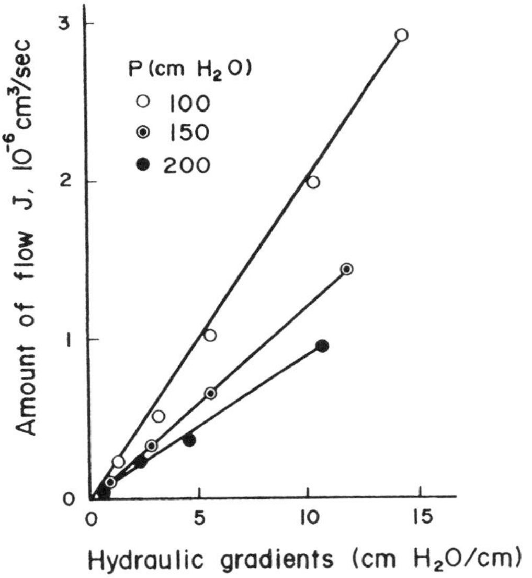

Fig. 6.2. Flow versus hydraulic gradient at constant pressure potential for a clay loam. (From Hasegawa and Maeda, 1977)

Table 6.1. Tests of the applicability of Darcy's law to water movement in soil.

State of water flow	Range of pressure, P (H_2O, cm)	Range of percent saturation, Sr, or volumetric water content, θ	Range of hydraulic gradient, J, or moisture gradient, W	Sample	Result	Reference
Percolation	36.4 < P	9.6 < Sr < 100	0.5 < J < 1	Sand (1-0.5 mm)	Darcy behavior	Childs and Collis-George (1950)
		0.30 < θ < 0.45	0 < W < 0.5	Silty clay loam	Darcy behavior	
Horizontal infiltration		0.15 < θ < 0.30	0 < W < 1.5	Silty clay loam	Non-Darcy behavior	Swartzendruber (1963)
Percolation	10 < P < 70		0.5 < J < 14	Sand (0.21-0.24 mm)	Non-Darcy behavior	Hadas (1964)
Drainage	40 < P < 50		J < 0.6	Sand (0.15-0.3 mm)	Darcy behavior	Watson (1966)
Percolation		10 < Sr < 88	1 < W < 28	Sandy loam	Non-Darcy behavior	Thames and Evans (1968)
Percolation	P < 200		0 < J < 20	Sand, volcanic ash soil, clay loam	Darcy behavior	Hasegawa and Maeda (1977)

Table 6.1. *Continued*

State of water flow	Range of pressure, P (H_2O, cm)	Range of percent saturation, Sr, or volumetric water content, θ	Range of hydraulic gradient, J, or moisture gradient, W	Sample	Result	Reference
Horizontal infiltration at constant negative head (5-100 cm)		$0.18 < \theta < 0.45$ $0.13 < \theta < 0.35$		Silt loam Sandy loam	Non-Darcy behavior (diffusivity is not a unique function of water content)	Nielsen et al. (1962)
Horizontal infiltration at constant negative head (53 cm)		$0.35 < \theta < 0.55$		Silty clay loam	Non-Darcy behavior (diffusivity is not a unique function of water content)	Rawlins and Gardner (1963)
Horizontal infiltration at constant negative head (3 cm)		$0.3 < \theta < 0.5$		Silty clay loam	Non-Darcy behavior (existence of a minimum gradient necessary to produce flow)	Ferguson and Gardner (1963)
Drying at constant negative head (30 cm)	$52 < P < 536$			Loam, sandy loam	Darcy behavior (diffusivity is a unique function of water content)	Weeks and Richards (1967)
Percolation under centrifugal field	around 200	$\theta = 0.0878$	$216 < J < 1650$	Sand	Darcy behavior	Nimmo et al. (1987)

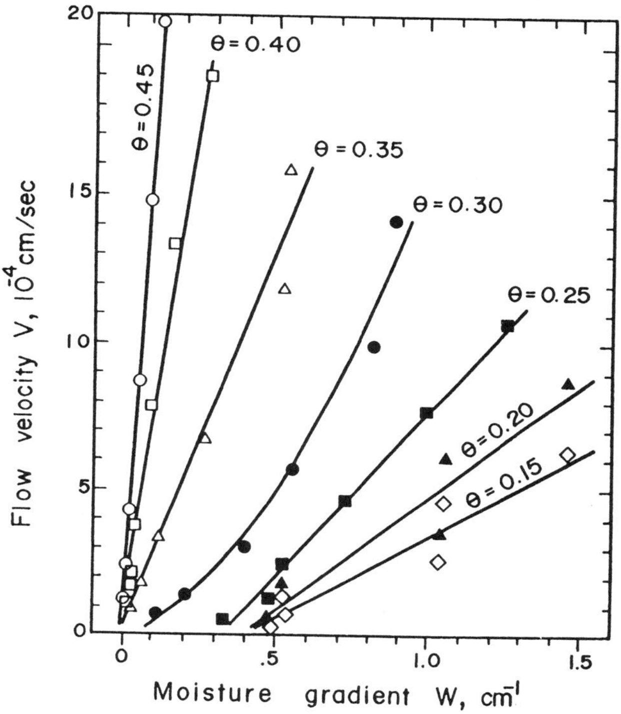

Fig. 6.3. Flow at different water contents. (From Swartzendruber, 1963)

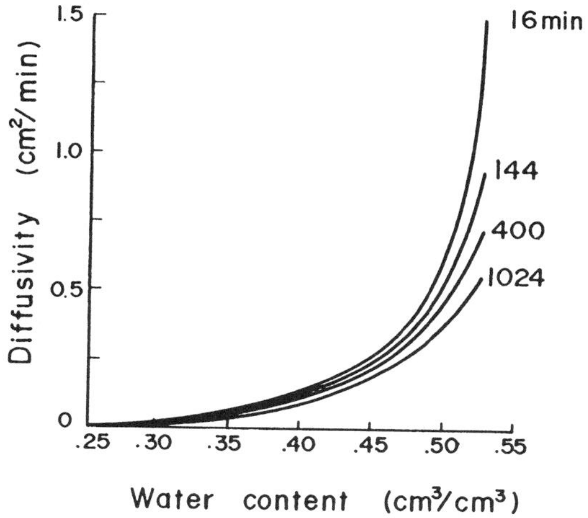

Fig. 6.4. The diffusivity relationship. (From Rawlins and Gardner, 1963)

et al. (1987) tested Darcy's law for unsaturated flow by measuring hydraulic conductivities for different centrifugal driving forces at constant water content. For the sand at a bulk density of 1.82 Mg/m^3 and 27% saturation, results were consistent with Darcy's law for hydraulic conductivity of 5.22 x 10^{-10} m/s and forces ranging from 216 to 1650 times gravity.

From the results in Table 6.1 it is difficult to draw an unqualified conclusion. However, there is a tendency for Darcy's law, or the diffusion equation, to be invalid when the water content becomes low or when the moisture gradient increases beyond certain limits.

6.1.3 Other Theories of Water Flow

Unsaturated Flow Theory Including Interfacial Phenomena

Gray and Hassanizadeh (1991a,b) introduced an unsaturated flow theory derived from basic principles of continuum mechanics with a thermodynamic analysis. They took explicit account of the interfaces. From a practical point of view the interface between air and water is an important attribute of unsaturated flow. This boundary does not exist in single phase flow. The theory adds new terms in the Darcy equation which must be properly accounted for in modeling unsaturated flow. In addition, the theory eliminates the unrealistic concept of negative absolute pressure of the water phase.

Non-Newtonian and Bingham Flow

Figure 6.5 shows the relationships between flow velocity and hydraulic gradient for three materials: a Newtonian liquid, a non-Newtonian liquid, and a Bingham material (Swartzendruber, 1962). Darcy flow is equivalent to the Newtonian flow. King (1898), Von Englehardt and Tunn (1955), Lutz and Kempter (1959), and Miller and Low (1963) studied the relation between hydraulic gradient and flow velocity for saturated clays and sandstones, and recognized that Darcy's law is not valid for saturated flows in their samples. They observed non-Newtonian or Bingham flow of water.

Swartzendruber (1962) was able to describe these results with a modified flow equation,

$$v = M\left[i - I\left(1 - e^{-i/I}\right)\right] \tag{6.3}$$

where M and I are constants and i is the hydraulic gradient. Swartzendruber (1962) and Miller and Low (1963) have explained the departure from Darcy's

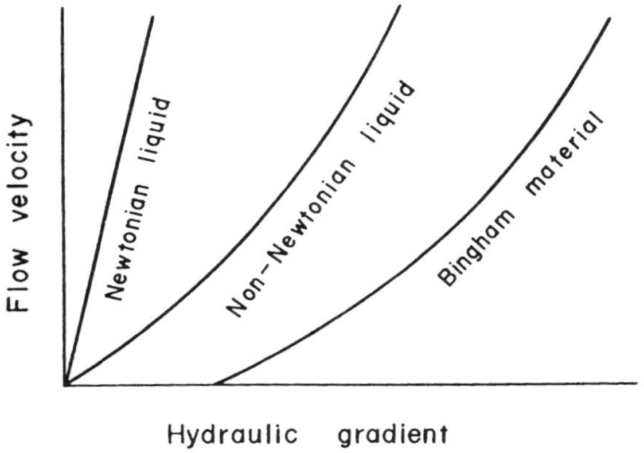

Fig. 6.5. Flow velocity versus gradient for materials flowing in a capillary tube. (From Swartzendruber, 1962)

law with the concept of quasi-crystalline water structure as a result of water-clay surface interaction (Low, 1961).

Tada (1965) found non-Newtonian flow in compacted samples of fresh Kantō loam, a volcanic ash soil (Fig. 6.6). He concluded that the deviation from Darcy's law could be due to interaction between particles and water. He calculated the dependence of hydraulic conductivity on hydraulic gradient in region (ii) in Fig. 6.6 by combining the Hagen-Poiseuille law and the Buckingham-Reiner equation. The values obtained agreed reasonably well with the experimental results. The Buckingham-Reiner equation, which expresses Bingham flow in a capillary tube, is given by

$$v = \frac{R^2 \rho g\, i}{8\eta}\left[1 - \frac{4}{3}\frac{i_0}{i} + \frac{1}{3}\left(\frac{i_0}{i}\right)^4\right] \tag{6.4}$$

where v is the velocity of fluid in the tube, R the radius of the tube, η the viscosity of the fluid, i the hydraulic gradient, and i_0 a threshold gradient. Nerpin and Kotov (1960) have also applied this equation to water flow in porous media.

Film Water Flow

Nerpin and Chudnovskii (1967) introduced basic equations for film water flow using the equation of continuity, the Navier-Stokes equations for the three

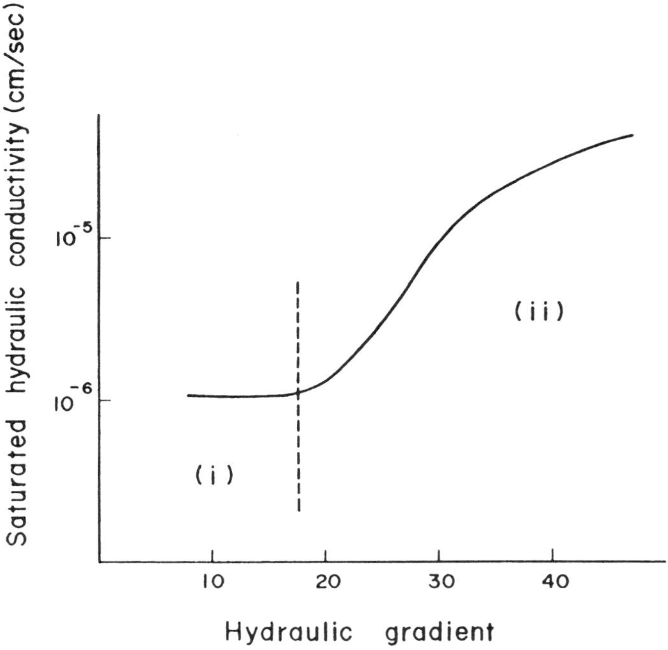

Fig. 6.6. Relation between conductivity and hydraulic gradient for a volcanic ash soil. (From Tada, 1965)

velocity components, and the heat transfer equation. They integrated the basic equations under given boundary conditions, and obtained equations representing the average velocities in the z and y directions. The equations show that Darcy's law is not valid for film water flow under usual conditions.

The ratio of the amount of water under the influence of force fields to the total amount of water in a soil increases as water content becomes low. Consequently, the probability of deviation from Darcy's law increases as water content decreases, and the diffusion equation would not be valid for unsaturated flow at low water contents. However, we do not have an equation of motion for unsaturated flow whose validity has been confirmed by a range of experimental results. Therefore, at the present stage, we are compelled to use Darcy's law and the diffusion equation as a first approximation at any water content. We must leave this for future study.

6.2 HETEROGENEITY OF HYDRAULIC CONDUCTIVITY IN SOIL

6.2.1 Introduction

Researchers who choose the natural world as the subject of study, soil scientists, hydrologists, geologists, and meteorologists, have wrestled with the spatial heterogeneities of the physical and chemical quantities they measure. Because hydraulic conductivity in a field soil varies greatly with position in the field, analyses of water and solute movement become difficult. Such difficulties cannot be resolved without considering the statistical distribution of measured physical, chemical, and biological properties.

It may be reasonable to assume that a physical or chemical characteristic such as bulk density, hydraulic conductivity, cation exchange capacity or pH in a natural soil under identical soil formation processes belongs to the same parent population, that is, the soil is statistically homogeneous relating to that characteristic. If this were not true, an astronomical number of measurements would be required to analyze a phenomenon related to the characteristic, and prediction of the phenomenon would be impossible. If the supposition is valid, the statistical distribution of that characteristic in the soil can be estimated from measurements on samples which are independent of each other, and prediction becomes possible. Two questions need to be answered: how many samples are required, and what is the relationship between adjacent samples.

What sample size is required to obtain the average of the physical quantity in the field to the desired accuracy? Small and large circles in Fig. 6.7 are distributed regularly over a wide area, but the diameters of the circles deviate from the respective averages, which are unknown. We will assume that there exists a physical quantity proportional to the fourth power of the diameter of each circle. Then, the physical quantity, K, corresponding to the total area is given by

$$K = B\left[\sum_{i=1}^{m} r_i^4 + \sum_{\ell=1}^{t} R_\ell^4\right] \tag{6.5}$$

where B is the proportionality constant, r_i and R_ℓ indicate the diameters of the ith small circle and the ℓth large circle, respectively, and m is the total number of small circles and t that of large circles in the total area S. Consequently, the average value of K per unit area is

$$\bar{K} = \frac{B\left[\sum_{i=1}^{m} r_i^4 + \sum_{\ell=1}^{t} R_\ell^4\right]}{S} \tag{6.6}$$

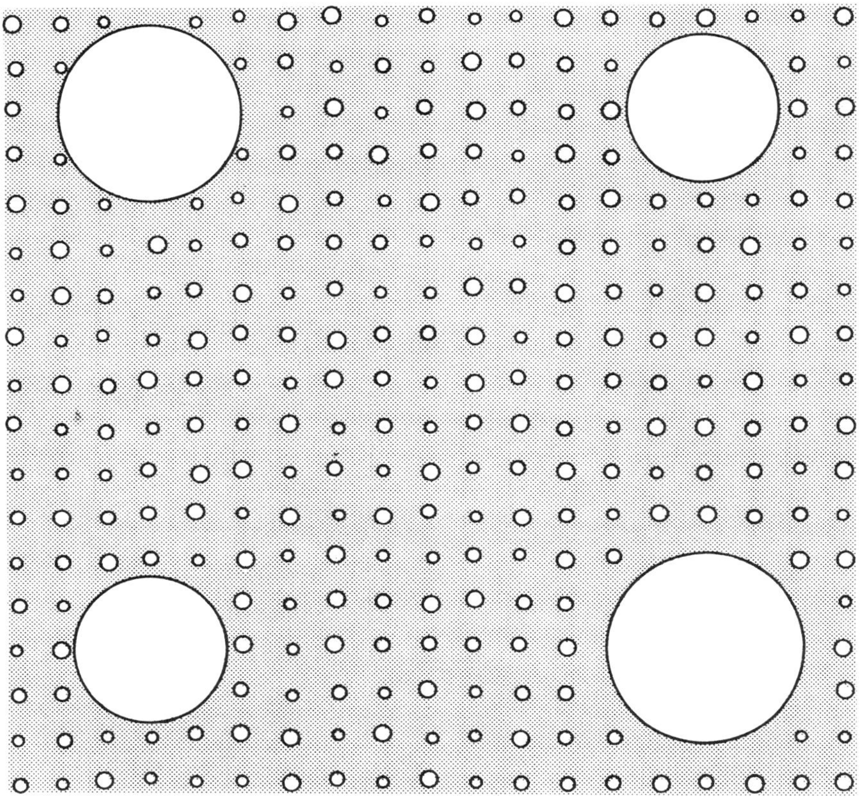

Fig. 6.7. Graphical representation to explain scale dependence of a physical property.

Let us try to estimate $\bar{K}$ by choosing n sections with an area s, and calculating the average value of the physical quantity per unit area in the respective sections, $\bar{k}_i(s)(1 \leq i \leq n)$, following Eqs. (6.5) and (6.6). Can s have any value, in other words, can $\bar{k}_i(s)$ calculated from any magnitude of s all belong to the same parent population?

We will explain the statistical meaning of these questions using Figure 6.7. Suppose we choose random square sections of number n with sides c. When c $= c_1$ and $\bar{r} << c_1 < \bar{R}$ where $\bar{r}$ and $\bar{R}$ are the average values of diameters of small and large circles, the contribution of small circles to the physical quantity is almost perfectly evaluated in the calculated $\bar{k}(c_1)$. However, if the physical quantity relating to a section is assumed to be proportional to the second power of the area of a large circle cut off by the section, the contribution of the large circle to the quantity cannot be evaluated closely. Therefore, this section size is unsuitable to estimate $\bar{K}$.

Assuming that $c = c_2$ and c_2 is larger than $\bar{R}$ but considerably smaller than the distance between neighboring large circles, all the sections include many small circles while only a few sections include a large circle. Consequently, the frequency distribution histogram of $\bar{k}(c_2)$ has two peaks of very high probability. As described later in detail, Bouma et al. (1989) have recognized such a phenomenon to occur in simulation of a soil with macropores.

In a case where $c = c_3 > 4\text{-}5$ times the distance between neighboring large circles, each section includes about 20 large circles in addition to many small circles. In this case, the physical quantity calculated from each section can be supposed to belong to an identical parent population. As a result, the statistical parameters such as mean and variance can be estimated (remember that the diameters of large circles are not the same). The smallest line, area, or volume among such lines, areas, or volumes from which the statistical parameters of a physical quantity in a field soil can be predicted is called the representative elementary line (REL), area (REA), or volume (REV). According to Bear (1972) a REV is defined as the smallest volume of soil that contains a representation of microscopic variation in all the forms and proportions present in a system. This concept is explained in more detail in Section 6.3. In addition, since the REV is for a homogeneous porous medium, its magnitude is supposed to be much smaller than that for a soil in the field. The REV depends not only on the physical quantity investigated but also on the soil under consideration.

Suppose we take a sample volume n times the REV to estimate the statistical properties of a physical quantity in a field soil. If the REVs contained in the sample volume are independent of each other, according to statistics the variance becomes 1/n of that predicted from samples with the REV. As described below, since neighboring samples are generally not independent, the variance is larger than 1/n. However, the variance becomes smaller as n increases. In practice, variability among replicate samples has been found to increase as the sample volume decreases (e.g. Reeve and Kirkham, 1951; Anderson and Bouma, 1973; Nagahori and Sato, 1971; Baker, 1977; Hawley et al., 1982). Sisson and Wierenga (1981) discussed the dependance of the standard deviation on sample size in a case where infiltration rate is log-normally distributed.

The second question is the dependance recognized between samples existing a short distance from each other. A soil has formed under the influence of primary soil forming factors such as temperature and rainfall. Many secondary factors have also operated locally on the soil to produce deviations in measured values of a physical quantity. An example of a secondary factor is the trampling by animals forming a small basin. Fine particles would flow into the basin and the soil would become finer compared with the surrounding soil.

Intuition would indicate that measurements on samples close together will give values of approximately the same magnitude. Measured values of a quantity in a small area may be regarded as a secondary parent population, included in the larger parent population determined by primary soil forming factors. The statistical properties of the quantity predicted from only measured values in the small area must deviate from the true statistical properties. Only the measured values on samples representing the different secondary effects can give the true statistical properties. In other words, only samples which have been under different secondary effects are independent of each other, while soil samples under the same secondary effect are not independent.

The distance between two sampling sites at which respective samples are judged to be independent has to be determined. This distance, called the "range of influence," characterizes the spatial variability of the soil. Ranges of influence have been predicted using the geostatistics developed by Journel and Huijbregts (1978). The ranges of influence for hydraulic properties tend to decrease with depth (Russo and Bresler, 1981) which seems to confirm the validity of the above discussion about secondary factors.

Therefore, statistical properties such as mean and variance of a physical quantity of a soil in the field cannot be predicted correctly without estimating the REV and the range of influence. However, only a few papers have discussed statistical properties for a physical quantity on the basis of the procedure described above.

6.2.2 Scale Effect and Representative Elementary Volume in Soil

Scale Effect

The dominant effect of small pores ("necks") on the hydraulic conductivity of a three-dimensional porous medium having a wide range of pore sizes has been widely recognized. The probability of a small plane forming a "planar neck" in the flow system increases, and the probability of a large planar pore being continuous decreases, as sample size increases. On the basis of this idea, Anderson and Bouma (1973) investigated the effect of different soil core lengths on measured values of saturated hydraulic conductivity of a silt loam soil. For a series of 10 cores with a diameter of 7.5 cm and heights of 5, 7.5, 10, and 17 cm (Fig. 6.8) both the mean and standard deviation for each set of measurements decreased as the cores became longer. The 17-cm samples were closest to the values measured *in situ* with the double tube method, which was considered the reference method. Values were estimated from an equation based on the physical relation between width of a planar pore and its capacity to transmit water at a given gradient, and a pore interaction model. The calculated values

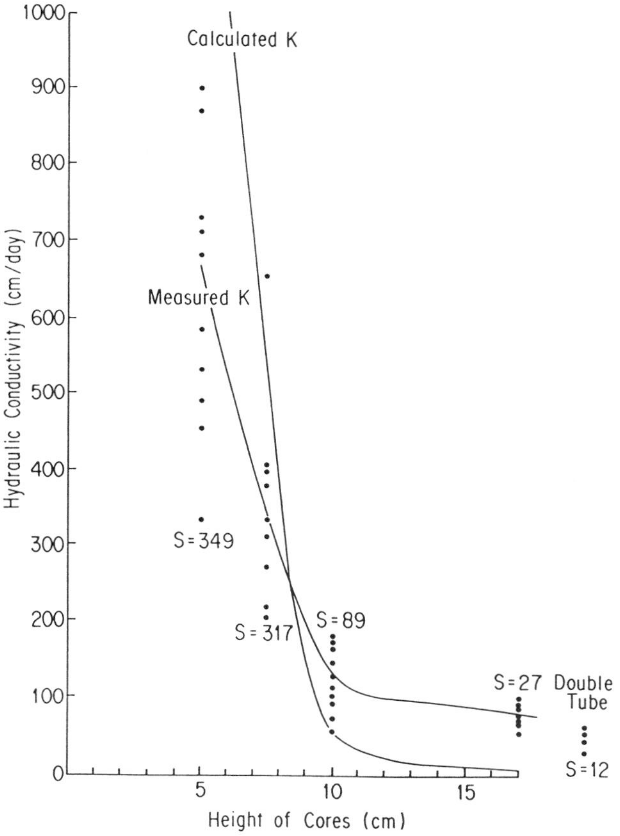

Fig. 6.8. Measured and calculated hydraulic conductivities and standard deviations (s) for cores of different height. K values measured with the double-tube method are included as a reference. (From Anderson and Bouma, 1973)

corresponding to cores with heights of 5, 7.5, 10, and 17 cm were 2,000, 500, 60, and 4 cm/day, respectively. The trend is similar to that in the measured values, although somewhat more extreme. This study clearly demonstrated the scale dependence of saturated hydraulic conductivity.

Estimation of REV

The procedure usually adopted to estimate the REV of a physical quantity is: 1) The quantity is measured for samples of different sizes. 2) Statistical properties such as mean, variance and median are calculated. 3) A smallest sample size whose means and medians show stable and approximately identical values is chosen as the REV. In practice, sample size is determined not only by the magnitude of the variance; operational convenience in measurement must be considered. An operationally cumbersome method can yield erroneous data.

Frequency distribution curves (histograms) of measured values from two different sample sizes may differ even though the means may be equal, for example, the histograms may have one peak or two. For this reason medians are used in addition to means.

Lauren et al. (1988) determined the REV for saturated hydraulic conductivity, K_s, of a clayey soil with macropores. The soil is classified in soil taxonomy (Soil Survey Staff, 1975) as a fine, illitic, mesic Glossaquic Hapludalf, and its structure is moderate to strong, subangular blocky, with medium prisms, firm, and very plastic. The prism sizes are 700 to 800 cm^3. K_s was measured at 37 equally spaced transect sampling locations using five different sizes of columns *in situ*. Each column, starting with the largest, was constructed within the previous column. Excavation of a block of soil on four sides, with the block remaining naturally attached at the bottom, defined a volume of soil to be used for K_s determination. The volumes used were 2.4×10^5 (A), 1.2×10^5 (B), 0.5×10^5 (C), 6.28×10^3 (D), and 884 cm^3 (E).

Columns B and C (Table 6.2) produced sample populations of K_s with statistically identical means and standard deviations (SD), while columns D and E had much larger, but statistically identical, means and SDs. Differences between B, C, and D are statistically significant at the 0.05 probability level. The largest column, A, produced a sample population with mean and SD intermediate between B, C, and D sizes and also different from each of the other sizes at the $\alpha = 0.05$ level. The questionable result for column A was explained as experimental error, in that this column was apparently not ponded long enough to establish true steady state flow conditions. The high values of K_s measured using small columns D and E were regarded to be greatly biased by the presence of a single crack or conducting pore, which dominated the hydraulic regime and produced extreme K_s estimates. Finally, the authors chose the C column as the practical sample size, considering labor, time, and materials in addition to the results obtained.

Once field K_s measurements had been made, columns D and E were detached and transported to the laboratory for K_s measurements. The K_s values for detached columns were at least two orders of magnitude greater than those for attached columns.

Bouma et al. (1989) studied the heterogeneity of a clay material consisting of a mixture of a dense ground mass and sandy soil material around frost fissures. They obtained the proportions of each component from drawings made of an exposed horizontal surface, 30 x 40 cm, in the 2 Btg horizon at a depth of

Table 6.2. Statistical parameters for K_s, saturated hydraulic conductivity (cm/day), measured on different sample volumes. (From Lauren et al., 1988)

Size	Mean	Mode	Median	SD	CV %	No. of samples
A	21.3	10.3	16.6	16.9	79	37
B	13.7	6.4	10.7	11.0	81	36
C	14.4	6.3	10.9	12.5	96	37
D	36.6	6.3	20.3	54.9	150	37
E	34.5	4.8	16.3	64.0	186	35

100 cm below the surface at two sites (Fig. 6.9). Using the proportions and the average measured K_s values for each component, they calculated values for K_s of 0.5 and 0.83 m/day, respectively. The values measured *in situ* for the two sites were 0.5 and 0.7 m/day, respectively, indicating satisfactory agreement between estimated and measured values.

They then evaluated effects of various sample volumes on K_s. The horizontal section shown in Fig. 6.9 with a real surface of 6,300 cm^2 was used to generate calculated values for different sample areas that represent cores of different sizes. The sample volumes were distinguished with a circular surface with diameters of 1 cm, 5 cm, 7.5 cm, 10 cm, 30 cm, and 60 cm, respectively.

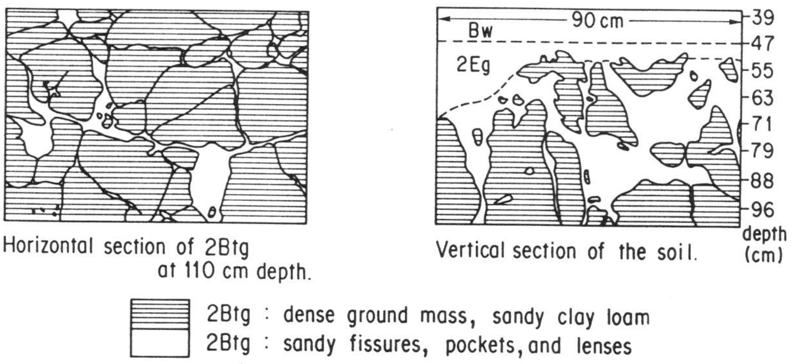

Fig. 6.9. Morphological features of horizontal and vertical sections of a Chromic Luvisol. (From Bouma et al., 1989)

Fifty random points were allocated for each sample volume in the area. Table 6.3 shows descriptive statistics for calculated K_s values. The results indicated that the variance is reduced when the sample diameter is increased. The histogram for the 1 cm core showed a bimodal distribution centered around 6.9 m/day for the sandy fissures, and 0.3 m/day for the groundmass. The distribution changed to a single peak for diameters larger than 20 cm. No significant differences were found in mean values of K_s among core sizes. The medians, however, were significantly different at the $\alpha = 0.05$ level. Data obtained for samples with a diameter of 20 cm were still highly skewed. A 30 cm diameter sample was, therefore, regarded to be the REV.

Lauren et al. (1988) concluded that a sampling REV for infiltration in the argillic horizon in a Typic Hapludalf should contain at least 20 peds or soil structure units. Espeby (1990) determined the spatial variability of K_s of 100 soil core samples taken from six soil pits in a forested glacial till soil. Samples of core length 50 or 100 mm with a diameter of 72 mm could not be distinguished statistically, and therefore the main analysis included all K_s values irrespective of core length. The frequency histogram of the data obtained by log-transforming the K_s values was close to a normal distribution, but showed a slight tendency to a bimodal distribution (Fig. 6.10). The diameter of samples may have been smaller than the REV.

The magnitude of REV depends on both the physical quantity measured and on the soil. Shigematsu and Iwata (1991) studied the effect of soil column length on unsaturated hydraulic conductivity, K_u, in a volcanic ash soil. An un-

Table 6.3. Descriptive statistics for calculated K_s values (m/day) using hypothetical samples with different diameters. (From Bouma et al., 1989)

Diameter (cm)	Mean	Std. dev.	Median	Skewness
1	1.28	1.98	0.30	2.0
5	1.33	1.40	0.98	2.2
7.5	1.30	1.51	0.76	2.5
10	0.99	0.78	0.76	1.5
20	1.43	1.05	1.22	1.5
30	1.47	0.61	1.49	0.1
60	1.43	0.35	1.36	0.5

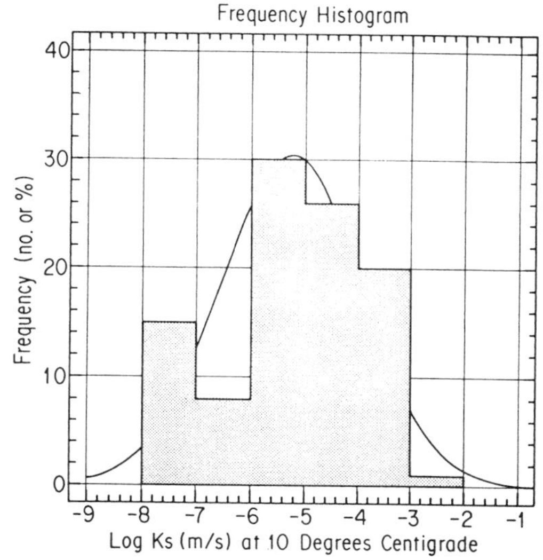

Fig. 6.10. Frequency distribution of saturated hydraulic conductivity compared with a normal distribution (six classes between 10^{-2} and 10^{-8} ms^{-1}). (From Espeby, 1990)

disturbed sample 10 cm in diameter and 30 cm long was taken vertically from depths of 60 to 90 cm in the subsoil. No roots or traces of roots were found below 60 cm. The subsoil has been formed on volcanic ash from the eruption of Mt. Fuji about 20,000 years ago. The soil has a well-developed structure, with porosity of 82% and saturated hydraulic conductivity of 1.6 x 10^{-2} cm/sec. The K_u was measured on drying as a function of suction, using a steady-flow method. Then the sample was carefully cut in half and K_u measured in the same way on the two samples of 15 cm height. The samples were then again cut in half to a length of 7.5 cm, and the measurements repeated. The black circles (Fig. 6.11) represent the mean K_u values of three samples with a diameter of 10 cm and a height of 4 cm taken separately from the same layer. Only the maximum and minimum values among the four 7.5 cm samples are shown in the figure. There is no effect of soil column length on K_u in the range of column length used here. This might suggest that the REV of K_u of a soil with a well-developed structure is small when the soil has formed under a condition where only the soil formation factors of rainfall and temperature have determined soil formation.

Keller et al. (1989) compared permeability results from two field methods with results from conventional laboratory consolidation and permeameter tests on a clayey till deposit. The field methods involved downward propagation of the seasonal cycle of water table fluctuation, or flow in the vicinity of a cavity

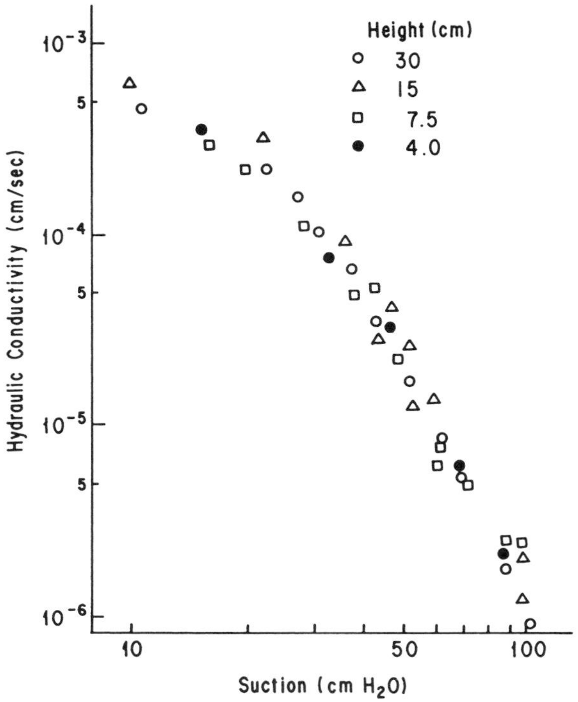

Fig. 6.11. Unsaturated hydraulic conductivity of samples with different height for a volcanic ash subsoil. (From Shigematsu and Iwata, 1991)

excavated in the till. The sample size for the laboratory test was 65 mm diameter and 32 mm thickness. The downward propagation method yields values of 10^{-10} m/sec, in close agreement with the results of the laboratory tests and the slug tests. Bjerg et al. (1992) found that the value of hydraulic conductivity determined by a mini slug test method corresponded quite well to the results obtained by two large-scale tracer experiments performed with chloride and tritium in a sandy aquifer.

The results of these limited measurements indicate that the magnitude of the REV for saturated hydraulic conductivity is approximately the same as the sample sizes used conventionally. This is true for a soil having an apparently uniform structure, such as sand or a clayey soil without macropores, and for unsaturated hydraulic conductivity in a soil with well-developed structure.

6.2.3 Determination of Range of Influence

The Semi-Variogram

The range of influence for a property in a field soil is estimated using a semi-variogram. Suppose we have random variable values of a property of the soil z(x) and z(x+h) at positions x and x+h, where h is the distance separating them. The value h is called the lag. If we have m pairs of observations, separated by the same lag, and z(x) is regarded to be a second order stationary random variable, the semi-variogram, $\gamma(h)$, which expresses the semi-variance as a function of h, is given by

$$\gamma(h) = \frac{1}{2m} \sum_{i=1}^{m} \left[z(x_i) - z(x_i + h)\right]^2 \tag{6.7}$$

This second order stationary random variable has the following properties: 1) the expected value of z at any place x is the mean, and 2) covariance of $z(x_1)$ and $z(x_2)$ is a function only of the distance, h, between x_1 and x_2.

In the semi-variogram for measured hydraulic conductivity in Fig. 6.12, the points are the values of semi-variance calculated from measured values. The solid line shows the semi-variogram function estimated from an exponential model, described in the next paragraph. In general, the semi-variogram for a soil property has the principal features represented by the solid line: 1) Although the semi-variance at zero lag (h=0) is itself zero by definition (refer to Eq. 6.7), a smooth curve approximating the sample semi-variances approaches a positive finite intercept on the coordinate h=0. This intercept is known as the nugget variance, and the phenomenon is called the nugget effect. The nugget effect results because it is impossible to take h to be nearly zero, since the sample has a finite size. (2) The semi-variogram, $\gamma(h)$, increases to a constant value as h becomes larger. That value is called the sill. The sill is equal to the population variance of the property, σ^2, in the field. The distance over which $\gamma(h)$ increases is called the range of influence, and is a very important concept. Within the range, the values of the random variable z(x) are correlated with each other. That is, two samples whose distance is larger than the range are regarded to be independent of one another.

Several models are used to describe a semi-variogram function $\gamma(h)$. The spherical model is given mathematically by

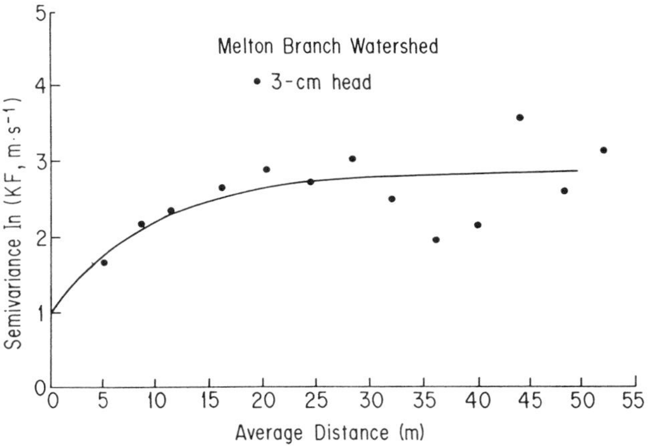

Fig. 6.12. Semivariogram for typical hydraulic conductivity measurements. (From Wilson et al., 1989)

$$\gamma(h) = n + c\left[\frac{3}{2}\frac{h}{\lambda} - \frac{1}{2}\frac{h^3}{\lambda^3}\right] \quad \textit{for } 0 < h \leq \lambda \tag{6.8}$$

$$\gamma(h) = n + c = \sigma^2 \qquad \textit{for } h > \lambda$$

where n is the nugget variance, c the sill value, and λ is the range of influence.

The exponential model is expressed by

$$\gamma(h) = n + c\left[1 - \exp(\frac{-h}{r})\right] \tag{6.9}$$

where r is the distance parameter controlling the spatial extent of the function and is often called the correlation length. For practical purposes $\lambda' = 3r$, where $\gamma(\lambda')$ is equal to approximately (n+0.95c), is used as the effective or practical range. This theory is developed in detail in several studies, e.g. Journel and Huijbregts (1978), Clark (1979), and Webster (1985).

Estimation of Parameters in the Semi-Variogram Function

We can easily draw up a semi-variogram for a hydraulic property in a field soil from a set of measured values. However, it is usually difficult to infer exact values of the sill, σ^2, the range of influence, λ, and the nugget variance, n, from the semi-variogram. The semi-variances calculated from the measurements usually do not lie smoothly on a curve, as shown in Fig. 6.12 or Fig. 6.13.

Knowledge of statistically described spatial structures for hydraulic properties is very important in practical applications such as estimating point or spatially averaged values of properties using kriging techniques, designing sampling networks and improving their efficiency, or stochastic modeling to understand the overall response of heterogeneous flow systems (Kitanidis and Lane, 1985). As an example, the range of influence for hydraulic conductivity is an important parameter in discussing macrodispersion in an aquifer. The relation between macrodispersivities and the three-dimensional spatial correlation structure of an aquifer has been developed for several special cases by Gelhar and Axness (1983). In a statistically isotropic medium, the longitudinal macrodispersivity, A_{11}, is expressed by

$$A_{11} \propto \sigma^2_{\ln k}\, \lambda \tag{6.10}$$

where $\sigma^2_{\ln k}$ is the variance of ln k where k is the hydraulic conductivity. Equation 6.10 shows that spatial dependence of dispersivity is proportional to λ. Many studies have attempted to evaluate the variance σ^2 and the range λ as exactly as possible (Jury et al., 1987; Russo and Jury, 1987; Samper and Neuman, 1989; Ünlü et al., 1990).

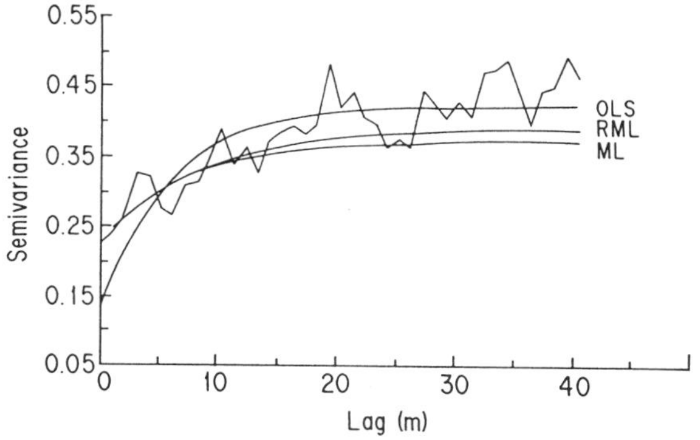

Fig. 6.13. Semivariogram for log saturated hydraulic conductivity. Comparison of experimental semivariograms with those estimated by the methods of ordinary least square (OLS), maximum likelihood (ML), and restricted maximum likelihood (RML). (From Ünlü et al., 1990)

Ünlü et al. (1990) evaluated the sensitivity of σ^2 and λ for three hydraulic properties to three estimation methods. The three properties were the log of saturated hydraulic conductivity, a pore size distribution parameter, and the specific water capacity. The statistical parameters were estimated assuming a second order stationary and an exponential semi-variogram model with nugget effect. Methods of ordinary least squares (OLS), maximum likelihood (ML), and restricted maximum likelihood (RML) were used to estimate σ^2 and λ, while methods of cross-validation (kriging) and uncorrelated residuals were used to validate the semi-variogram model with estimated σ^2 and λ (Fig. 6.13). For all three soil hydraulic properties, the OLS method consistently produced higher σ^2 and lower λ values than ML and RML methods. In the majority of cases, the OLS estimates of σ^2 and λ failed the validation test. In general, for a given data set, RML estimates of σ^2 and λ were more accurate and consistent, in comparison with those of OLS and ML. The semi-variogram for saturated hydraulic conductivity (Fig. 6.13) shows that estimated values of σ^2 and λ are highly dependent on the estimation methods.

Effect of Sample Size on Range of Influence

Lauren et al. (1988) investigated the range of influence (spatial structure) of silty loam soil with macropores using five different sample sizes (see Table 6.2). Semi-variograms of log-transformed hydraulic conductivity measured in columns A, D, and E showed no spatial structure, only random variation. Semi-variograms for columns B and C indicated spatial dependence that could be described by either a linear or spherical model. The range of influence was 60-80 m. As to the reason why spatial structure would not be expected for columns D and E (the two smallest), the authors pointed out that they are smaller than the REV and unrepresentative of the true water flow regime in the field. If sample size is small compared with the scale of heterogeneity in the field (REV), spatial structure cannot be recognized. The authors concluded that the largest column, A, was unsaturated when the measurements were performed. This indicates that it is desirable to investigate the REV for a soil before sampling to estimate the range of influence, especially for a soil expected to show large heterogeneity.

6.2.4 Spatial Dependence of Hydraulic Conductivity in the Field

Sudicky (1986) measured spatial variability of hydraulic conductivity of 1,279 samples in a sand aquifer of horizontal discontinuous lenses of medium-grained, fine-grained and silty fine grained sand, with infrequent silt, silty-clay, and coarse sand layers. The measurements were made in the laboratory using the standard falling head procedure, on samples 0.05 m diameter and 0.05 m long. The geometric mean of hydraulic conductivity for the samples was 9.75 x 10^{-5} m/sec. The three-dimensional covariance structure of the aquifer from the log-transformed data indicates that use of an exponential autocorrelation function model (n = 0) with variance equal to 0.29, an isotropic horizontal correlation length equal to around 2.8 m, and a vertical correlation length equal to 0.12 m is representative. That is, the practical horizontal and vertical ranges of influence are about 8.5 m and 0.36 m, respectively.

The spatial variability of hydraulic conductivity in a sandy aquifer has been determined using mini slug test methods by Bjerg et al. (1992). The hydraulic conductivity of the aquifer had a geometric mean of 5.05 x 10^{-4} m/sec. The semi-variogram function for log-transformed hydraulic conductivity, ln k, was expressed by an exponential model without nugget effect. The variance of ln k was equal to 0.37, and the horizontal correlation length was found to be small (1-2.5 m). Consequently, the practical horizontal range was 3-7.5 m. In the above two papers, the ranges were obtained without investigation of REV for each soil. However, REVs for sand are regarded to be small, as described before.

Wilson et al. (1989) quantified the spatial dependence of hydraulic conductivity in the subsoils of two forested watersheds using the Guelph Permeameter (Reynolds and Elrick, 1985; 1986). The two watersheds have contrasting morphological and mineralogical properties; one is deeply weathered from dolomitic shale, and the second has shallow soils weathered from limey shale. The practical horizontal range was 30 m for the first watershed and less than 4.2 m for the second. The hydraulic conductivity of soils in the first watershed had a greater arithmetic mean (1.50 vs. 0.76 x 10^{-5} m/sec) and greater random variability.

Lauren et al. (1988) obtained 60-80 m as the range of influence in hydraulic conductivity for a silty clay loam with macropores. Russo and Bresler (1981) estimated the integral scales (corresponding to the ranges of influence) for several hydraulic properties using the data obtained by Nielsen et al. (1973). According to their results, the calculated integral scales of saturated hydraulic conductivities at depths of 0, 30, 60, and 90 cm are 34, 39, 21, and 14 m, respectively.

Moutonnet et al. (1988) investigated spatial variability of unsaturated hydraulic conductivity in a calcareous brown soil using a geostatistical procedure by Vauclin (1982). Soil internal drainage was measured along a 100-m transect with 20 neutron probe access tubes at 5-m intervals using an automatic neutron moisture gauge and a tensiometer data logger. The geostatistical analysis of measurements of conductivity vs. water content indicated the independence of unsaturated hydraulic conductivities along the transect, even at the minimum lag of 5 m. Viera et al. (1981) studied the spatial variability of infiltration rate measured in the field for a Typic Xerorthent soil and found that samples separated by 50 m or more were not correlated.

No general conclusions can be drawn from these few experimental results. However, it may be said that 1) even for a sand having a considerably complicated structure, the range of influence for hydraulic conductivity is smaller than for fine-grained soils, and 2) the spatial structure of unsaturated conductivity may be considerably different from that of saturated hydraulic conductivity.

6.2.5 Stochastic Theory for Water Flow in Heterogeneous Soil

What is the relationship between flow modeled in a fictitious homogeneous field and flow in an actual spatially variable field? (Dagan and Bresler, 1983).

Even if the statistical distribution function (probability density function) of hydraulic conductivity in a field soil is obtained using REV and range of influence, it is very difficult to predict water flow in the field. In general, the variance is so large that a deterministic partial differential equation such as the Richards equation cannot be used. The equation is valid only for an approximately homogeneous medium, while field observations have shown that hydraulic properties of soils vary significantly with location even within a given soil type (Warrick and Nielsen, 1980). The standard deviation of the natural logarithm of saturated hydraulic conductivity has been observed to range from 0.3 to 3. Statistical distributions have been recognized by many researchers to be approximately lognormal (e.g. Nielsen, et al., 1973; Freeze, 1975; Baker and Bouma, 1967; Warrick et al., 1977; Sudicky, 1986; Lauren et al., 1988).

Nielsen and Biggar (1982) have discussed the limitations of the standard deterministic approach, and emphasized the need for statistical techniques which incorporate the effects of natural variability. A rational approach, related to field variability, has been set forth by stochastic modeling since the 1980s (e.g. Anderson and Shapiro, 1983; Dagan and Bresler, 1983; Mualem, 1984; Yeh et al., 1985a,b,c; Mantoglou and Gelhar, 1987a,b,c; Polmann, 1991). Most stochastic theories proposed for water flow, especially unsaturated water flow, in soil have included so many assumptions and been so mathematically complex that they cannot be summarized briefly. This reflects the stage of development of the theories. Only their basic features will be introduced here.

The fundamental features are: 1) It is assumed that a classical deterministic partial differential equation, such as the Richards equation or the diffusion equation, is valid for water flow at a local scale. 2) The hydraulic properties appearing in the equation are assumed to be stationary, i.e. each property is defined to fluctuate randomly around its mean depending on statistics such as the mean, variance and correlation scale or range of influence. A stochastic partial differential equation is derived by substituting the defined random functions of the properties into the classical equation. 3) The equation is solved using a finite element method under the initial and boundary conditions.

As an example to explain these features, we take a stochastic, two-dimensional flow simulation by Polmann et al. (1991) based on the Mantoglou and Gelhar theory (1987a,b,c):

1) Unsaturated flow at a local scale (of the order of tens of centimeters) is assumed to be adequately described by the classical Richards equation.

$$C(\psi)\,\frac{\partial\psi}{\partial t} = \frac{\partial}{\partial x_i}\left[K(\psi)\,\frac{\partial(\psi + x_1)}{\partial x_i}\right] \tag{6.11}$$

where ψ is the (positive) soil-water tension at location x and time t, x_1 is the depth (increasing downward), $C(\psi) = -\partial\theta/\partial\psi$ is the specific moisture capacity, θ the volumetric moisture content, and $K(\psi)$ is an isotropic unsaturated hydraulic conductivity function. Spatial vectors are assumed and repeated subscripts are understood to be summed from 1 to 3. Though the local soil property functions, $C(\psi)$ and $K(\psi)$, can vary greatly over the distance of interest in the field, the specific capacity is assumed to be a known spatially-invariant function of tension, and only the hydraulic conductivity variability is taken into consideration.

2) Following Gardner (1958), the hydraulic conductivity function is presumed to have a log linear form.

$$\ln K(\psi) = \ln K_s - \alpha\psi \tag{6.12}$$

where K_s is the saturated hydraulic conductivity, and α is the slope of the unsaturated log conductivity function. The effects of spatial variability are accounted for by assuming that ln Ks and α are stationary random fields with known statistical properties, described later. That is,

$$\alpha = A + a \tag{6.13}$$

$$\ln K_s = F + f \tag{6.14}$$

where A and F are ensemble means, a and f are random fluctuations.

3) Substituting Eqs. 6.12, 6.13 and 6.14 into Eq. 6.11 gives a stochastic partial differential equation which depends on the random fields, ln K_s and α. In other words, the dependent variable $\psi(x,t)$ is a random field which can only be described in terms of its statistical properties. This also applies to a variable such as $\theta(x,t)$ directly related to ψ. A small perturbation approach is used to derive approximate expressions for the mean and variance of ψ. Their mean

tension satisfies a deterministic differential equation having the same form as the Richards equation.

$$C(H)\ \frac{\partial H}{\partial t} = \frac{\partial}{\partial x_i}\left[\hat{K}_{ij}\frac{\partial(H + x_i)}{\partial x_i}\right] \tag{6.15}$$

where H is the mean tension and $\hat{K}_{ij}$ is an anisotropic effective hydraulic conductivity function which depends on the mean tension, the time and space derivatives of the mean tension, and the statistical properties of f and a.

It should be noted that an isotropic unsaturated conductivity function in the Richards equation is changed to an anisotropic function depending on $\partial H/\partial t$, $\partial H/\partial x$, f and a. Theoretical stochastic researches for unsaturated soils suggested that the hydraulic conductivity anisotropy in a heterogeneous soil, and especially for a layered soil, will vary as the hydraulic state (mean pressure head, and spatial and temporal head gradients) of the medium vary (state-dependent anisotropy). That is, as the mean pressure head decreases below the air-entry value, anisotropy may increase by more than several orders of magnitude. Fig. 6.14 (McCord et al., 1991) shows anisotropy as a function of pressure head for a dune sand calculated using the stochastic estimator of Yeh et al. (1985b) for three different correlation lengths. Mualem (1984) has suggested that for a horizontally stratified soil, anisotropy at a given pressure head may be estimated as the ratio of the effective mean hydraulic conductivity in the horizontal and vertical directions at the same pressure. Stephens and Heermann (1988) and McCord et al. (1991) have confirmed experimentally the suggestion obtained from the theoretical stochastic studies. Variable, state-dependent anisotropy in unsaturated soils is a large-scale (macroscopic) flow property which results from media with texture heterogeneities at a small scale.

Equation 6.15 may be solved for the mean tension distribution under initial and boundary conditions for H. In addition, a similar approximation for the mean moisture content Θ yields

$$\Theta = \theta\ (H) \tag{6.16}$$

The mean tension H is related to the random tension field ψ by a relation analogous to Eq. 6.14.

$$\psi = H + h \tag{6.17}$$

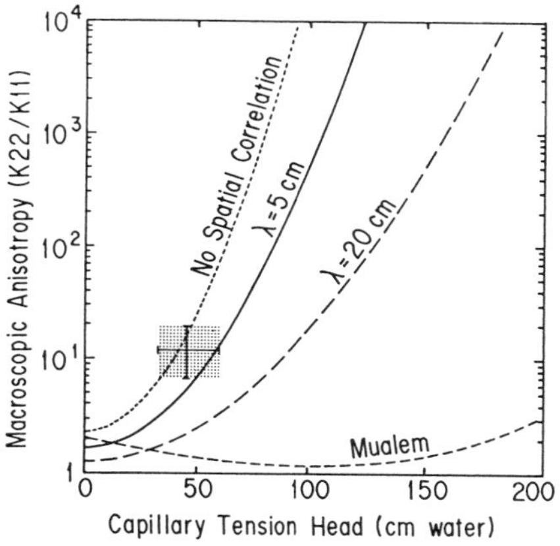

Fig. 6.14. Anisotropy as a function of pressure head for the Sevilleta dune sands calculated using the stochastic estimator of Yeh et al. (1985b) for three different correlation lengths. Dots show value of field estimated anisotropies (mean ± one standard deviation) at *in situ* tensions (mean ± one standard deviation). (From McCord et al., 1991)

4) The function, $\theta(\psi)$, proposed by van Genuchten (1980) is used as the moisture retention curve. All parameters in the function are dealt with as deterministic spatially-invariant. Consequently, C(H) also becomes a spatially invariant function of H.

5) K_{ij} and σ_h^2 (variance of tension) are obtained on assumptions: First, the random fields of f and a are expressed by exponential autocorrelation functions with a common vertical correlation scale λ_1, and with variances σ_f^2 and σ_a^2, respectively. The horizontal correlation scales are much larger than λ_1, representing a situation where the soil is highly stratified. Second, the random fields f and a are uncorrelated. Third, K_{ij} and σ_h^2 depend on the mean tension gradient, and $\partial H/\partial x_i$ is neglected.

When these assumptions are introduced, the Mantoglou and Gelhar (1987a,b,c) expressions for $\hat{K}_{ij}$ and σ_h^2 become

$$\hat{K}_{11} = K_G \exp\left[-AH - E(ah) - \frac{\sigma_E^2}{2} \right]$$

$$\hat{K}_{22} = K_G \exp\left[-AH - E(ah) + \frac{\sigma_E^2}{2} \right]$$

$$\hat{K}_{12} = \hat{K}_{21} = 0 \qquad (6.18)$$

$$\sigma_h^2 = 2 \frac{\sigma_f^2 \lambda_1}{\pi} I_{hh}$$

$$E(ah) = 2 \frac{\sigma_f^2 \lambda_1}{\pi} I_{ah}$$

The parameter K_G is the geometric mean of the saturated hydraulic conductivity, defined as $K_G = \exp(F)$. σ_E^2, I_{hh}, and I_{ah} are given as complex functions of A, H, C(H), $\partial H/\partial t$ and σ_a^2, not shown here.

6) The left-hand side of Eq. 6.15 is rewritten in terms of mean moisture content ($\partial\Theta/\partial t$), since the mixed formulation (moisture content-tension) assures conservation of mass (Bouloutas, 1989). The governing nonlinear flow equation is linearized with a modified Picard approximation and is discretized over space with a Galerkin finite element method. Thus, the mean moisture content and the variance of tension in addition to the mean tension are computed using K_{ij} and σ_h^2 defined above.

Figure 6.15 shows a tension distribution in the profile from this two-dimensional mean flow simulation, under a condition where water is supplied continuously at the soil surface at 2 cm/day to a strip 4 m wide and 15 m long. The profile is on the plane vertical to the strip centerline through the midpoint of the centerline, and the distribution is ten days from the beginning of water supply. Figure 6.15b represents the vertical tension profile along line A B shown in Fig. 6.15a together with the ±2 standard deviation confidence intervals. It should be borne in mind that one of the most attractive features of a stochastic approach is its ability to quantify not only mean behavior but also to put approximate bounds on deviations from the mean.

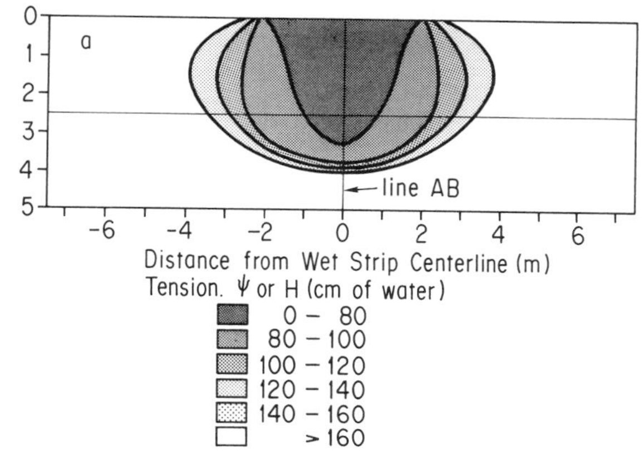

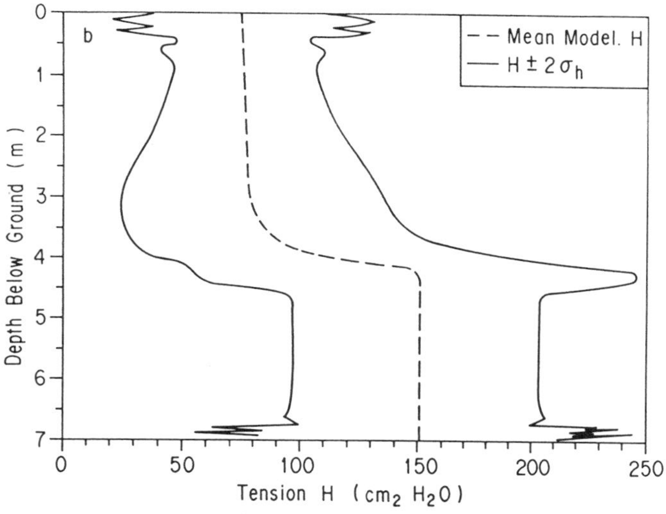

Fig. 6.15. Tension distribution after 10 days and profile estimated by two-dimensional mean flow model. (From Polmann et al., 1991)

6.3 SOLUTE MOVEMENT IN FIELD SOILS

6.3.1 Historical Review

The convection-dispersion equation for solute movement under water-saturated conditions is

$$\theta \frac{\partial C}{\partial t} = \nabla \cdot (\theta \bar{D} \cdot \nabla C) - J \cdot \nabla C \tag{6.19}$$

where C is the concentration of solute in an aqueous solution whose volumetric content in a porous medium is θ and whose volumetric flux density through the medium is J, the elements of the second tensor $\bar{D}$ are the solute dispersion coefficients, and t is the time. The parameter $\bar{D}$ in Eq. (6.19) has often been expressed mathematically in model applications with the constitutive relationship (e.g. Biggar and Nielsen, 1976; Bear, 1972; Anderson, 1979; Gelhar and Axness, 1983; Beese and Wierenga, 1983)

$$\bar{D} \sim \bar{D}_m + \bar{\lambda} U \tag{6.20}$$

where $\bar{D}_m = D_M \bar{I}$, D_M is the scalar coefficient of molecular diffusion of the solute in the porous medium, $\bar{I}$ is an identity matrix, and $\bar{\lambda}$ is the laboratory scale dispersivity. The symbol ~ represents the approximation that holds only for sufficiently large values of pore water velocity, U = J/θ. The correctness of (6.20) as a strictly empirical law to describe solute transport through water-saturated porous media in laboratory columns seems generally accepted (Sposito et al., 1986). Recently, this equation is referred to as the classical convection-dispersion equation.

Only a few early papers related to solute movement in soil (e.g. Slichter, 1905; Burd and Martin, 1923). After 1950 studies relating to solute transfer in porous media, especially theoretical ones, began to appear (e.g. Lapidus and Amundson, 1952; Taylor, 1953; Scheidegger, 1954, 1958; Rifai et al., 1956; de Josselin de Jong, 1958).

Lapidus and Amundson (1952) proposed an equation similar to the convection-dispersion equation, without any explanation.

$$D \frac{\partial^2 C}{\partial z^2} = v \frac{\partial C}{\partial z} + \frac{\partial C}{\partial t} + \frac{1}{\sigma} \frac{\partial n}{\partial t} \tag{6.21}$$

where D is the diffusion coefficient of the adsorbate in solution, C is the concentration of the adsorbate in the fluid stream, z is the distance, v is the pore velocity of fluid, t is the time, σ is the fractional void volume, and n is the amount of adsorbate on the absorbent. In the equation it should be noted that D denotes not the dispersion coefficient but the diffusion coefficient of the adsorbate. They simplified solute transport as the sum of only two fluxes, due to water movement and longitudinal diffusion, in order to solve the partial differential equation. They stated that Eq. (6.21) is in error not only in neglecting the hydrodynamic variations but also the consequent concentration effect.

Scheidegger (1954) employed random walk theory, extending it to three dimensions to derive the probability density function of the displacement. He assumed stationary flow and a statistically homogeneous porous medium. He also neglected molecular diffusion, so only the velocity distribution was considered to cause spreading of the moving boundary. The probability function obtained was Gaussian (normal) with a variance proportional to time. This work was probably the first attempt to treat hydrodynamic dispersion as a phenomenon in three-dimensional space (Bear, 1969, 1972). However, because Scheidegger did not take molecular diffusion into account, his coefficient of dispersion is a scalar with no distinction between the longitudinal and transvere directions.

Rifai et al. (1956) and de Josselin de Jong (1958) also proposed statistical models. The former suggested that transverse molecular diffusion may introduce variation in the dispersion of the front. The latter used a theory which combines both transverse and longitudinal dispersion into one formula. Experimental studies were reported by several researchers (e.g. Kaufman and Orlob, 1956; Day, 1956; Day and Forsythe, 1957; Handy, 1959). Day and Forsythe (1957) suggested that $\alpha = D/U$ in one-dimensional flow be considered a soil characteristic, where α is the dispersivity.

The 1960s saw the establishment of a convective and dispersive model and the theoretical derivation of the convection-dispersion equation. The model expressed by the convection-dispersion equation was established by Nielsen and Biggar (1962). They compared measured breakthrough curves with results calculated from the three theoretical models of Scheidegger (1954), Rifai et al. (1956), and Lapidus and Amundson (1952). The results suggested that the last model was the most satisfactory for predicting spreading of a noninteracting solute in porous media where spreading results from dispersion diffusion. They changed the model by replacing D, indicating the molecular diffusion coefficient, with a new parameter characterized by both diffusion and dispersion. Thus, the convective and dispersive model had been used before the theoretical derivation of the convection-dispersion equation.

Bear and Bachmat (1966, 1967) derived the convection-dispersion equation. The continuum approach was adopted as it was not practical to treat transport phenomena in porous media by referring only to the fluid continuum filling the void space. The detailed geometry of the fluid-solid interfaces is not known. In the continuum approach, the multiphase porous medium composed of a network of interconnected random narrow passages of varying length, cross-section, and orientation, is replaced by a fictitious continuum. Values of several variables and parameters are assigned to any mathematical point in the continuum. These are averaged values for any particular variable over the representative elementary volume (REV) of the medium around the point considered.

Thus, the foundation of the convective and dispersive model established by Nielsen and Biggar (1962) was solidified not only by experimental results but also by the theoretical result. Application of the convection-dispersion equation successfully described solute transport in laboratory column experiments or homogeneous soils. However, experimental results inconsistent with the convection-dispersion equation soon began to appear. The problem was considered to be due to heterogeneities in field soils.

In 1976, Biggar and Nielsen published a report on a pioneering field-scale leaching experiment and its analysis in conjunction with the soil water experiment of Nielsen et al. (1973) (Sposito and Jury, 1986). Their analysis indicated that 1) pronounced spatial variability was observed for the dispersion coefficient as well as for pore water velocity and hydraulic conductivity, and the measured values had lognormal distributions (Fig. 6.16); and 2) the values of the pore water velocity varied not only among field plots but also among different depths in a single plot, where corresponding soil porosity data showed little variation with depth. This suggested that solute transport would show both lateral and vertical variability even when the solute application rate and soil porosity are relatively uniformly distributed. The recognition of extreme spatial variability in field transport parameters established by this report led to many studies showing the convection-dispersion equation not valid in field soils, and to stochastic modelling for solute transport in the field.

Experimental results that questioned the validity of the convection-dispersion equation were: 1) dispersivities (or dispersion coefficients) measured in the laboratory were found to be much smaller than those estimated from field studies (e.g. Martin, 1971; Fried, 1975; Anderson, 1979; Pickens and Grisak, 1981a; Schulin et al., 1987), and 2) field-scale dispersion coefficients were recognized to increase with distance and time (e.g. Warrick et al., 1971; Lawson and Elrick, 1972; Sauty, 1980; Sudicky et al., 1985; Butters and Jury, 1989). Theoretical studies had also shown the same results (e.g. Warren and Skiba, 1964; Mercado, 1967; Gelhar et al., 1979; Sauty, 1980; Smith and Schwartz, 1980; Matheron and Marsily, 1980; Pickens and Grisak, 1981a,b; Dagan, 1982; Gelhar and Axness, 1983; Güven et al., 1984).

The modeling approaches that have been used in field-scale solute transport studies may be divided into two categories (Butters and Jury, 1989). One is the application of the convection-dispersion equation with random variable coefficients (e.g. Dagan and Bresler, 1979; Simmons, 1982; Dagan, 1984, 1986, 1987, 1988; Neuman et al., 1987; Naff et al., 1988; Naff, 1990). These

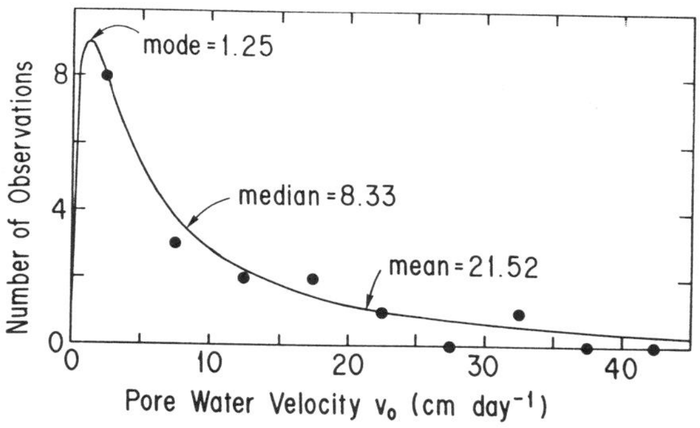

Fig. 6.16. Frequency distribution of pore water velocity. (From Nielsen et al., 1973)

stochastic convection-dispersion models have been built on an assumption that the classical convection-dispersion equation is valid at some local scale. The stochastic analyses have focused on representation of the hydraulic conductivity tensor as realizations of a random field, and its influence on the groundwater velocity variability and dispersion parameters. These stochastic approaches, as well as new continuum approaches for dispersion in heterogeneous porous media (e.g. Chu and Sposito, 1981; Cushman, 1984; Tompson and Gray, 1986; Plumb and Whitaker, 1988) have been very useful in examining mechanisms of solute transport in heterogeneous porous media and the dependance of dispersivity on scale and time. At the present stage, they do not seem to be effective for practical analyses and predictions of actual solute movement in field soils, because they use too many assumptions. For a summary and a critical review of stochastic methods to derive transport equations the reader is referred to Sposito et al. (1986) and Cushman (1987).

Deterministic approaches that account for the scale dependence of dispersion parameters are also beginning to appear (Pickens and Grisak, 1981b; Gupta and Bhatthacharya, 1986; Barry and Sposito, 1989; Yates, 1990). However, since the dependance of the parameters on scale in field soils has not yet been clarified, the approaches cannot be adopted as practical tools.

A second category includes stochastic-convective transfer function models (Jury, 1982; Jury et al., 1986). This is a "black box" approach. The pore system of a natural soil is very complicated and the detailed behavior of a solute moving in the system is not well understood. Instead of modeling the processes, the solute flux at different depths can be estimated from measurements of the flux in outflow at one depth as a function of cumulative net applied water with a given concentration of solute. Considering the stage of development of the theories and the accumulation of experimental results for solute transport in heterogeneous field soils, this approach may be the most practical available now to estimate solute transport. For further details, refer to books by Jury et al. (1991) and Jury and Roth (1990).

In addition, a mobile-immobile water model has been proposed, in which the liquid water is divided into two pore size groups as mobile and immobile phases. It includes preferential flow of solute, and considers the interaction between solutes in the two phases (van Genuchten and Wierenga, 1976, 1977; van Genuchten et al., 1977; Skopp et al., 1981). This model is an improvement over the classical convection-dispersion model; the idea of two water phases is more realistic for soils than a model with solute transport in a medium consisting of noninteracting elementary particles such as a sand. The idea will surely be used in field-scale solute transport models.

6.3.2 Limitations on the Validity of the Convection-Dispersion Equation

Since the 1970s, experimental results both confirming and rejecting the validity of the convection-dispersion equation have been reported. It is most important that limitations on the validity of the classical convection-dispersion equation be clarified.

Experimental Results

Experiments with undisturbed soils usually show tracer breakthrough occurring much earlier than predicted by the convection-dispersion equation (e.g. Kissel et al., 1973; Bouma and Anderson, 1977; White et al., 1984; Ishiguro, 1992). Ishiguro (1992) studied Br transport through undisturbed hard pans of paddy field soils having vertical tubular pores with diameters less than 1 mm (macropores) made by rice roots. The densities of the macropores calculated from the differential water capacity curves for the gray lowland and the gley soil were 2.1 and 2.9 cm^{-2}. The breakthrough curve for the gray lowland soil was predicted approximately by the analytical solution of the one-dimensional convection-dispersion equation. The curve for the gley soil could not be approximated by the equation but could be predicted by a coaxial cylindrical model proposed by the author.

The physical properties of the soil columns, the time for convection in the macropores and the time for radial molecular diffusion in each soil column are shown in Table 6.4. The mean velocities in vertical macropores were calculated from an equation containing the measured water flux. The main mechanisms of Br transport in the soil column are convection in the macropores and radial molecular diffusion into the soil matrix away from the macropores. The dominant mechanism can be evaluated by comparing the time, t_1, necessary for

Table 6.4. Convection and diffusion measured for soil columns. (From Ishiguro, 1992)

	Hydraulic conductivity cm/s (measured)	Porosity % (measured)	Vertical macroporosity % (measured)	Mean velocity in vertical macropores cm/s	Time for Convection t_i sec	Radial diffusion t_2 sec	t_1 / t_2
Gray lowland soil, compost applied	1.1×10^{-5}	49.8	2.04	1.1×10^{-3}	3600	1300	2.8
Gley soil	4.8×10^{-4}	53.6	0.99	9.7×10^{-2}	41	940	0.044

convection to make an appreciable change in concentration to the time, t_2, necessary for the radial variation of concentration in the soil matrix to decrease to about 1/3 of its value. The values of t_1 and t_2 were estimated using the method of Taylor (1953) for evaluation of hydrodynamic dispersion in a tube. The following inferences were drawn from the results: 1) when $t_1 > t_2$, the breakthrough curve is steep at the initial stage and is not predicted by the convection-dispersion equation, and 2) if $t_1 < t_2$, the curve becomes a skewed sigmoid curve, and is approximated by the convection-dispersion equation.

Fried (1972) measured longitudinal dispersivities at several sites and obtained 0.1 to 0.6 m for the aquifer stratum, 5 to 11 m for the aquifer thickness, and 12.2 m for the regional scale. Pickens and Grisak (1981a) compared published experimental data on dispersivities obtained using various methods, and showed that 1) the longitudinal dispersivities obtained from computer modeling studies of contamination zones in granular geologic media range from 12 to 61 m, 2) those obtained from analysis of laboratory breakthrough data on repacked granular materials are the order of 0.01-1 cm, and 3) intermediate are those obtained from analysis of field tracer tests, which range between 0.012 and 15 m.

Lawson and Elrick (1972), in measurements on solute transport in glass beads, found that the dispersivity showed a marked increase over a distance of 0.8 m. Molinari and Peaudecref (1977) also observed a statistically significant increase of dispersivity with distance from the source for a two-dimensional uniform flow field (Fig. 6.17). Souty (1980) reviewed the physics of mass transport in groundwater systems and then developed an approach for slug injection and continuous injection of tracers in porous media for both one- and two-dimensional systems. The results were developed for both uniform and radial (converging and diverging) flow conditions. Applying this approach to results of a series of field tests performed in France, he found that dispersivity

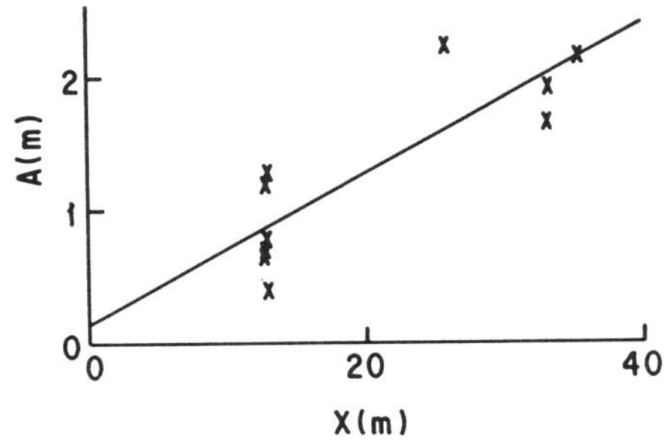

Fig. 6.17. Dispersivity (A) vs. distance (x) given by Molinari et al. (1977) for field experiments at the Bonnaud site in France. The linear regression line has a slope of 0.0566 with a 90% confidence interval (0.0211, 0.0921). (From Gelhar et al., 1979)

at first increases with distance until a characteristic value is reached, and then remains constant.

Pickens and Grisak (1981a) investigated the magnitude of longitudinal dispersivity in a sandy stratified aquifer using laboratory column and field tracer tests. The field investigations included two single-well injection-withdrawal tracer tests using ^{131}I and a two-well recirculating withdrawal-injection tracer test using ^{51}Cr-EDTA. Tracer movement in the aquifer was monitored in great detail with multilevel point sampling instrumentation. The results were: 1) The concentration profile measured within the aquifer, or the breakthrough curve at a sampling point, is strongly dependent on the type of monitoring system, i.e. the groundwater sampling scale. The monitoring system (e.g. multilevel point sampling devices, piezometer, wells) can result in some apparent spreading, giving larger estimates of dispersivity. 2) The mean longitudinal dispersivity obtained from analysis of transport at the scale of individual levels in the aquifer for the single-well tests was 0.7 cm. 3) The average dispersivity value obtained from three laboratory tracer tests with a repacked column of sand was 0.035 cm. 4) The full-aquifer dispersivity from analysis of the withdrawal concentrations for the injection-withdrawal test depended largely on the effect and extent of transverse migration between layers in response to hydraulic and concentration gradients. Dispersivity values of 3 and 9 cm were obtained from tracer tests with average radial front positions of 313 and 499 cm at the end of the injection phase. The full-aquifer dispersivity exhibits a scale effect, increasing approximately in proportion to the square of the average radial front position. The scale-dependent nature of the full-aquifer dispersivity is also evident by comparison to the relatively constant value of 0.7 cm obtained for the individual levels. 5) The full-aquifer longitudinal dispersivity obtained from analysis of the withdrawal-well breakthrough curve of a two-well test is also scale dependent. The dispersivity (50 cm) obtained in this test describes transport in the aquifer only for fully penetrating wells and with the same distance between input and sample locations (8 m). 6) Detailed relative vertical hydraulic conductivity distributions were obtained by monitoring tracer movement at various levels within the aquifer. For simulating solute transport at the scale of the full-aquifer thickness, the following scale-dependent dispersivity expression is representative for the study site aquifer: $\alpha_L = 0.1\ L$, where α_L is the longitudinal dispersivity and L is the mean travel distance. The effect of transverse migration between aquifer layers would become more significant with increasing travel distance, resulting in a reduction in the rate of increase of dispersivity with mean travel distance.

Sudicky et al. (1983) conducted natural gradient chloride tracer tests. These were the first field tests that permitted estimation of three principal values of the dispersion coefficients for anisotropic (laminated) porous media. The results demonstrated the scale dependence of dispersion. Table 6.5 shows longitudinal and transverse dispersivity values in relation to travel distance. Because of local

Table 6.5. Longitudinal and transverse dispersivity values in relation to travel distance. (From Sudicky, 1983)

Fast zone			Slow zone		
Distance from injection wells (m)	α_l (m)	α_t (m)	Distance from injection wells (m)	α_l (m)	α_t (m)
1.15	0.032	0.01	0.075	0.0125	0.005
1.25	0.034	0.01	0.90	0.0125	0.005
1.75	0.043	0.015	1.10	0.02	0.0075
1.80	0.041	0.015	1.25	0.025	0.0075
2.50	0.058	0.02	1.50	0.025	0.01
2.70	0.060	0.02	1.70	0.025	0.01
3.50	0.075	0.02	1.75	0.025	0.01
3.80	0.08	0.02	2.50	0.06	0.025
			3.50	0.08	0.03
			11.00	0.08	0.03

heterogeneity in the aquifer, the tracer slug gradually split into two halves, one moving horizontally at an average velocity of 2.9 x 10^{-6} m/s and the other horizontally at 8.2 x 10^{-7} m/s. The fast and slow zones indicated in the table correspond to the two flows. Both the longitudinal and transverse dispersivities in the two zones initially increase with distance and then become constant.

Butters and Jury (1989), using data measured in field experiments by Butters et al. (1989), compared two models of vadose zone solute transport: the deterministic one-dimensional convection-dispersion equation model (CDE) and the stochastic-convective lognormal transfer function model (CLT) with an assumed lognormal distribution of travel times. The parameters used to test model predictions were calculated for normalized solute breakthrough curves measured at a depth of 0.3 m. In Fig. 6.18 the averaged and normalized solute breakthrough curves at depths of 0.3 m and 1.8 m are plotted as solid curves, and the distributions predicted by the CDE and CLT models are plotted as dashed and dotted curves, respectively. The distributions calculated by the CDE and CLT at 0.3 m coincide well with the observed curve. At 1.8 m, the distribution estimated by the CLT agrees well with the measured curve but the CDE deviates greatly. The deviation of the CDE curve is due to an assumption that the dispersion coefficient is constant irrespective of the distance from the solute input. The lognormal probability density function used in the CLT has been shown to produce a longitudinal dispersion of the solute pulse which grows linearly with distance (Jury and Sposito, 1985).

Dyson and White (1989) studied solute transport through layered columns of repacked aggregates overlying sand, under steady flow conditions. The predictions by the convection-dispersion model and the transfer function model with an assumed lognormal distribution of travel times were tested against experimental values of drainage effluent concentration and solute concentration with depth in the columns. The parameter values in each model were obtained

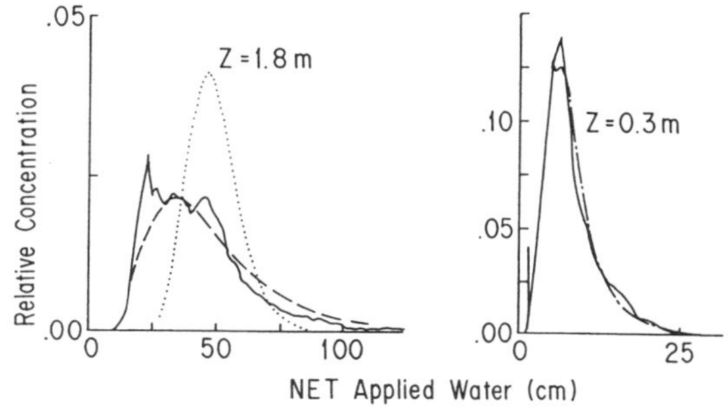

Fig. 6.18. Calibration of the CLT and CDE models and observed area-averaged breakthrough curves. (From Butters and Jury, 1989)

from experiments on columns containing only aggregate or sand. The predictions by the transfer function model and the convection-dispersion model did not differ significantly, and both predictions agreed with the experimental results.

Güven et al. (1985) carried out simulations of the field experiments by Pickens and Grisak (1981a) using the single-well advection-dispersion model, SWADM. The agreement between the simulated and experimental values (Fig. 6.19) is very good. The authors concluded "the results show that the movement of an injected tracer in a stratified aquifer may be accurately simulated without resorting to the use of a scale-dependent dispersivity if the flow field and local dispersion coefficients are known in sufficient detail. When the advection process is simulated accurately, the values of local dispersivity will be small, constant, and on the order of those measured at individual levels in the aquifer."

Theoretical Results

Many reports describe the dependence of dispersivity on distance from the point of solute input. Pickens and Grisak (1981a) derived scale-dependent full-aquifer dispersivity expressions relating dispersivity to the statistical properties of a stratified geologic system where the hydraulic conductivity distribution is normal, lognormal, or arbitrary. Dispersivity is a linear function of mean travel distance (Table 6.6).

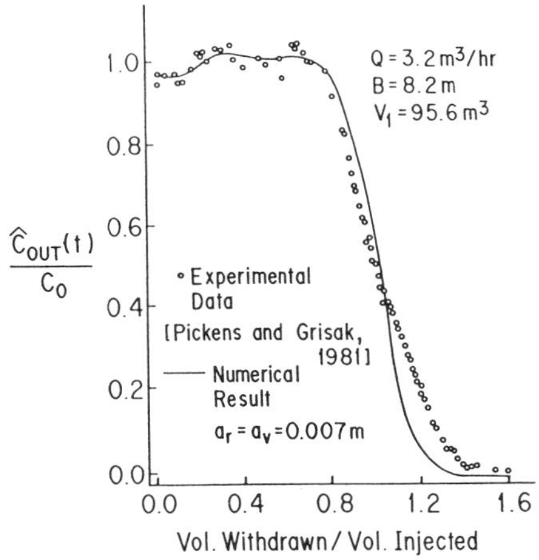

Fig. 6.19. Comparison of SWADM results with field data for the concentration leaving an injection-withdrawal well. (From Güven et al., 1985)

Table 6.6. Scale-dependent dispersivity expressions obtained from stratification results for tracer tests. (From Pickens and Grisak, 1981a)

		Two-well test	
		Distance from injection well	
Stratification	Single-well test	4 m	7 m
Normal	$\alpha_L = 0.256$ L	$\alpha_L = 0.095$ L	$\alpha_L = 0.093$ L
Log normal	$\alpha_L = 0.178$ L	$\alpha_L = 0.041$ L	$\alpha_L = 0.051$ L
Arbitrary	$\alpha_L = 0.117$ L	$\alpha_L = 0.059$ L	$\alpha_L = 0.043$ L

Gelhar et al. (1979) analyzed the longitudinal dispersion produced as a result of vertical variations of hydraulic conductivity in a stratified aquifer by treating the variability of conductivity and concentration as homogeneous stochastic processes. The mass transport process is described using a first-order approximation which is analogous to that of Taylor (1953) for flow in tubes. The resulting stochastic differential equation describing the concentration field is solved using spectral representations. The results of the analysis demonstrated: 1) the mean transport process becomes Fickian for large times, 2) the dispersion coefficient for large times is in the form of a product of mean velocity and a dispersivity, 3) the asymptotic value of this dispersivity is related to statistical properties of the medium, 4) the approach to the asymptotic dispersive condition is slow and significant non-Fickian transport can occur early in the process, and 5) during development of the dispersion process, there are significant departures from the classical normal concentration distribution of the Fickian process.

Most theoretical studies have arrived at the conclusion in 5) above. However, there are two theoretical predictions about the asymptotic behavior of dispersivity described in 1) to 4). This will be discussed in detail in the next section.

Serrano (1992) attempted to describe field-scale transport parameters in terms of regional hydraulic and aquifer properties such as recharge rate, transmissivity, hydraulic gradient, aquifer thickness, and soil properties. He found that hydrologic and hydraulic aquifer properties, in particular recharge, have a strong effect on the magnitude and distribution of the groundwater velocity field. Aquifers subject to natural recharge, which generate variable velocity fields, exhibit increasing dispersion coefficients with distance, even when constant (laboratory scale) values of dispersivities and homogeneity assumptions are adopted. This may be one of the reasons, along with the aquifer heterogeneities at the field scale, for the increasing values so frequently reported in the literature of dispersion coefficient with spatial and temporal scales. It would appear that constant, laboratory scale, dispersivities may be sufficient for modeling field-scale concentrations if an equation which accounts for the effects of hydrologic and hydraulic variables is used.

Investigation of Conditions in the Theoretical Derivation of the Convection-Dispersion Equation

Limitations on the validity of any physical law need to be examined. Theoretically derived physical laws are generally based on some assumptions. The law can become invalid under conditions where the assumptions are not satisfied. When the convection-dispersion equation fails, the phenomenon does not satisfy at least one of the assumptions adopted in the derivation of the equation. We will discuss the limitations on the validity of the convection-dispersion equation through investigations of the assumptions.

Taylor (1953) introduced the convection-dispersion equation in his investigations of the displacement of one liquid by another containing a miscible solute in a straight capillary tube. One assumption playing a very important role in the derivation was "The time necessary for appreciable effects to appear, owing to convection transport, is long compared with the time decay during which radial variations of concentration are reduced to a fraction of their initial value through the action of molecular diffusion." In other words, the time required for radial concentration differences to be appreciably reduced by radial diffusion is short relative to the time required for longitudinal convection to cause appreciable radial concentration changes (Bear, 1972).

This condition is given by

$$L \,/\, \bar{v} \gg R^2 \,/\, 14.4\, D \tag{6.22}$$

where L is the length of tube, $\bar{v}$ is the average velocity, R is the tube radius, and D is the coefficient of molecular diffusion. Assuming that the Hagen-Poiseuille law is valid for the flow, D is 10^{-5} cm^2/s, and L is 1 m, we obtain

$$0.1\ mm \gg R \tag{6.23}$$

Since the measured value of hydraulic conductivity of soil is in general much smaller than that calculated using the Hagen-Poiseuille law, the numerical value on the left-hand side must be considerably larger than 0.1 mm.

Bear and Bachmat (1966, 1967) and Bear (1972) introduced the convection-dispersion equation for porous media on the basis of a continuum approach. In the derivation, the concept of "representative elementary volume (REV)" plays a very important role. Bear explained the REV taking porosity as an example. P is a mathematical point inside the domain with a volume, ΔU_i, much larger than a single pore or grain for which P is the centroid. For this volume, we may determine the ratio:

$$n_i \equiv n_i(\Delta U_i) = (\Delta U_v)_i \;/\; \Delta U_i \tag{6.24}$$

where n_i is the porosity as a function of ΔU_i, and $(\Delta U_v)_i$ is the volume of void space within U_i. Figure 6.20 is a schematic representation illustrating the definition of porosity and its REV. The abscissa indicates ΔU_i, and the ordinate shows n_i. A sequence of values n_i (ΔU_i) may be obtained by gradually reducing the size of ΔU_i around P as a centroid; n_i may change gradually as ΔU_i decreases, especially if the domain under consideration is not homogenous. Homogeneity is defined by Plumb and Whitaker (1988) as follows: a porous medium is homogeneous with respect to a given process and a given averaging volume when the effective transport coefficients (such as the Darcy's law permeability tensor or the hydraulic dispersion tensor) in the volume-averaged transport equation are independent of position. Otherwise the porous medium is heterogeneous.

After the value of ΔU_i decreases below a certain value, the changes or fluctuations may decay, leaving only small amplitude fluctuations that are due to the random distribution of pore sizes in the neighborhood of P. However, when ΔU_i falls below a certain value, ΔU_o, n_i begins to fluctuate because the dimensions of ΔU_i approach those of a single pore. Finally, as ΔU_i becomes zero, converging with P, n_i will take a value of one if P is inside a pore.

Thus, the porosity n(P) of the medium at point P is defined by

$$n(P) = \lim_{\Delta U_i \to \Delta U_o} n_i \,\{\Delta U_i(P)\} = \lim_{\Delta U_i \to \Delta U_o} (\Delta U_v)_i(P) \;/\; \Delta U_i \tag{6.25}$$

Since both ΔU_o and ΔU_v are assumed to vary smoothly in the vicinity of P, n is a continuous function of the position of P within the porous medium. The REV for any physical property other than porosity can also be defined, while the size of the REV depends on the property. It should be noted that in introducing the concept of REV, the actual medium has been replaced by a fictitious continuum in which the value of any property may be assigned to any mathematical point in it.

As a matter of course, REV should be much smaller than the size of the entire flow domain; if not, the resulting average cannot represent what happens at P. On the other hand, it must be sufficiently larger than the size of a single pore, including a sufficient number of pores to permit the meaningful statistical average required in the continuum concept. In addition, the concept of REV

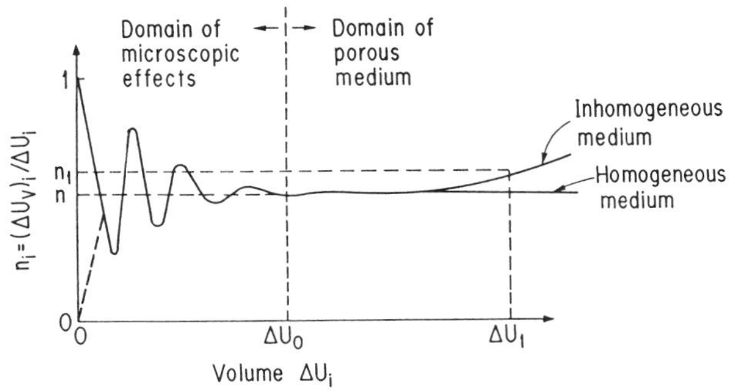

Fig. 6.20. Definition of porosity and representative elementary volume. (From Bear, 1972)

seems to be useful only for approximately homogeneous or statistically homogeneous media, as will be described later. In this case, the size of REV depends on the physical characteristics of a porous medium, i.e. the size of REV for a clayey soil with cracks is larger than for a sand.

The main conditions assumed for an REV in deriving the convection-dispersion equation are: 1) The void space of REV in a porous medium is composed of a spatial network of interconnected random passages, i.e. channels or tubes of varying length, cross-section and orientation, and junctions where at least three channels meet. In addition, the channels and the junctions have a more or less uniform spatial distribution. A spatial network extended like the mesh of a net has a tendency to even out a physical quantity such as velocity or solute concentration. 2) The flow regime is laminar, i.e. the Darcy law is valid for the flow. 3) The fluid loses energy only during passage through the narrow channels and not while passing from one channel to the next through the junction whose volume is much smaller than that of a channel. 4) Thus, the network of channels connected to junctions produces average gradients of pressure, density, viscosity and solute concentration in any REV that includes a sufficiently large number of channels and junctions. These average gradients are practically independent of the geometric shape of a single channel within the REV. 5) Local deviations form the average at points within the void space, which depend strongly on the local geometry of the solid matrix, and are assumed to be much smaller than the average itself.

The convection-dispersion equation was derived by the following procedure: 1) The general mass conservation for chemically inactive solute is taken as the starting point. 2) The equation is first averaged over a channel and next over the void space of an REV. 3) The convection-dispersion equation is obtained after some transformations for respective terms. It is assumed that $(\partial \rho_\alpha / \partial t) \approx 0$, i.e. the solute density distribution, ρ_α, in an REV may be taken instantaneously as constant, and $(\partial n_L / \partial x_i) = 0$ where n_L is the linear porosity.

The conditions assumed for the void space of an REV and the flow, suggest that in the REV the value of a physical quantity such as velocity or solute concentration averaged over each channel across a plane perpendicular to the direction of its average gradient shows an approximately uniform value. On the other hand, a porous medium for which the convection-dispersion equation is valid must be approximately homogeneous in the concept defined by Plumb and Whitaker (1988), described before. If the effective transport coefficients such as dispersivity are dependent on position, the classical convection-dispersion equation becomes invalid.

We know this intuitively also from the fact that the convection-dispersion equation is valid not only for porous media but also for a capillary tube. As emphasized before, the most substantial condition for the tube is Eq. (6.22). Similarly, the convection-dispersion equation may not be applicable to a porous medium where rapid mixing of solutes in the direction vertical to the flow is not assured.

The above conditions may be equivalent to those for the validity of the convection-dispersion equation defined by Jury et al. (1991): "the medium is homogeneous through the volume in which solute transport occurs, and the time required for solutes in stream tubes of different velocity to mix by diffusion or transverse dispersion along a direction normal to the direction of mean convection is short compared to the time required for solutes to move through the volume by mean convection." This condition seems to be looser than that introduced in the derivation by Bear (1972).

It is difficult to judge, based on the above conditions, whether the equation is applicable to a field soil. The difficulties include the interrelation between the magnitude of the REV and the size of the entire flow domain under consideration. This is discussed in the following section. However, it is not difficult to make the evaluation when the sample size under consideration is known. The equation is generally valid for a sand or sandy loam. On the other hand, two criteria are needed for the evaluation of a soil with macropores, including cracks, or a well-structured soil. First, it is necessary to investigate whether the macropores or interaggregate pores satisfy the condition in Eq. (6.22) proposed by Taylor (1953). If they do not satisfy the condition, the equation is not used for the soil. Thus, knowledge of the morphological features of the soil is very important. Second, when the condition is satisfied, the value of the interaction coefficient between flow in the soil matrix and that in macropores or interaggregate pores becomes a key in the evaluation. If the interaction coefficient is large, the model approaches the convection-dispersion model (e.g. Skopp et al., 1981; Steenhuis et al., 1990). A similar conclusion is reached for solute movement in a stratified soil where the hydraulic conductivities of neighboring layers differ and the flow in each layer is parallel to the stratification. In this case, the magnitude of the transverse dispersion, or the ratio of the velocity to the diffusion coefficient, play a decisive role (e.g. Silliman and Simpson, 1987; Matheron and de Marsily, 1980).

6.3.3 Physical Meaning of REV in Solute Movement

In Section 6.2, it was emphasized that water movement in a field soil cannot be predicted without knowing the REV for hydraulic conductivity. The same it true for solute movement. However, choosing the REV for dispersivity is more difficult because the dispersivity in a field soil usually increases with distance from the solute input. If the dispersivity increases infinitely with distance in a field soil, the REV is not defined. If the dispersivity reaches a constant value in a definite distance, the REV may be chosen and the classical or stochastic convection-dispersion equation may predict solute transport in the soil. Therefore, whether or not asymptotic behavior of dispersivity exists will be investigated first.

Does Asymptotic Behavior Exist?

Experimental results: Martin (1971), in a series of laboratory experiments on different lengths of a sandstone, observed that the asymptotic state is reached only when the travel distance of the tracer is large compared with the heterogeneity scale. As stated before, Sauty (1980) found that dispersivity first increases with distance and then reaches a constant value. At travel distances on the order of 1 m, irregular concentration patterns were observed, while at larger distances the effect of heterogeneity was much less distinct.

Pickens and Grisak (1981a) recognized that the individual-level dispersivity obtained from analysis of breakthrough curves from sampling points for single-well tracer tests was not scale dependent for various travel distances (Table 6.7). They inferred from their results that a constant or asymptotic dispersivity had been reached at a travel distance closer than the nearest sampling point, 0.36 m. The aquifer had laminations of the order of 0.1-0.5 cm. On the other hand, the full aquifer dispersivity obtained from the analysis of the withdrawal-well breakthrough curve of a two-well test showed a scale dependence. As shown in Table 6.5, Sudicky et al. (1983) observed that the dispersivity becomes constant at a distance of 3-4 m from the injection well.

Silliman and Simpson (1987) carried out tests under controlled laboratory conditions to investigate the changes in dispersivity caused by heterogeneities. The experiments were conducted in a 2.4 x 1.07 x 0.1 m "sandbox" with measurements of tracer concentration taken at five distances from the input. Packing arrangements were: a) uniform, coarse 20-mesh sand, b) a three-layer arrangement of coarse over fine 90-mesh sand over coarse sand, c) a layered-

Table 6.7. Parameter values obtained from analysis of breakthrough curves from sampling points for a single-well tracer test. (From Pickens and Grisak, 1981a)

Depth (m)	Radial distance (m)	$t_{0.5}$ (days)	(Q/b_i) (m^2/day)	α (m)
2.29	0.36	0.028	5.53	0.7
	0.66	0.094	5.53	0.7
	2.06	0.88	5.74	0.7
2.67	0.36	0.028	5.53	0.7
	0.66	0.085	6.12	0.7
	2.06	1.06	4.78	0.7
3.05	0.36	0.087	1.78	1.5
	0.66	0.213	2.44	0.7
	2.06	1.43	3.54	1.5
4.57	0.36	0.038	4.07	0.4
	0.66	0.132	3.94	0.4
	2.06	1.22	4.15	0.4
8.08	0.66	0.053	9.81	0.7
	2.06	0.34	14.90	0.7

$t_{0.5}$ = time to reach relative concentration of 0.5
Q/b = recharge per unit aquifer thickness

block packing where the mid-depth fine sand layer in b) was replaced by small blocks of fine sand surrounded by the coarse sand, and d) uniform heterogeneous packing where small blocks of the fine sand were uniformly distributed through the entire coarse sand matrix.

They transformed the solution of the one-dimensional convection-dispersion equation (Ogata and Banks, 1961) into an equation showing the relation between the relative concentration C/C_o and V, where C is the concentration of a chemical within the pore fluid, C_o is the source concentration, and V is expressed by

$(x\text{-}vt)/(2t)^{1/2}$ (x is distance, v is average velocity, and t is time). The relative concentration data from one-dimensional experiments were plotted as a function of v on normal probability paper. A linear relationship indicated the convection-dispersion equation was valid, and the slope of the line provided a measure of the dispersion coefficient or dispersivity. Any nonlinearities in such a plot would indicate deviation from the theory, and changes in slope at measuring positions would indicate a nonconstant dispersivity or a "scale effect."

The results for various packing arrangements were as follows. 1) The breakthrough curves for arrangement a) showed a constant dispersivity of approximately 0.02 m. 2) The breakthrough curves for arrangement b) deviated from the homogeneous results only in the late-time tails. The breakthrough curves had two components with different apparent dispersivities—tracer movement through the coarse sand layers before the stage of tailing and that through the fine sand layer in the late-time tails. 3) The two distinct portions of the breakthrough curves at the shortest flow distance (0.15 m from the input) for arrangement c) could be identified—one for the coarse sand matrix and one for the heterogeneous layer. The portion for the heterogeneous layer appeared in the late-time, and the relationship between C/C_0 and v at that time was nearly linear. This indicates that the central region (the heterogeneous layer) is acting as a homogeneous medium with a constant dispersivity within a distance of 0.15 m from the input. 4) Results for arrangement d) showed an increase in dispersivity with distance from the input. The results were not clear on whether the dispersivity reached a constant value.

The most interesting finding was the difference in the results of arrangements c) and d). Arrangement c) has the same structure as in arrangement d), but the former acts as a homogeneous medium within a distance of 0.15 m while the latter does not. This means that the scale at which a heterogeneous medium may be regarded as a homogeneous continuum depends on the structure and scale of the heterogeneity. It appears that the central layer of arrangement c) could be modeled as a homogeneous unit within the length scale of this laboratory experiment, whereas arrangement d), composed of a similar type of heterogeneity at large scale, was still subject to a nonconstant dispersivity even at the final sampling location. In the former case, the scale effect may be interpreted as a result of averaging breakthrough curves from multiple layers, each of which could be defined by constant parameters.

Sudicky et al. (1985) studied the migration of a nonactive solute in layered porous media under controlled laboratory conditions by injecting a tracer into a thin sand layer bounded by silt layers. The experimental results showed a delay in the breakthrough of the tracer in the effluent of the permeable sand layer along with a substantial decrease in concentration after a 1 m travel distance. They emphasized that a transient redistribution of the tracer across the strata by transverse diffusion plays a decisive role, while local longitudinal dispersion is only of secondary importance as a spreading process in such systems.

Theoretical results: Gelhar et al. (1979) represented the flow and mass transport processes in a stratified aquifer through stochastic equations by assuming that variations in hydraulic conductivity and concentration are statistically homogeneous. Their results indicated that 1) the approach to a constant or asymptotic dispersivity is slow, and 2) the constant value is achieved by vertical integration and consequently the dispersivity is averaged over numerous layers.

Smith and Schwartz (1980) used a hybrid deterministic-probabilistic approach to investigate the transport of a population of tracer particles in a hypothetical heterogeneous medium where the two-dimensional, spatially autocorrelated hydraulic conductivity field is generated as a first-order nearest-neighbor stochastic process. Their conclusion was that a constant dispersivity value for a geologic unit can theoretically exist if the system is long enough to provide sufficient spatial averaging of the tracer particles in the velocity field.

Dagan (1982) studied the problem of solute transport in heterogeneous geologic formations whose transmissivity or hydraulic conductivity are subject to uncertainty for two- and three-dimensional flows. The results showed that the concentration obeys a diffusion-type equation for two-dimensional flows only when the solute body has traveled a distance larger than a few tens of the transmissivity integral scale, which may be very large in many conceivable applications. Where the ergodic assumption is bound to apply to three-dimensional structures, longitudinal dispersion can be presented asymptotically by a Fickian equation, with dispersivity much larger than pore scale dispersivity.

Gelhar and Axness (1983) analyzed the dispersive mixing resulting from complex flow in three-dimensionally heterogeneous porous media using a stochastic continuum theory, and demonstrated that a classical gradient transport (Fickian) relationship is valid for large-scale displacement. Güven et al. (1984) calculated the travel distance to be the order of 5-100 km. Taking a typical mean velocity of the order of 10^{-5} m/s, the time to reach a Fickian approximation was found to range from 16 to 300 years. Freyberg (1986) investigated the three-dimensional movement of a tracer plume containing bromide and chloride using data from a large-scale natural gradient field experiment on groundwater solute transport. Though the asymptotic longitudinal dispersivity for the first 647 days of transport obtained from the calibration was 0.49 m, asymptotic conditions were apparently not reached. Schwartz (1977), using a Monte Carlo technique, demonstrated that a unique value of dispersivity may not exist for certain hypothetical geologic media.

In summary, the theoretical analyses seem to show the convection-dispersion equation is not valid within a practical time scale. However, experimental results suggest that a Fickian approximation can be reached much faster than expected by theoretical analyses (e.g. Sauty, 1980; Sudicky et al., 1983). The description by Sudicky et al. (1983) related to the discrepancy may be instructive. The concentrations were measured at the scale of many or most of the heterogeneities. The smoothing of the concentration patterns with time appears to be the result of local vertical intra-heterogeneous mixing that allows an infilling of zones of lower concentration within the tracer pulse without causing a gross increase in its vertical extent. This observation is inconsistent with the highly irregular concentration patterns generally predicted by stochastic theories of dispersion in heterogeneous media, that require a vertical averaging of concentrations with depth to achieve regularity. The results suggest that the smoothing process is the result of a mechanism occurring within the heterogeneous medium itself.

Matheron and de Marsily (1980) used a random motion model to solve solute transport in stratified media, with flow parallel to the stratification. They suggested that although Fickian behavior will in general not occur for flow parallel to the stratification, diffuse behavior is much more likely to appear when the flow is not exactly parallel to the stratification.

The extent to which pre-asymptotic behavior manifests itself in natural aquifers needs to be investigated more thoroughly. Published results from systematic, field-scale experiments in aquifers are generally lacking (Sposito et al., 1986).

REV for Solute Transport

In Bear's (1972) definition of the concept of REV (Eq. 6.25), an averaged quantity has to be independent of the magnitude, shape, and orientation of the averaging region used for its evaluation. The smallest averaging region satisfying the condition is called the representative elementary volume (REV).

As described in Section 6.2.1, this definition is somewhat different from that used for an actual field soil because of the difference in static properties of physical quantities in the porous media. The porous media considered by Bear seem to be homogeneous in the concept defined by Plumb and Whitaker (1988), though this is not stated specifically. On the other hand, since a physical quantity in a field soil is expected to be statistically homogeneous because of the soil formation process, the values of the quantity in the REV belong to a statistical distribution instead of having a unique value. Consequently, the REV is determined by investigating changes of the statistical parameters accompanying the change of sample volume. That is, REV is defined as the smallest volume among such volumes that the statistical distribution for the property in the field can be correctly estimated.

The concept of REV lacks experimental verification, which casts serious doubt on the existence of spatial invariance of the REV for real porous media. Baveye and Sposito (1984) used an averaging via a weight function. They held that a weight function may be associated with the window of the measuring instruments. The transport equations, averaged with a weight function, then take on meaning only when measurements of physical properties used in the transport equations are taken by an instrument with the corresponding weight function. In other words, they got around the concept of scale by associating the concept of a weight function with that of an instrument. Cushman (1984) gave a high value to this approach as being a very general extension of the original REV. Certainly, this approach is attractive when the existence of REV has been confirmed in field soils, although the definition of the REV is somewhat different from that of Bear (e.g. Lauren et al., 1988; Bouma et al., 1989).

An REV for solute transport in a porous medium cannot be defined in the absence of asymptotic behavior of dispersivity in the medium. In this case, the prediction of solute transport in the medium becomes possible only when we know in detail the spatial distributions of physical quantities related to solute transport. This would require laborious investigations.

Sposito et al. (1986) discussed conditions under which asymptotic behavior appears. In general, asymptotic Fickian behavior cannot be expected to hold under a condition where the spatial variability of U(x) (pore water velocity) is arbitrary. "Instead, U(x) must have a somewhat repetitive structure for the asymptotic Fickian approximation to be expected. Examples of this repetitive structure include U(x) as a periodic function, or an almost periodic function, or a sample realization of a homogeneous, ergodic random function." However, the conditions required to produce asymptotic behavior have not yet been clarified.

Even if the dispersivity shows asymptotic behavior, the REV is not necessarily defined. If the distance at which the dispersivity becomes constant is large compared with the magnitude of the entire flow domain under investigation, the REV may become meaningless.

As described above, experimental results suggest that a Fickian behavior can be reached much faster than generally expected from theoretical analyses. The following discussion assumes that the experimental results reflect the process correctly. Then how is the REV determined under conditions where asymptotic behavior of dispersivity is found in a porous medium?

Sauty (1980) stated that dispersivity apparently increases with distance until a characteristic value is reached where no further increases are observed. At this distance, the experimental data fit the result for a monolayer-type curve, and one has a measure of the scale of the controlling heterogeneity. At smaller distances the consequences of local inhomogeneities are equivalent to that of a multilayered system, but as distance increases, the effects of transverse dispersivity become more and more dominant and the macroscopic result is that of an equivalent homogeneous aquifer. This is somewhat analogous to the concept that the strong heterogeneities that one can visualize on the scale of pores can be averaged into a homogeneous porous medium in the REV. That is, if the dispersivity becomes constant at a distance from the solute input in a given direction, the distance is to be called the REV in that direction.

Van Wesenbeeck and Kachanoski (1991) proposed a method for estimating the horizontal scale dependence of dispersion, and related the scale dependence to the REV. They conducted field solute transport experiments on two plots approximately 50 m apart in a well-drained, coarse-textured Typic Hapludalf exhibiting well-developed horizonation. One site had been under cultivation at least 50 years, while the forested site had never been cultivated. Soil solution samplers were installed in a transect at 0.4 m depth and 0.2 m spacing. A uniform pulse of KCl was applied under conditions of constant surface flux density of water supplied through a trickle irrigation system. Samples were collected from all solution samplers simultaneously at regular time intervals using an electric vacuum pump and manifold. Solution samples from selected sites were monitored in the field for bulk electrical conductivity using a portable meter to determine when the pulse had passed. A total of 800 samples for the forested site and 960 samples for the cultivated site were analyzed for chloride. The variance of solute travel time at different spatial scales was calculated from moment analysis of breakthrough curves obtained by averaging local breakthrough curves across different spatial scales. A graph showing the relationship between solute travel time variance and spatial scale was used to decide the minimum horizontal scale which was regraded to correspond to the REV for the dispersion process. The presence of a sill value in the graph was considered to indicate the occurrence of a true field-scale variance estimate, i.e. an estimate that is no longer a function of scale, in a manner similar to the semivariogram.

The calculated scale dependence of solute travel time variance (Fig. 6.21) indicated that distances of at least 2.8 m and 3.8 m were needed to reach an effective field variance for the uncultivated and the cultivated sites. The larger value in the cultivated site was due to an increase in horizontal correlation length

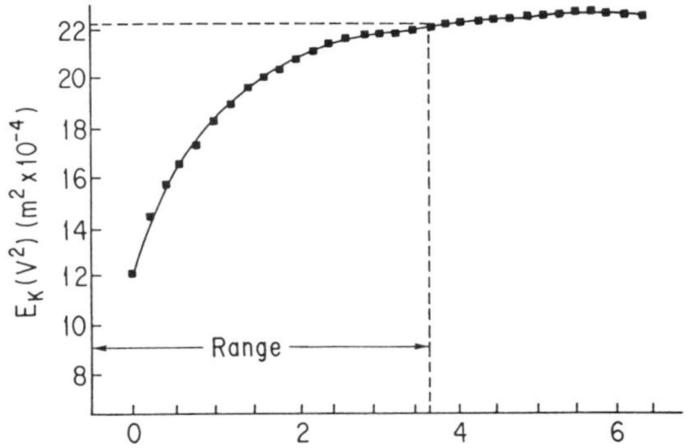

Fig. 6.21. Solute travel-time variance, $E_k(V^2)$, as a function of spatial scale for a cultivated soil. (From Van Wesenbeeck and Kachanoski, 1991)

scales of soil properties caused by tillage mixing. The larger scale might also be considered to be due to decrease in mixing for moving solute because of destruction of pore continuity by tillage. More such studies of spatial dependance of solute transport in the field are needed.

6.4 APPLICATION OF FRACTAL THEORY TO SOLUTE AND WATER FLOWS IN FIELD SOIL

Many recent studies have analyzed physical properties of soils using concepts of fractal geometry (e.g., Armstrong, 1986 for spatial variations of soil properties; Wheatcraft and Tyler, 1988 for scale-dependent dispersivity in heterogeneous aquifers; Tyler and Wheatcraft, 1990 for soil water retention; Pedro et al., 1990 for soil hydraulic conductivity; Rieu and Sposito, 1991a,b for soil porosity and soil water properties; Tyler and Wheatcraft, 1992 for soil particle-size distribution; Hatano et al., 1992 for flow paths in solute transport; Hatano and Booltink, 1992 for bypass flow). This section will describe some applications to field soils.

6.4.1 Concepts of Fractal Geometry

It is easy to say which of the curves in Fig. 6.22 is more irregular. However, it is very difficult to answer the question of how much more irregular the lower curve is, without the help of fractal concepts. This section briefly introduces fundamentals of fractal geometry.

The degree of irregularity of a fractal object is independent of scale. When we cut up the upper curve shown in Fig. 6.23 with a unit of length which is one-third of the distance across of the curve, L, the curve is divided into four parts.

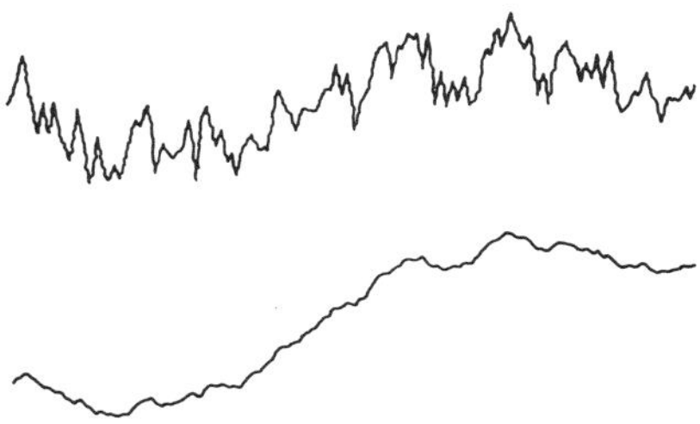

Fig. 6.22. Two fractal curves with different fractal dimensions. (From Takayasu and Takayasu, 1988)

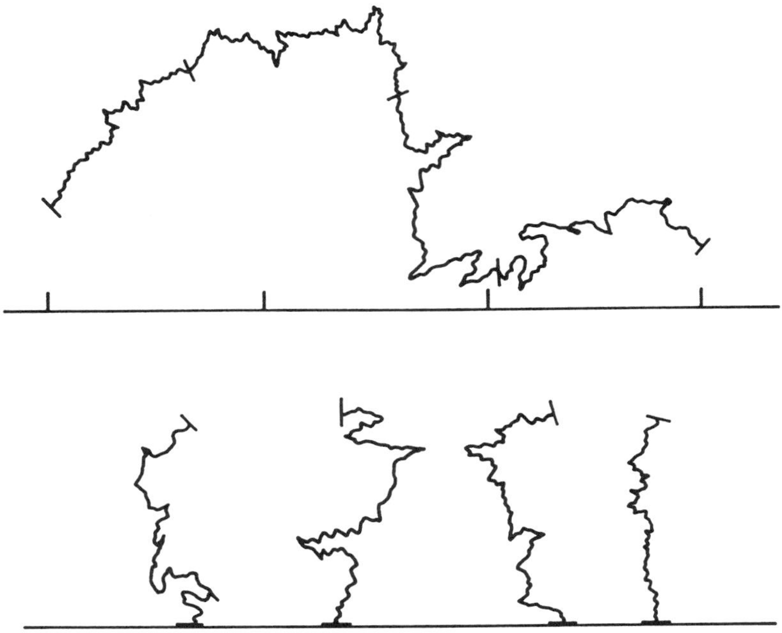

Fig. 6.23. A fractal curve for explaining self-similarity and fractal dimension. (From Takayasu and Takayasu, 1988)

Comparison of the original curve and the four curves produced by cutting (see lower part of Fig. 6.23) indicates no difference in degree of irregularity. Where the original curve is divided by a unit length of 1/9L, the same conclusion can be obtained for the sixteen curves, that is, the irregularity is statistically the same at any scale of viewing the original curve. Such a curve is defined to be statistically "self-similar." On the other hand, if the circumference of a circle is divided equally into four, the degree of irregularity of each arc produced by the division is clearly simpler than that of the circumference.

The curve has one more interesting feature. In order to explain, suppose that we divide each side of a square into n equal parts where n is a natural number. Then, the number of small squares produced by the division, N, is given by

$$N = n^2 \tag{6.26a}$$

Performing similar procedures for a straight line and a cube, we obtain

$$N = n^1 \qquad \textit{for the straight line} \tag{6.26b}$$

$$N = n^3 \qquad \textit{for the cube} \tag{6.26c}$$

Considering that the dimensions of a straight line, a square, and a cube are one, two, and three, respectively, and generalizing the above equation, we get

$$N = n^{D_T} \tag{6.27}$$

where D_T is the dimension, a natural number.

As described before, if the curve shown in Fig. 6.23 is cut up with a unit length of 1/3 L, four curves with the same degree of irregularity are obtained. When divided with a unit length of 1/27 L, 64 curves are obtained. Since $4 = 3^{\log_3 4}$ and $64 = 27^{\log_3 4}$, we get

$$N = n^{\log_3 4} = n^{1.26} \tag{6.28}$$

Extending Eq. (6.27), the dimension of the curve may be defined to be 1.26. The concept of dimension is extended to a fractional number. Eq. (6.29), generalizing Eq. (6.28), has been recognized to be valid for many natural objects.

$$N = n^D \tag{6.29}$$

where D was named the fractal dimension by Mandelbrot (1967).

For examples, the values of D for pore size distributions of a sandy loam, clay loam, silty clay loam, and silt were 1.011, 1.017, 1.404, and 1.485 (Tyler and Wheatcraft, 1989), and for the size distribution of flocs of silica particles with a radius of ~25 Å the D value was 2.12 ± 0.05 (Schaefer et al., 1984). The values of D for the upper and lower curves in Fig. 6.22 are 1.10 and 1.80, respectively. The degree of irregularity increases with increase in the D value.

In addition, the measuring unit of length can be as small as one wishes for artificial fractal curves such as the Koch curve, but there is a lower limit for a natural fractal object. The reader is referred to Mandelbrot (1983) and Feder (1988) for more information on the fundamental concepts of fractal dimension.

6.4.2 Estimation of Fractal Dimension

Of the various methods for estimating fractal dimension of a natural object, we will introduce only methods often used in the field of soil science.

Fractal Dimension of a Curve

The value n in Eq. (6.29) is related to a measuring unit of length, r, by the relation n = L/r. Substituting L/r for n in Eq. (6.29), we get

$$N(r) = (L/r)^D \tag{6.30}$$

Transforming Eq. (6.30), we get

$$\log N(r) = -D \log r + C \; (=\log L^D) \tag{6.31}$$

N(r) is measured at different r values and displayed as a log-log plot of N(r) versus r. The slope, D, in this case ranges from 1 to 2. Again the degree of irregularity of the curve increases as D increases.

Another method exists for estimating the roughness (or smoothness) of pore walls or circumferences of islands. The dimension characterizing the roughness, D_{pe}, also ranges from 1 to 2. If the log-log plots for the perimeters (Pe) and the areas (A) of pores or islands are correlated to a straight line, D_{pe} can be estimated from its slope. In other words, the relation is described by

$$\log A = (2/D_{pe}) \log Pe + C \tag{6.32}$$

where C is a constant.

Fractal Dimension of the Distribution Pattern of an Object Distributed in Space

Figure 6.24 shows images of the distributions of stains in horizontal cross-sections of soil cores obtained by percolating methylene blue solution (Hatano et al., 1992). The two-dimensional fractal dimension of staining patterns, Ds(2), can be derived by fractionalizing a horizontal cross-section into squares (pixels) of side r, and then counting the number of stained pixels N(2,r). By repeating this process with increasing r values, a series of N(2,r) values is obtained. If the pattern can be expressed by fractal geometry, the relationship is given by

$$\log N(2,r) = -Ds(2) \log r + C \tag{6.33}$$

where C is a constant.

Staining patterns in soil satisfied this relationship (Hatano et al., 1992), with Ds(2) ranging from 0 to 2. In the range of $0 < D < 1$, the more linear the

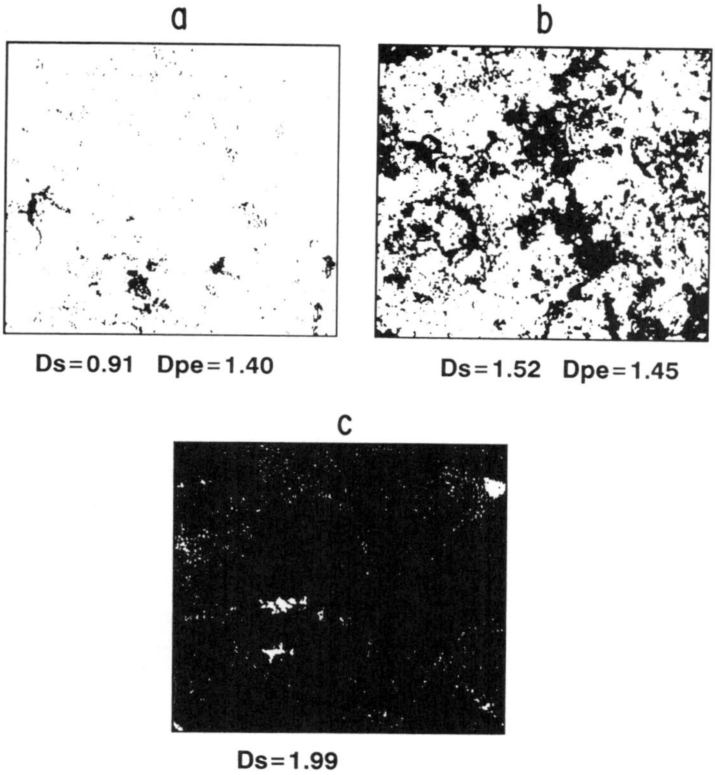

Fig. 6.24. Images of the distribution of stains in soil columns a) Humic Acrisol at depth of 11.5 cm, b) Dystric Gleysol at a depth of 3.5 cm, and c) Ochric Andisol at a depth of 3.5 cm. (From Hatano et al., 1992)

pattern, the larger is D, while in a range of $1 < D < 2$, the pattern becomes more planar with the increase of D.

The three-dimensional fractal dimension can be obtained in a similar way. In this case, cubes of side r are used, instead of squares. The relationship for a three-dimensional fractal is

$$\log N(3,r) = -Ds(3) \log r + C \tag{6.34}$$

where C is a constant.

6.4.3 Examples

Use of Fractal Dimension for Predicting Bypass Flow

Hatano and Booltink (1992) characterized methylene blue staining patterns of undisturbed unsaturated soil cores with a height and diameter of 200 mm, using the fractal dimension of each pattern to predict the measured bypass flows. Soil samples were taken at a depth of 15-35 cm at five randomly chosen locations in the field from a well-structured clay soil classified as a Hydric Fluvaquent. Aggregate size in the samples was 30-50 cm^3; this implies that 120-200 peds were present in a sample, which should ensure that the sample was representative (Lauren et al., 1988). A thin layer of methylene blue powder was added to the sample surface, followed by 10 mm water at an intensity of 13 mm/h. Outflow rates in the cores were almost equal to the irrigation rates after outflow had started. This indicates that nearly all the applied water contributed to the bypass flow.

The three-dimensional fractal dimension of the staining pattern in each soil sample, Ds(3), was calculated for the upper and lower halves of the core using the values of Ds(2) and the stained area in cross-sections. A statistically significant empirical equation for total outflow, O_m, whose correlation coefficient is 0.995 was obtained, using both upper and lower values of Ds(3) and the volume fraction of stained parts, Vs.

$$O_m = -230.6(Vs^{Ds(3)-1})_{upper} + 232.4(Vs^{Ds(3)-1})_{lower} + 12.6 \tag{6.35}$$

This equation indicates that the contribution to bypass flow in the lower half of the cores differs from that in the upper half; that is, greater values of Vs and smaller values of Ds(3) in the lower half of the column increase the total amount of outflow. A physical morphological explanation for this phenomenon is that stains in the upper half of the cores develop around relatively small peds and are mostly not continuous. In the lower half with fewer stains only vertically continuous macropores, which dominate bypass flow, remain. This tendency was also found in other experiments by Hatano et al. (1992).

Estimation of the Brenner Number Using the Fractal Dimension

The curves shown in Fig. 6.25 represent mathematical solutions to the classical one-dimensional convection-dispersion equation corresponding to initial and boundary conditions describing the experiment. The numbers in the figure denote the Brenner number defined by B = VL/D where V is the average pore velocity, L is the length of soil column, and D is the dispersion coefficient. The

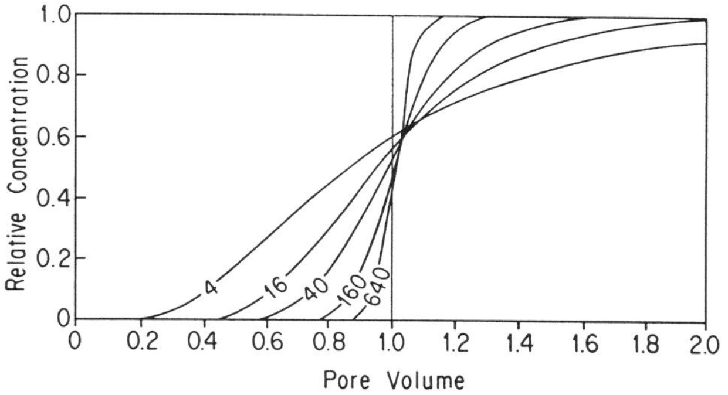

Fig. 6.25. Breakthrough curves whose parameter is the Brenner number. (From Rose and Passioura, 1971)

Brenner number has been widely used to characterize the shapes of breakthrough curves.

Hatano et al. (1992) used fractal dimensions of methylene blue staining patterns in five undisturbed columns with a diameter of 150 mm and a height of 150 mm to analyze solute transport in soil columns (refer to Fig. 6.24). The five soils were classified as Dystric Cambisol (DC), Humic Aridisol (HA), Dystric Gleysol (DG), Ochric Andisol (OA), and Andic Regosol (AR), respectively. The two-dimensional fractal dimension of the staining pattern, Ds(2), and the fractal dimension of the perimeter of stains, Dpe, at different depths of each column were estimated. Ds(2) decreased with increase in depth, ranging from 0.59 to 2.0. A value less than 1 was obtained for stains found in parts of cracks and in root channels in the deeper layers of HA and DG columns, while a value greater than 1.7 was obtained for large stains throughout the OA column. On the other hand, Dpe was estimated to be almost the same through each column.

An empirical equation relating the Brenner number, B, for chloride breakthrough as a function of macroporosity, θ_m, and the depth averages of $\bar{D}s(2)$ and $\bar{D}pe$ was obtained.

$$B = a\,\theta_m^{\bar{D}s(2)} + b\,\bar{D}pe + c \tag{6.36}$$

where a, b, and c are constants. When the average value of Ds(2) obtained in the lower half of each column was used, the correlation coefficient was 0.995. This equation suggests that macropore flow contributes significantly to miscible displacement and that the contribution depends on the internal structure of the pores and on the smoothness of their boundaries.

An Explanation of Scale-Dependent Dispersivity Based on Fractal Geometry

Field-measured dispersivities have been shown to be scale-dependent (see Section 6.3), that is, a tracer test conducted over a longer travel path will yield a larger dispersivity value than a tracer test conducted in the same geologic formation over a shorter travel path. Wheatcraft and Tyler (1988) explained scale-dependent dispersivities in heterogenous aquifers using concepts of fractal geometry. From fractal scaling relationships and expressions for concentration variance for the classical convection-dispersion theory, they developed fractal expressions for field-measured dispersivities for two cases: dispersion in a single fractal stream tube in which a particle moves along a fractal path, and dispersion in a set (or bundle) of stream tubes. The equations predict that the field-measured dispersivity will be scale-dependent. The strength of the scale-dependence is directly related to the fractal dimension of the stream tube, D. For the single stream tube model, the field-measured dispersivity is proportional to the straight-line travel distance raised to a power of (D-1). For the set of stream tubes it is proportional to the straight-line travel distance raised to a power of (2D-1). Consequently, in the case where the stream tube is not fractal (D=1), the field-measured dispersivity is constant and equivalent to the traditional laboratory dispersivity, and for a non-fractal set of stream tubes the dispersivity is shown to grow linearly with distance. Comparison with field data indicated that most field systems exhibit behaviors similar to a set of fractal stream tubes.

Wheatcraft and Tyler (1988) also developed a fractal random walk model to examine solute transport in a fractal Lagrangian system and to verify and elaborate the theoretical results. They concluded: 1) The computer model and the equations show that it is possible to obtain the fractal dimension from a tracer experiment in which the breakthrough curve (and hence the concentration variance) is obtained for at least two travel distances. 2) The fractal dimension obtained from these measurements and equations will be related directly to the fractal pathways taken by the tracer. 3) Although these pathways are obviously related to aquifer heterogeneity, it remains to be determined whether the fractal dimension of the hydraulic conductivity distribution is equivalent to the dimension of the fractal path distribution. 4) As long as the fractal dimension remains constant, it is possible to predict solute migration at scales larger than the scale of observation in which the fractal dimension and the dispersivity were obtained. This is the most important potential advantage to using a fractal approach to field-scale solute transport.

ACKNOWLEDGMENTS

Figures 6.3, 6.4, 6.8, 6.12, 6.13, and 6.21 are from *Soil Science Society of America Proceedings or Journal.* Figures 6.5, 6.10, and 6.25 are from *Soil Science* (© Williams and Wilkins). Figures 6.9 and 6.24 are from *Geoderma.* Figures 6.2 and 6.6 are from *JSIDRE* (© The Japanese Society for Irrigation, Drainage and Reclamation Engineering). Figures 6.14, 6.15, 6.17, 6.18, and 6.19 are from *Water Resources Research.* Figures 6.22 and 6.23 are from © Diamond, Inc., Tokyo. Figure 6.16 is from *Hilgardia.* Figure 6.20 is from "Dynamics of Fluids in Porous Media" © Elsevier, New York.

REFERENCES

Anderson, J., and J. Bouma, Relationships between saturated hydraulic conductivity and morphometric data of an argillic horizon, *Soil Sci. Soc. Am. Proc.* 37:408-413 (1973).

Anderson, J., and A. M. Shapiro, Stochastic analysis of one-dimensional steady state unsaturated flow: A comparison of Monte Carlo and perturbation methods, *Water Resour. Res.* 19:121-133 (1983).

Anderson, M., Modelling of groundwater flow systems as they relate to the movement of contaminants, *CRC Critical Rev. Environ. Control* 9:97-156 (1979).

Armstrong, A. C., On the fractal dimensions of some transient soil properties, *J. Soil Sci.* 37:641-652 (1986).

Baker, F. G., Factors influencing the crust test for *is situ* measurement of hydraulic conductivity, *Soil Sci. Soc. Am. J.* 41:1029-1032 (1977).

Baker, F. G., and J. Bouma, Variability of hydraulic conductivity in two subsurface horizons of two silt loam soils, *Soil Sci. Soc. Am. J.* 40:219-222 (1976).

Barry, D. A., and G. Sposito, Analytical solution of a convection-dispersion model with time-dependent transport coefficients, *Water Resour. Res.* 25:2407-2416 (1989).

Baveye, P., and G. Sposito, The operational significance of the continuum hypothesis in the theory of water movement through soils and aquifers, *Water Resour. Res.* 20:521-530 (1984).

Bear, J., Hydrodynamic dispersion, *In* R. J. M. De Wiest (ed.) *Flow Through Porous Media*, Academic Press, New York, pp. 109-199 (1969).

Bear, J., *Dynamics of Fluids in Porous Media*, Elsevier, New York (1972).

Bear, J., and Y. Bachmat, Hydrodynamic dispersion in non-uniform flow through porous media, taking into account density and viscosity differences, Hydraulic Lab, Technion, Haifa, Israel, *IASH, P.N.* 4/66 (1966). (Hebrew with English summary)

Bear, J., and Y. Bachmat, A generalized theory on hydraulic dispersion in porous media, IASH Symp. Artificial Recharge and Management of Aquifers, Haifa, Israel, *IASH, P.N.* 72:7-16 (1967).

Beese, F., and P. J. Wierenga, The variability of the apparent diffusion coefficient in undisturbed soil columns, *Z. Pflanzenernähr. Bodenkd.* 146:302-315 (1983).

Biggar, J. W., and D. R. Nielsen, Miscible displacement and leaching phenomena, *Adv. Agron.* 11:254-274 (1967).

Biggar, J. W., and D. R. Nielsen, Spatial variability of the leaching characteristics of a field soil, *Water Resour. Res.* 12:78-84 (1976).

Bjerg, P. L., K. Hinsby, T. H. Christensen, and P. Gravesen, Spatial variability of hydraulic conductivity of an unconfined sandy aquifer determined by a mini slug test, *J. Hydrology* 136:107-122 (1992).

Bouloutas, E. T., Improved numerical methods for modeling flow and transport processes in partially saturated porous media, Ph.D. thesis, Dep. of Civ. Eng., Mass. Inst. of Technol., Cambridge, MA (1989).

Bouma, J., and J. L. Anderson, Relationships between soil structure characteristics and hydraulic conductivity, *In* R. R. Bruce et al. (ed.) *Field Soil Water Regime, Soil Sci. Soc. Amer. Spec. Publ.* no. 5 (1973).

Bouma, J., and J. L. Anderson, Water and chloride movement through soil columns simulating pedal soils, *Soil Sci. Soc. Am. J.* 41:766-770 (1977).

Bouma, J., A. G. Jongmans, A. Stein, and G. Peek, Characterizing spatially variable hydraulic properties of a boulder clay deposit in The Netherlands, *Geoderma* 45:19-29 (1989).

Burd, J. S., and J. C. Martin, Water displacement of soil and the soil solution, *J. Agr. Sci.* 13:265-295 (1923).

Butters, G. L., and W. A. Jury, Field scale transport of bromide in an unsaturated soil, 2. Dispersion modeling, *Water Resour. Res.* 25:1582-1588 (1989).

Childs, E. C., and N. Collis-George, The permeability of porous materials, *Proc. Roy. Soc. London* A201:392-405 (1950).

Chu, S-Y., and G. Sposito, A derivation of the macroscopic solute transport equation for homogeneous saturated, porous media, *Water Resour. Res.* 16:542-546 (1980).

Clark, I., *Practical Geostatistics*, Applied Science Publishers, London (1979).

Cushman, J. H., On unifying concepts of scale, instrumentation, and stochastics in the development of multiphase transport theory, *Water Resour. Res.* 20:1668-1676 (1984).

Cushman, J. H., Development of stochastic partial differential equations for subsurface hydrology, *Stochastic Hydrol. Hydraul.* 1:241-262 (1987).

Dagan, G., Stochastic modeling of groundwater flow by unconditional and conditional probabilities, 2. The solute transport, *Water Resour. Res.* 18:835-848 (1982).

Dagan, G., Stochastic transport in heterogenous porous formations, *J. Fluid Mech.* 145:151-177 (1984).

Dagan, G., Statistical theory of groundwater flow and transport: pore to laboratory, laboratory to formation, and formation to regional scale, *Water Resour. Res.* 22:120-134 (1986).

Dagan, G., Theory of solute transport by groundwater, *Ann. Rev. Fluid Mech.* 19:183-215 (1987).

Dagan, G., Time-dependent macrodispersion for solute transport in anisotropic heterogenous aquifers, *Water Resour. Res.* 24:1491-1500 (1988).

Dagan, G., and E. Bresler, Solute dispersion in unsaturated heterogeneous soil at field scale, 1. Theory, *Soil Sci. Soc. Am. J.* 43:461-467 (1979).

Dagan, G., and E. Bresler, Unsaturated flow in spatially variable fields, I. Derivation of models of infiltration and redistribution, *Water Resour. Res.* 19:413-420 (1983).

Day, P. R., Dispersion of a moving salt-water boundary advancing through saturated sand, *Trans. Am. Geophys. Union* 37:595-601 (1956).

Day, P. R., and W. M. Forsythe, Hydrodynamic dispersion of solutes in the soil moisture stream, *Soil Sci. Soc. Am. Proc.* 21:477-480 (1957).

de Josselin de Jong, G., Longitudinal transverse diffusion in granular deposits, *Trans. Am. Geophys. Union* 39:67-74 (1958).

Dyson, J. S., and R. E. White, A simple predictive approach to solute transport in layered soil, *J. Soil Sci.* 40:525-542 (1989).

Espedy, B., An analysis of saturated hydraulic conductivity in a forested glacial till slope, *Soil Sci.* 150:485-494 (1990).

Feder, J., *Fractals*, Plenum Press, New York (1988).

Ferguson, H., and W. H. Gardner, Diffusion theory applied to water flow data obtained using gamma ray adsorption, *Soil Sci. Soc. Am. Proc.* 27:243-246 (1963).

Freeze, R. A., A stochastic-conceptual analysis of one-dimensional groundwater flow in nonuniform homogeneous media, *Water Resour. Res.* 11:725-741 (1975).

Freyberg, D. L., A natural gradient experiment on solute transport in a sand aquifer, 2. Spatial moments and the advection and dispersion of nonreactive tracers, *Water Resour. Res.* 22:2031-2046 (1986).

Fried, J. J., *Groundwater Pollution*, Elsevier, Amsterdam, 330 pp. (1975).

Gaillard, B., A. Lallemand-Barres, J. Molinari, and P. Peaudecerf, Etude méthodologique des caractéristiques de transfer des substances chimiques dans les nappes, 43 pp., Centre Natl. de la Rech. Sci., *Rep. SARR/FARTH1/ 76-07/JM-BG*, Orleans, France (1976).

Gardner, W. R., Some steady state solutions of the unsaturated moisture flow equation with application to evaporation from a water table, *Soil Sci.* 85:288-232 (1958).

Gelhar, L. W., and C. L. Axness, Three-dimensional stochastic analysis of macrodispersion in aquifers, *Water Resour. Res.* 19:161-170 (1983).

Gelhar, L. W., A. L. Gutjahr, and R. L. Naff, Stochastic analysis of macrodispersion in a stratified aquifer, *Water Resour. Res.* 15:1387-1397 (1979).

Gray, W. G., and S. M. Hassanizadeh, Paradoxes and realities in unsaturated flow theory, *Water Resour. Res.* 27:1847-1854 (1991a).

Gray, W. G., and S. M. Hassanizadeh, Unsaturated flow theory including interfacial phenomena, *Water Resour. Res.* 27:1855-1863 (1991b).

Gupta, V. K., and R. N. Bhatthacharya, Solute dispersion in multidimensional periodic saturated porous media, *Water Resour. Res.* 22:156-164 (1986).

Güven, O., F. J. Molz, and J. G. Melville, An analysis of dispersion in a stratified aquifer, *Water Resour. Res.* 20:1337-1354 (1984).

Güven, O., W. Falta, F. J. Molz, and J. G. Melville, Analysis and interpretation of single-well tracer tests in stratified aquifers, *Water Resour. Res.* 21:676-684 (1985).

Hadas, A., Deviation from Darcy's law for the flow of water in unsaturated soils, *Isr. J. Agric. Res.* 14:159-167 (1964).

Handy, L. L., An evaluation of diffusion effects in miscible displacement, *J. Petrol. Technol.* :61-63 (1959).

Harlan, R. L., An analysis of coupled heat-fluid transport in partially frozen soil, *Water Resour. Res.* 9:1314-1325 (1973).

Hasegawa, S., and T. Maeda, Applicability of Darcy's law for unsaturated flow in soils, *Trans. Jap. Soc. Irrig. Drain. Reclam. Eng.* 70:13-19 (1977). (Japanese)

Hatano, R., and H. W. G. Booltink, Using fractal dimensions of stained flow patterns in a clay soil to predict bypass flow, *J. Hydrol.* 135:121-131 (1992).

Hatano, R., N. Kawamura, J. Ikeda, and T. Sakuma, Evaluation of the effect of morphological features of flow paths on solute transport by using fractal dimensions of methylene blue staining pattern, *Geoderma* 53:31-44 (1992).

Hawley, M. E., A. A. McCuen, and T. J. Jackson, Volume:accuracy relationship in soil moisture sampling, *J. Irrig. Drain. Div., Proc. Am. Soc. Civ. Eng.* 108:1-11 (1982).

Ishiguro, M., Solute transport through hard pans: 1. Effect of vertical tubular pores made by rice roots on solute transport, *Soil Sci.* 152:432-439 (1991).

Journel, A. G., and Ch. J. Huijbregts, *Mining Geostatistics*, Academic Press, New York (1978).

Jury, W. A., Simulation of solute transport using a transfer function model, *Water Resour. Res.* 18:363-368 (1982).

Jury, W. A., W. R. Gardner, and W. H. Gardner, *Soil Physics* (5th edition), John Wiley & Sons, New York (1991).

Jury, W. A., and K. Roth, *Transfer Functions and Solute Transport Through Soil: Theory and Applications*, Berkhäuser, Basel (1990).

Jury, W. A., and G. Sposito, Field calibration and validation of solute transport models for the unsaturated zone, *Soil Sci. Soc. Am. J.* 49:1331-1341 (1985).

Jury, W. A., G. Sposito, and R. E. White, A transfer function model of solute movement through soil, I. Fundamental concepts, *Water Resour. Res.* 22:243-247 (1986).

Jury, W. A., D. Russo, G. Sposito, and H. Elabd, The spatial variability of water and solute transport properties in unsaturated soils, I. Analysis of property variation and spatial structure with statistical models, *Hilgardia* 55:(4)1-32 (1987).

Kaufman, W. J., and G. T. Orlob, An evaluation of ground-water tracers, *Trans. Am. Geophys. Union*, 37:297-306 (1956).

Keller, C. K., G. van der Kamp, and J. A. Cherry, A multiscale study of the permeability of a thick clayey till, *Water Resour. Res.* 24:2299-2317 (1989).

King, F. H., Principles and conditions of the movement of ground water, *U.S. Geol. Surv. 19th Ann. Rep., Part 2*, 59-294 (1898).

Kissel, D. E., J. T. Ritchie, and E. Burnett, Chloride movement in undisturbed swelling clay soil, *Soil Sci. Soc. Am. J.* 37:21-24 (1973).

Kitanidis, P. K. and R. W. Lane, Maximum likelihood parameter estimation of hydrologic spatial processes by the Gauss-Newton method, *J. Hydrol.* 79:53-71 (1985).

Krischer, O., Grundgesetze der Feuchtigkeitsbewegung in Trockengüten, *VDI Z.* 82:373-378 (1938).

Lapidus, L., and N. R. Amundson, Mathematics of adsorption in beds: IV. The effect of longitudinal diffusion in ion exchange and chromatographic columns, *J. Phys. Chem.* 56:984-988 (1952).

Lauren, J. G., R. J. Wagenet, J. Bouma, and J. H. M. Wosten, Variability of saturated hydraulic conductivity in a Glossaquic Hapludalf with macropores, *Soil Sci.* 145:20-28 (1988).

Lawson, D. E. and Elrick, D. E., A new method for determining and interpreting dispersion coefficients in porous media, *Proc. 2nd Symp. Fundamentals of Transport Phenomena in Porous Media*, Univ. Guelph, Guelph, Ont., pp. 753-777 (1972).

Lutz, J. F., and W. D. Kemper, Intrinsic permeability of clay as affected by clay-water interaction, *Soil Sci.* 88:83-90 (1959).

Mandelbrot, B. B., How long is the coastline of Great Britain? Statistical self-similarity and fractal dimension, *Science* 155:636-638 (1967).

Mandelbrot, B. B., *The Fractal Geometery of Nature*, W. H. Freeman, New York (1983).

Mantoglou, A., and L. W. Gelhar, Large-scale models of transient unsaturated flow systems, *Water Resour. Res.* 23:37-46 (1987a).

Mantoglou, A., and L. W. Gelhar, Capillary tension head variance, mean soil moisture content, and effective specific soil moisture capacity of transient unsaturated flow in stratified soils, *Water Resour. Res.* 23:47-56 (1987b).

Mantoglou, A., and L. W. Gelhar, Effective hydraulic conductivities of transient unsaturated flow in stratified soils, *Water Resour. Res.* 23:57-68 (1987c).

Martin, J. M., Deplacements miscibles dans les milieux poreux naturels de grande extension, *Rev. Inst. Fr. Pét.* 26:1065-1075 (1971).

Matheron, G., and G. de Marsily, Is transport in porous media always diffusive? A counter example, *Water Resour. Res.* 16:901-917 (1980).

McCord, J. T., D. B. Stephens, and J. L. Wilson, Hysteresis and state-dependent anisotropy in modeling unsaturated hillslope hydrologic processes, *Water Resour. Res.* 27:1501-1518 (1991).

Mercado, A., The spreading pattern of injected water in a permeability stratified aquifer, *Int. Assoc. Sci. Hydrol. Publ.* 72:23-36 (1967).

Miller, R. J., and P. F. Low, Threshold gradient for water flow in clay systems, *Soil Sci. Soc. Am. Proc.* 27:605-609 (1963).

Molinari, J., and P. Peaudecref, Essais conjoints en laboratoire et sur le terrain en vue d'une approche simplifée de la prévision de propagation des substances miscibles dans les aquifères réels, Paper presented at the Symposium on Hydrodynamic Diffusion and Dispersion in Porous Media, *Int. Assoc. Hydraul. Res.*, Pavia, Italy (1977).

Moutonnet, P., E. Pluyette, N. E. Mourabit, and P. Couchat, Measuring the spatial variability of soil hydraulic conductivity using an automatic neutron moisture gauge, *Soil Sci. Soc. Am. J.* 52:1521-1526 (1988).

Mualem, Y., Anisotropy of unsaturated soil, *Soil Sci. Soc. Am. J.* 48:505-509 (1984).

Naff, R. L., On the nature of the dispersive flux in saturated heterogeneous porous media, *Water Resour. Res.* 26:1013-1026 (1990).

Naff, R. L., T.-C. J. Yeh, and M. W. Kemblowski, A note on the recent natural gradient tracer test at the Borden site, *Water Resour. Res.* 24:2099-2103 (1988).

Nagahori, K., and K. Sato, On the nonuniformity of moisture content and dry density in gravelly sandy loam field, XI. The fundamental studies on the soil sampling in farm land, *Trans. Jap. Soc. Irrig. Drain. Reclam. Eng.* 36:67-73 (1971). (Japanese with English Summary)

Nerpin, S. V., and A. F. Chudnovskii, Physics of Soil, Chap. 3, Moisture films in soils, pp. 107-134, *Physico-Mathematic Lit.*, Moscow (1967); translation into English, Winter Bindery, Jerusalem (1970).

Nerpin, S. V., and A. I. Kotov, Report of Soviet Soil Science, *7th Int. Cong. Soil Sci.*, Madison, WI (1960).

Neuman, S. P., C. L. Winter, and C. M. Newman, Stochastic theory of field scale Fickian dispersion in anisotropic porous media, *Water Resour. Res.* 23:453-466 (1987).

Nielsen, D. R., and J. W. Biggar, Miscible displacement: III. Theoretical consideration, *Soil Sci. Soc. Am. Proc.* 26:216-221 (1962).

Nielsen, D. R., and J. W. Biggar, Implications of the vadose zone to water-resource management, *In* Scientific Basis of Water-Resource Management, pp. 41-50, National Academy Press, Washington, D.C. (1982).

Nielsen, D. R., J. W. Biggar, and J. M. Davidson, Experimental consideration of diffusion analysis in unsaturated flow problems, *Soil Sci. Soc. Am. Proc.* 37:522-527 (1962).

Nielsen, D. R., J. W. Biggar, and K. T. Erh, Spatial variability of field-measured soil-water properties, *Hilgardia* 42:215-259 (1973).

Nimmo, J. R., J. Rubin, and D. P. Hammermeister, Unsaturated flow in a centrifugal field: Measurement of hydraulic conductivity and testing of Darcy's law, *Water Resour. Res.* 23:124-134 (1987).

Ogata, A., and R. B. Banks, A solution of the differential equation of longitudinal dispersion in porous media, *U.S. Geol. Surv. Prof. Pap. 411A*, p. 34 (1961).

Otani, S., and S. Maeda, Mechanism of water movement in moist granular materials, *Chem. Ind.*, 28:362-367 (1964). (Japanese)

Pedro, G. T., R. A. Novy, H. T. Davis, and L. E. Scriven, Hydraulic conductivity of porous media at low water content, *Soil Sci. Soc. Am. J.* 54:673-679 (1990).

Pickens, J. F., and G. E. Grisak, Scale-dependent dispersion in a stratified granular aquifer, *Water Resour. Res.* 17:1191-1211 (1981a).

Pickens, J. F., and G. E. Grisak, Modeling of scale-dependent dispersion in hydrogeologic systems, *Water Resour. Res.* 17:1701-1711 (1981b).

Plumb, O. A., and S. Whitaker, Dispersion in heterogeneous porous media, 1. Local volume averaging and large-scale averaging, *Water Resour. Res.* 24:913-926 (1988).

Polmann, D. J., D. McLaughlin, S. Luis, L. W. Gelhar, and R. Ababou, Stochastic modeling of large-scale flow in heterogeneous unsaturated soil, *Water Resour. Res.* 27:1477-1458 (1991).

Rawlins, S. L., and W. H. Gardner, A test of the validity of the diffusion equation for unsaturated flow of soil water, *Soil Sci. Soc. Am. Proc.* 27:507-511 (1963).

Reeve, R. C. A., and D. Kirkham, Soil anisotropy and some field methods for measuring permeability, *Trans. Am. Geophys. Union* 32:582-590 (1951).

Reynolds, W. D., and D. E. Elrick, *In situ* measurement of field-saturated hydraulic conductivity, sorptivity, and the α-parameter using the Guelph Permeameter, *Soil Sci.* 140:292-302 (1985).

Reynolds, W. D., and D. E. Elrick, A method for simultaneous *in situ* measurement in the vadose zone of field-saturated hydraulic conductivity, sorptivity, and the conductivity-pressure head relationship, *Ground Water Monit. Rev.* 16:84-95 (1986).

Rieu, M., and G. Sposito, Fractal fragmentation, soil porosity, and water properties: I. Theory, *Soil Sci. Soc. Am. J.* 55:1231-1238 (1991a).

Rieu, M., and G. Sposito, Fractal fragmentation, soil porosity, and water properties: II. Applications, *Soil Sci. Soc. Am. J.* 55:1239-1244 (1991b).

Rifai, M. N. E., W. J. Kaufman, and D. K. Todd, Dispersion phenomenon in laminar flow through porous media, *Prog. Rept. 2*, Canal Seepage Research, Univ. Calif., Berkeley, 157 pp. (1956).

Rose, D. A., and J. B. Passioura, The analysis of experiments on hydrodynamic dispersion, *Soil Sci.* 111:252-257 (1971).

Russo, D., and E. Bresler, Soil hydraulic properties as stochastic processes, I. An analysis of spatial variability, *Soil Sci. Soc. Am. J.* 45:682-687 (1981).

Russo, D., and W. A. Jury, A theoretical study of the esimation of the correlation scale in spatially variable fields, I. Stationary fields, *Water Resour. Res.* 23:1257-1268 (1987).

Samper, F. J., and S. P. Neuman, Estimation of spatial covariance structures by adjoint state maximum likelihood cross validation, 1. Theory, *Water Resour. Res.* 25:351-362 (1989).

Sauty, J., An analysis of hydrodispersive transfer in aquifers, *Water Resour. Res.* 16:145-158 (1980).

Schaeffer, D. W., J. E. Martin, P. Wiltzius, and D. S. Cannell, Fractal geometry of colloidal aggregates, *Phy. Rev. Lett.* 52:2371-2374 (1984).

Scheidegger, A. E., Statistical hydrodynamics in porous media, *J. Appl. Phys.* 25:994-1001 (1954).

Scheidegger, A. E., On the theory of flow of miscible phases in porous media, *Extrait des Comptes Rendus et Rapports*, Assemblee Generale de Toronto, Tome II, pp. 236-242, Gentbrugge (1958).

Schulin, R., M. Th. van Genuchten, H. Flühler, and P. Ferlin, An experimental study of solute transport in a stony field soil, *Water Resour. Res.* 23:1785-1794 (1987).

Schwartz, F. W., Macroscopic dispersion in porous media: The controlling factors, *Water Resour. Res.* 13:743-752 (1977).

Serrano, S. E., The form of the dispersion equation under recharge and variable velocity, and its analytical solution, *Water Resour. Res.* 28:1801-1808 (1992).

Shigematsu, M., and S. Iwata, Scale dependence of unsaturated hydraulic conductivity, *Proc. Ann. Meet., Jap. Soc. Irrig. Drain. Reclam. Eng.* pp. 78-79 (1991). (Japanese)

Silliman, S. E., and E. S. Simpson, Laboratory evidence of the scale effect in dispersion of solutes in porous media, *Water Resour. Res.* 23:1667-1673 (1987).

Simmons, C. S., A stochastic-convective transport representative of dispersion in one-dimensional porous media systems, *Water Resour. Res.* 18:1193-1214 (1982).

Sisson, J. B., and P. J. Wierenga, Spatial variability of steady-state infiltration rates as a stochastic process, *Soil Sci. Soc. Am. J.* 45:699-704 (1981).

Skopp, J., W. R. Gardner, and E. J. Tyler, Solute movement in structured soils: Two-region model with small interactions, *Soil Sci. Soc. Am. J.* 45:837-842 (1981).

Slichter, C. S., Field measurements of the rate of movement of underground water, *U.S. Geol. Survey, Water Supply and Irrigation Paper 140*, 122 pp. (1905).

Smith, L., and F. W. Schartz, Mass transport, 1. A stochastic analysis of macroscopic dispersion, *Water Resour. Res.* 16:303-313 (1980).

Soil Survey Staff, *Soil Taxonomy*. U.S. Dept. Agric. Handbook 436, Washington, D.C. (1975).

Sposito, G., and W. A. Jury, Group invariance and field-scale transport, *Water Resour. Res.* 22:1743-1748 (1986).

Sposito, G., W. A. Jury, and V. K. Gupta, Fundamental problems in the stochastic convection-dispersion model of solute transport in aquifers and field soils, *Water Resour. Res.* 22:77-88 (1986).

Steenhuis, T. S., J.-Y. Parlange, and M. S. Andreini, A numerical model for preferential solute movement in saturated soils, *Geoderma* 46:193-208 (1990).

Stephens, B. D., and S. Heerman, Dependence of anisotropy on saturation in a stratified sand, *Water Resour. Res.* 24:770-778 (1988).

Sudicky, E. A., A natural gradient experiment on solute transport in a sandy aquifer: Spatial variability of hydraulic conductivity and its role in dispersion processes, *Water Resour. Res.* 22:2069-2082 (1986).

Sudicky, E. A., J. A. Cherry, and E. O. Frind, Migration of contaminants in ground water at a landfill: A case study, 4. A natural-gradient dispersion test, *J. Hydrol.* 63:81-108 (1983).

Sudicky, E. A., R. W. Gillham, and E. O. Frind, Experimental investigation of solute transport in stratified porous media, 1. The nonreactive case, *Water Resour. Res.* 21:1035-1041 (1985).
Swartzendruber, D., Modification of Darcy's law for the flow of water in soils, *Soil Sci.* 93:22-29 (1962).
Swartzendruber, D., Non-Darcy behavior and the flow of water in unsaturated soils, *Soil Sci. Soc. Am. Proc.* 27:491-495 (1963).
Takayasu, H., and M. Takayasu, *What Is Fractal?*, Diamond, Inc., Tokyo (1988). (Japanese)
Taylor, G. I., The dispersion of matter in solvent flowing slowly through a tube, *Proc. Roy. Soc. London, Ser. A* 219:189-203 (1953).
Thames, J. L., and D. D. Evans, An analysis of the vertical infiltration of water into soil columns, *Water Resour. Res.* 14:817-828 (1968).
Thompson, A. F. B., and W. G. Gary, A second-order approach for modeling dispersive transport in porous media, 1. Theoretical development, *Water Resour. Res.* 22:591-599 (1986).
Tyler, S. W., and S. W. Wheatcraft, Fractal processes in soil water retention, *Water Resour. Res* 26:1047-1054 (1990).
Tyler, S. W., and S. W. Wheatcraft, Fractal scaling of soil particle-size distributions: Analysis and limitations, *Soil Sci. Soc. Am. J.* 56:362-369 (1992).
Ünlü, K., D. R. Nielsen, J. W. Biggar, and F. Morkoc, Statistical parameters characterizing the spatial variability of selected soil hydraulic properties, *Soil Sci. Soc. Am. J.* 54:1537-1547 (1990).
van Genuchten, M. Th., A closed-form equation for predicting the hydraulic conductivity of unsaturated soils, *Soil Sci. Soc. Am. J.* 44:892-898 (1980).
van Genuchten, M. Th., and P. J. Wierenga, Mass transfer studies in sorbing porous media, I. Analytical solutions, *Soil Sci. Soc. Am. J.* 40:473-480 (1976).
van Genuchten, M. Th., and P. J. Wierenga, Mass transfer studies in sorbing porous media, II. Experimental evaluation with tritium, *Soil Sci. Soc. Am. J.* 41:272-278 (1977).
van Genuchten, M. Th., P. J. Wierenga, and G. A. O'Connor, Mass transfer studies in sorbing porous media, III. Experimental evaluation with 2,4,5-T, *Soil Sci. Soc. Am. J.* 41:278-285 (1977).
Van Wesenbeeck, I. J., and R. G. Kachanoski, Spatial scale dependence of *in situ* solute transport, *Soil Sci. Soc. Am. J.* 55:3-7 (1991).
Vauclin, M., Méthodes d'étude de la variabilité spatiale des propriétés d'un sol, *In* Variabilité Spatiale des Processus de Transfert dans les Sols, Ed. INRA, Paris (1982).
Viera, S. R., D. R. Nielsen, and J. W. Biggar, Spatial variability of field measured infiltration rates, *Soil Sci. Soc. Am. J.* 45:1040-1048 (1981).

Von Engelhardt, W., and W. L. Tunn, The flow of fluids through sandstones, III., *State Geol. Surv. Circ. 194* (1955).

Warren, J, E., and F. F. Skiba, Macroscopic dispersion, *Soc. Pet. Eng. J.* 4:215-230 (1964).

Warrick, A. W., J. W. Biggar, and D. R. Nielsen, Simultaneous solute and water transfer for an unsaturated soil, *Water Resour. Res.* 7:1216-1225 (1971).

Warrick, A. W., G. J. Mullen, and D. R. Nielsen, Scaling field measured soil hydraulic properties using a similar media concept, *Water Resour. Res.* 13:355-362 (1977).

Warrick, A. W., and D. R. Nielsen, Spatial variability of soil physical properties in the field, *Applications of Soil Physics*, D. Hillel (ed.), Academic Press, London, pp. 319-344 (1980).

Watson, K. K., An instantaneous profile method for determining the hydraulic conductivity of unsaturated porous materials, *Water Resour. Res.* 12:709-715 (1966).

Webster, R., Quantitative spatial analysis of soil in the field, *In* Advances in Soil Science, Vol. 3, B. A. Stewart (ed.), Springer-Verlag, New York, pp. 1-71 (1985).

Weeks, L. V., and S. J. Richards, Soil water properties computed from transient flow data, *Soil Sci. Soc. Am. Proc.* 31:721-725 (1967).

Wheatcraft, S. W. and S. W. Tyler, An explanation of scale-dependent dispersivity in heterogenous aquifers using concepts of fractal geometry, *Water Resour. Res.* 24:566-578 (1988).

White, R. E., G. W. Thomas, and M. S. Smith, Modelling water flow through undisturbed soil cores using a transfer function model derived from ^{3}HOH and Cl transport, *J. Soil Sci.* 35:159-168 (1984).

Wilson, G. V., J. M. Alfonsi, and P. M. Jardine, Spatial variability of saturated hydraulic conductivity of the subsoil of two forested watersheds, *Soil Sci. Soc. Am. J.* 53:679-685 (1989).

Yates, S. R., An analytical solution for one-dimensional transport in heterogeneous porous media, *Water Resour. Res.* 26:2331-1338 (1990). [Correction, *Water Resour. Res.* 27:2167 (1991).]

Yeh, T.-C., L. W. Gelhar, and A. L. Gutjahr, Stochastic analysis of unsaturated flow in heterogeneous soils, 1. Statistically isotropic media, *Water Resour. Res.* 21:447-456 (1985a).

Yeh, T.-C., L. W. Gelhar, and A. L. Gutjahr, Stochastic analysis of unsaturated flow in heterogeneous soils, 2. Statistically anisotropic media with variable α, *Water Resour. Res.* 21:457-464 (1985b).

Yeh, T.-C., L. W. Gelhar, and A. L. Gutjahr, Stochastic analysis of unsaturated flow in heterogeneous soils, 3. Observations and applications, *Water Resour. Res.* 21:465-471 (1985c).

7

FIELD WATER REGIMES

7.1 WATER IN UPLAND SOILS

Volcanic ash soils are distributed widely in Japan, occupying about 60% of the total area of upland soils. The unique physical and chemical properties that influence the field water regime of volcanic ash soils are due mainly to the presence of allophane. Many measurements have been made of water movement in volcanic ash soils, especially in Kantō loam, a typical volcanic ash soil in Japan.

Kantō loam is described as (Ministry of Agriculture and Forests, Government of Japan, 1964; Yamazaki et al., 1963): a surface soil horizon usually 30 to 50 cm thick and very rich in humus (15 to 20%); its color is close to pitch black; the surface soil has a granular structure; the B-C horizon (subsoil) is comparatively light brown, and the humus content is 2 to 3%; the thickness of the subsoil is 70 to 100 cm, and the structure is massive.

7.1.1 Physical Properties

Table 7.1 shows representative grain size and specific surface areas of Utsunomiya volcanic ash soil (Kantō loam) and Iwatagahra red yellow soil (Diluvial) (Tada et al., 1963). The dominant clay minerals in the Iwatagahara soil are 14 Å and kaolin minerals. Volcanic ash soils have large specific surface areas due to the presence of allophane.

The total porosity of nonvolcanic ash soils ranges from 30 to 60%, decreasing slightly with depth. The total porosity of Kantō loam is about 80%, both in the surface soil and in the subsoil. The subsoil has a very high total porosity in spite of its low humus content of 2 to 3%.

Tabuchi (1963) has studied the distribution of large pores (2.0 to 0.01 mm) in these soils by microscopic observation. Some of the results are shown in Table 7.2. Both the surface soil and subsoil of the Kantō loam have a greater

Table 7.1. Grain size and specific surface area of Utsunomiya and Iwatagahara soils. (From Tada et al., 1963)

Soil	Depth (cm)	Grain size, mm, %					Specific surface area ($m^2\ g^{-1}$)
		2.0-0.2	0.2-0.1	0.1-0.02	0.02-0.002	<0.002	
Utsunomiya	15	17.9	6.9	7.8	28.7	38.9	320
(Kantō loam)	100	4.4	6.2	10.4	38.1	40.9	330
Iwatagahara	10	10.1	3.4	31.8	22.8	31.9	76
(red-yellow soil)	40	10.4	2.3	32.3	23.9	31.1	86

Table 7.2. Pore distribution of soils from microscopic observations. (From Tabuchi et al., 1963)

Soil	Depth (cm)	Number of pores, cm^{-2} 2.0-1.0 mm	1.0-0.5 mm	0.5-0.25 mm	0.25-0.01 mm	Structure	Interconnection of pores
Utsunomiya	15	0	14	81	94	Granular	Good
(Kantō loam)	100	0	11	85	112	Massive	Poor
Iwatagahara	5	2	5	18	90	Massive	Poor
(red-yellow soil)	30	0	5	23	58	Massive	Poor

number of large pores than are found in the red-yellow soil. Most of the large pores in the surface horizon of the Kantō loam are interconnected.

Misono et al. (1953), Misono and Terasawa (1957), and Terasawa (1963) have clarified the water retention characteristics of volcanic ash soils (Figs. 7.1 and 7.2). The high water retention above pF 4.2 is due to the large specific surface area, while the irreversibility of water retention on air drying is due to physicochemical properties of allophane.

The values of saturated hydraulic conductivity of some soils in Japan are shown in Table 7.3. Although light clay soils generally have conductivities of less than 10^{-4} cm/sec, volcanic ash soils have conductivities greater than 10^{-2} cm/sec. Water moves downward easily in volcanic ash soils. This difference in hydraulic conductivity shows the importance of measuring hydraulic conductivity of soils in planning irrigation and drainage. Sometimes it becomes necessary to decrease the hydraulic conductivity of volcanic ash soils.

Nonuniformity of hydraulic conductivity has been investigated by Nagata (1971). The coefficient of variability is over 40% and is larger for paddy fields than for upland fields. It would require over 100 samples for an accuracy of ±10%. The unsaturated hydraulic conductivity of two representative soils is shown in Fig. 7.3.

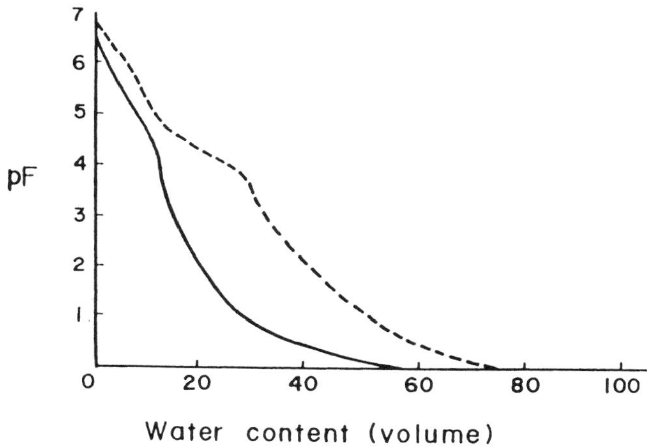

Fig. 7.1. Water retention curves for a volcanic ash soil (dashed line) and a red-yellow soil (solid line). (From Terasawa, 1963)

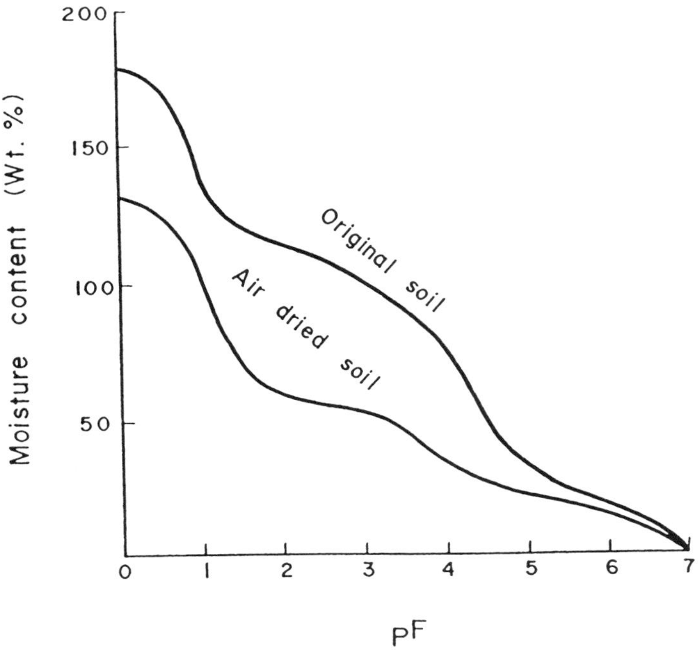

Fig. 7.2. Change of water retention due to air-drying of a volcanic ash soil. (From Misono et al., 1953)

Table 7.3. Values of saturated hydraulic conductivity of some soils in Japan.

Soil	Texture	Saturated hydraulic conductivity (cm/sec)
Volcanic ash soils		
Utsunomiya	LiC	$4 - 8 \times 10^{-2}$
Nodai	LiC	$6 - 13 \times 10^{-2}$
Non-volcanic-ash soils		
Matsuyama	SL	$1 - 4 \times 10^{-4}$
Nonoichi	CL	$5 - 10 \times 10^{-6}$
Kumamoto	LiC	1×10^{-6}
Akita	LiC	1×10^{-4}
Aomori	HC	6×10^{-6}
Akita	HC	$3 - 5 \times 10^{-6}$

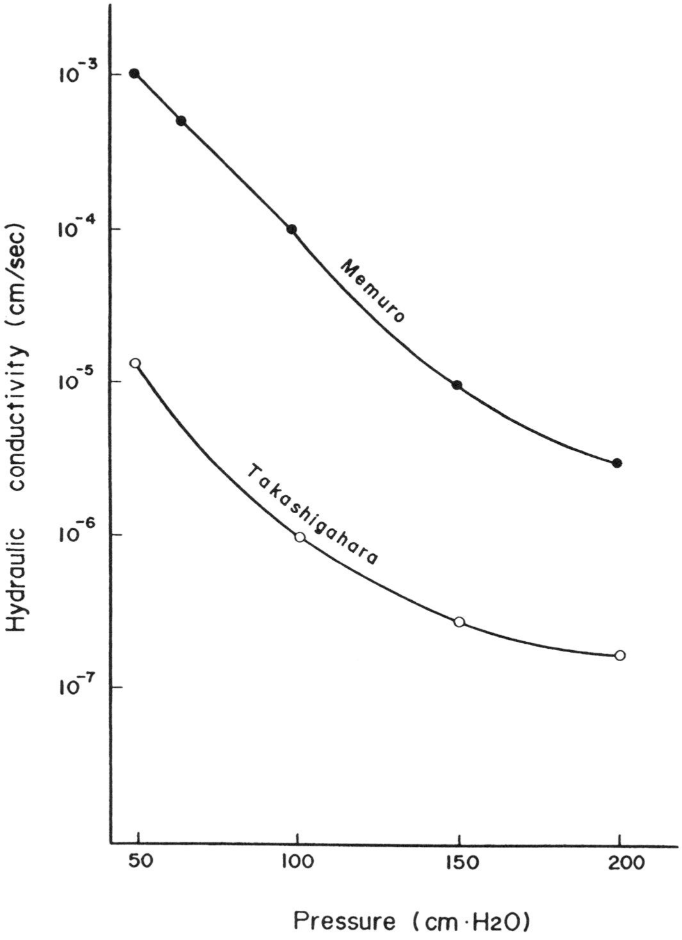

Fig. 7.3. Unsaturated hydraulic conductivity of Takashigahara, a red-yellow soil, and Memuro, a volcanic ash soil. (From Hasegawa, 1963)

7.1.2 Water Regimes in Upland Soils in Japan

Hayasaka (1978) measured the distribution of water content and soil suction in a volcanic ash soil (a mulberry field) at Kumamoto on Kyūshū Island over a period of 2 years. The porosity and texture of the soil are shown in Fig. 7.4. Monthly rainfall at Kumamoto in 1971 is given in Table 7.4. Figure 7.5 shows

Table 7.4. Monthly rainfall at Kumamoto in 1971.

Month	Rainfall (mm)	Month	Rainfall (mm)
1	41	7	336
2	61	8	287
3	73	9	148
4	63	10	110
5	252	11	22
6	494	12	54

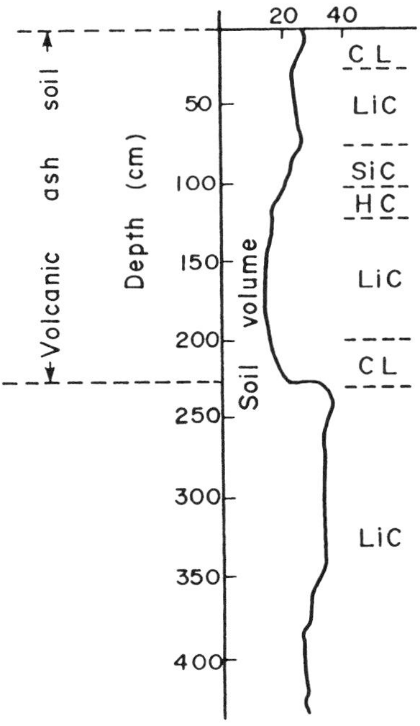

Fig. 7.4. Changes of soil volume and texture with depth for a volcanic ash soil. (From Hayasaka, 1978)

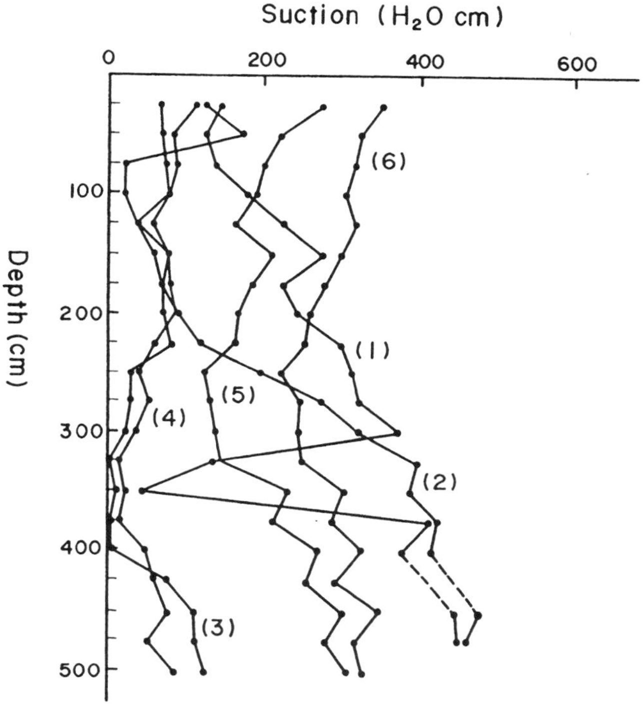

Fig. 7.5. Water regime of a volcanic ash soil in 1971. Water suction on: (1) Feb. 10; (2) April 22; (3) June 14; (4) Aug. 10; (5) Oct. 2; and (6) Dec. 10. (From Hayasaka, 1978)

some of the observed results in 1971. The water suction at 100 cm did not exceed 200 cm H_2O except in December. The same result was obtained by Kira et al. (1963) for a Kantō loam.

In 1967, Kyūshū Island experienced a one in 73 year drought. Kitajima and Hashimoto (1967) measured the distribution of water content and suction in a volcanic ash soil at Kumamoto during that period (Fig. 7.6). There was no rain during the period from August 14 to September 27, but water suction at a depth of 90 cm on September 27 did not exceed 400 cm H_2O.

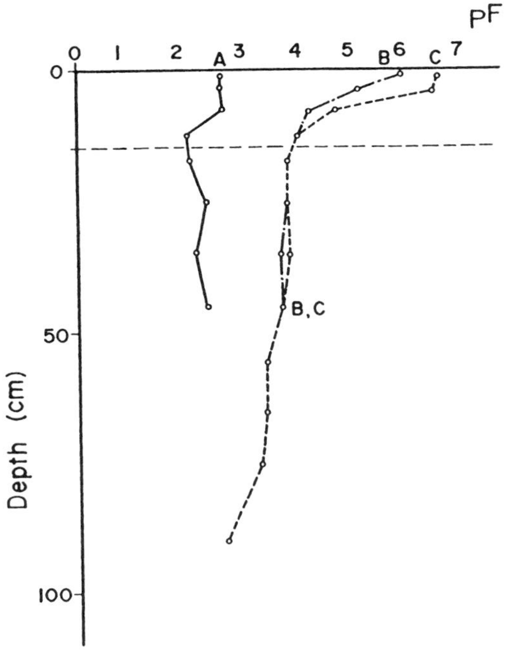

Fig. 7.6. Water regime of a volcanic ash soil during a drought in 1967. (A) Aug. 14; (B) Sept. 21; and (C) Sept. 27. (From Kitajima and Hashimoto, 1967)

7.1.3 Water Movement in Upland Soils

Field moisture capacity, the water content after natural drainage, is determined not only by the water characteristics of the soil and the layers of the soil profile, but also by the water condition of the subsoil. Table 7.5 lists the values of soil suction corresponding to field capacities of several soils in Japan. These values are lower than those obtained by researchers in arid regions, probably due to the wet subsoils in Japan.

Terasawa (1962), using soil columns of a volcanic ash soil, showed that a considerable amount of water was transferred from the subsoil to the surface soil during a dry season. Kira et al. (1963) obtained the same result. They placed a sheet of iron horizontally into a test plot at a depth of 45 cm to disrupt water transfer from the subsoil to the surface soil, and measured the changes of water suction in this plot (plot A) and in another test plot (plot B) without the cutoff. Both plots were protected from rainfall. The results are shown in Fig. 7.7.

Table 7.5. pF values corresponding to field moisture capacity of several soils in Japan.

Volcanic ash soils	Non-volcanic-ash soils
pF 2.0 for Kumamoto (Kitajima and Hashimoto, 1967)	pF 1.6 for Taketoyo (Yokoi, 1965)
pF 1.8 for Saitama (Arai, 1976)	pF 1.5 for Aichi (Iwata, 1966)
	pF 1.5 for Takashigahara (Iwata, 1966)
	pF 1.4 for Kashima (Hasegawa, 1963)

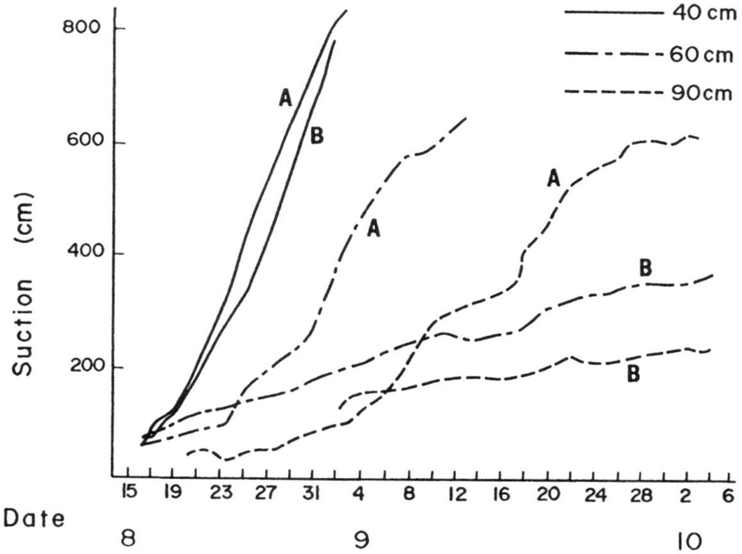

Fig. 7.7. Effect of access to subsoil moisture on water suction in surface soil. (A) subsoil moisture cut off; (B) subsoil moisture available. (From Kira et al., 1963)

Although the upland rice in plot A exhibited signs of wilting by mid-September, plot B did not show signs of wilting until harvesting began at the end of October. The suction at 60 and 90 cm increased less rapidly in plot A than in plot B. Water must have been transferred from the subsoil to the surface soil in considerable amounts.

Nakano (1970) found that the amount of water moving upward from the subsoil to the surface layer was about one-third of evapotranspiration. Nakajimada (1976) observed that water suction at a depth of 45 cm in a bare red-yellow soil did not exceed 100 cm H_2O even with low rainfall. All these results indicate that a considerable quantity of water moves up from the subsoil.

7.2 WATER BALANCE IN RICE PADDY FIELDS

Water balance in a paddy field (Fig. 7.8) has inputs of surface inflow S_i and rainfall R, and outputs by evapotranspiration ET, surface outflow S_o, and percolation P:

$$\Delta H = (S_i - S_o) + R - ET - P \tag{7.1}$$

Figure 7.9 shows the seasonal changes measured in two paddy fields. Surface inflow from irrigation water continued from the seedling transplanting period in May to the drainage period in September. Rainfall and evapotranspiration are seasonal. Percolation in paddy field (a) is low due to the clayey soil. But in field (b) percolation is large during the first stage of irrigation in May due to the pervious soil. Surface outflow in May is due to drainage for rice planting and that in July is due to midsummer drainage.

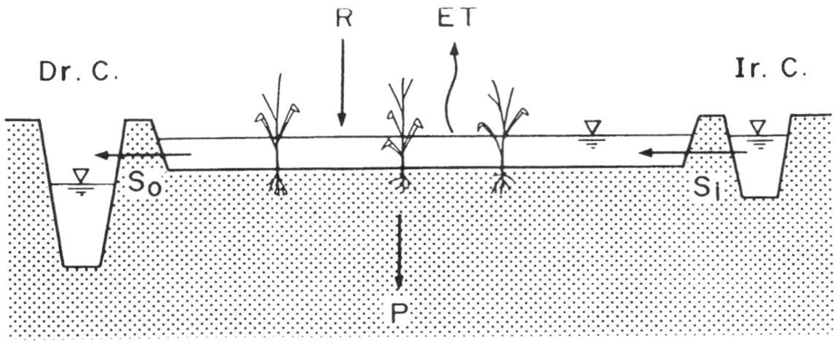

Fig. 7.8. Input and output of water in a paddy field, showing irrigation and drainage canals.

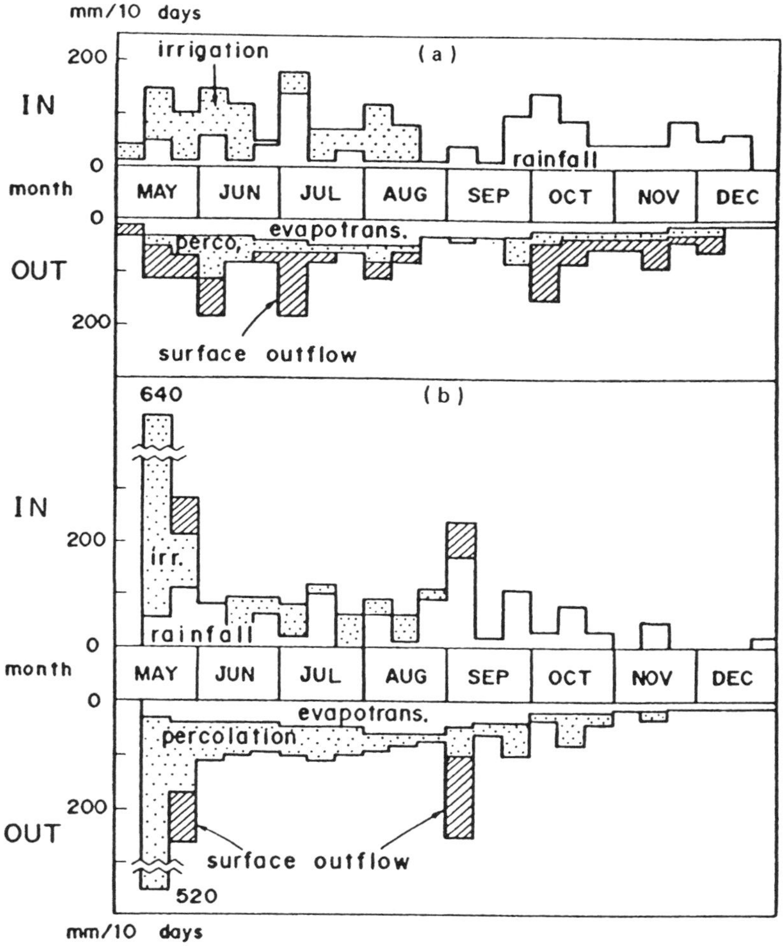

Fig. 7.9. Examples of the seasonal changes of input and output of water in two types of paddy fields: (a) high groundwater table; (b) low groundwater table.

7.2.1 Types of Paddy Fields

A standard paddy field in Japan is surrounded by borders; on one side there is an irrigation canal with an inlet into the paddy, and on the other side a drainage canal with an outlet from the paddy. Other types of paddy fields include those with underdrains, those in which water is irrigated from plot to plot without a canal, and those with one canal that is used for both irrigation and drainage (Yamazaki, 1988).

The drainage of paddy fields is determined by their position in the landscape. On low alluvial plains or in valley bottoms the groundwater level is close to the soil surface (Fig. 7.10). Percolation rates are low. High groundwater also results where the water level in the drainage canals is raised to decrease seepage losses.

In paddy fields on hills or alluvial cones, the groundwater level is much lower than the surface of the paddy fields (Fig. 7.11). In these cases, water flows through the subsoil in an unsaturated condition with negative water pressure. In Japan, there are many paddy fields on sloping land. Such paddy fields are called "Tanada." Water flow is controlled by the slope of the impervious layer; water from the upper part of the slope is often reused in paddy fields in the lower part.

7.2.2 Water Requirement in Paddy Fields

Water requirement in a paddy field, W_{ri}, is defined as

$$W_{ri} = ET + P + S_o \tag{7.2}$$

The total water requirement in a given area of paddy fields, W_r, is

$$W_r = \Sigma W_{ri} - R_u + L_o \tag{7.3}$$

where R_u is reused water and L_o is loss in canals. Substituting Eq. (7.2) into Eq. (7.3), we get

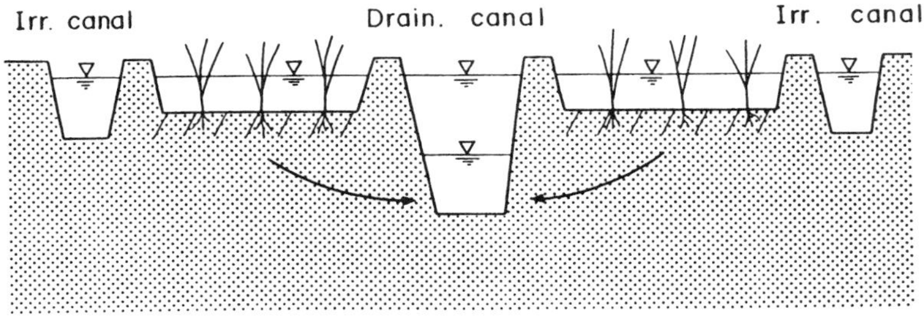

Fig. 7.10. Paddy fields with a high groundwater level.

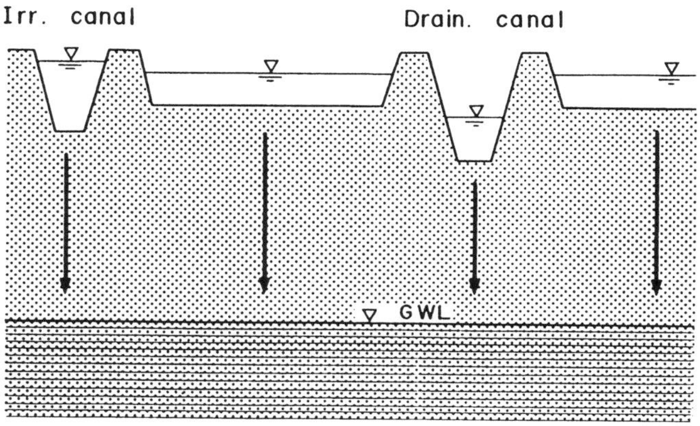

Fig. 7.11. Paddy fields with a low groundwater level.

$$W_r = \Sigma ET + \Sigma P + \Sigma S_o - R_u + L_o \quad (7.4)$$

In Japan, P and S_o are the main variable factors determining W_r, because the value of ET is almost constant during the irrigation period. The factor of reused water is important for the water requirement in an area. If water is perfectly reused by pumping in a high groundwater district, W_r will be obtained approximately by

$$W_r \doteqdot \Sigma\ ET \quad (7.5)$$

Therefore,

$$\Sigma\ P + \Sigma\ S_o + L_o \doteqdot R_u \quad (7.6)$$

The irrigation requirement is calculated as

$$I_{ri} = W_r - R_e \quad (7.7)$$

where R_e is effective rainfall.

In addition to these factors, water is required to saturate the soil for puddling before rice planting. Puddling-water requirement W_p can be calculated as

$$W_p = W_{ri} + \Delta H_p + \Delta S \quad (7.8)$$

$$\Delta S = L \times n \times (1 - S_w) \tag{7.9}$$

where ΔH_p is the water depth for puddling work, ΔS the water volume required to saturate the soil, L the depth of the surface soil layer, n the porosity, and S_w the saturation ratio before puddling. The amount of water required for puddling is about 100 to 150 mm in Japan (Nakagawa, 1967).

Water loss from percolation and evapotranspiration in a paddy field is usually measured daily as water depth, and is called the "daily decreased water depth" (mm/day or cm/day). Evapotranspiration is fairly constant in Japan, both by region and over the season. It ranges from 4 to 6 mm/day, and averages 4.9 mm/day (Nakagawa, 1967). ET is somewhat larger in August. In southeast Asia, values of ET range from 4 to 9 mm/day. Percolation rate varies widely and is affected by factors such as texture of the soil, depth of the topsoil, and groundwater level. Sato et al. (1975), in their investigation of 530 paddy fields, found an average water depth decrease of 7.9 mm/day, with a large coefficient of variation of 42%.

Puddling has a large effect in decreasing water loss (Masujima, 1966; Kawasaki, 1975). Underdrains increase water loss. After soil reformation into well-drained paddy fields, water loss increases (Kanō et al., 1961).

7.2.3 Percolation and Rice Growth

Percolation of water through the subsoil in rice paddy fields is an important management factor for rice culture in Japan. Excess percolation causes a large water requirement, injury due to low water temperature, and fertilizer loss. But adequate percolation is said to have some benefits, such as ensuring air supply to the soil and eliminating harmful materials (Ishihara, 1967). Isozaki (1957) has shown that maximum rice yields occur at percolation loss values in the range of 10 to 20 mm/day. Terasawa and Ueda (1970) found much lower optimum values in their experiment with various types of paddy field soils in lysimeters. No difference in plant growth was observed where the percolation rate was more than 1 mm/day. The poorest plant growth was at a percolation rate of 0.2 mm/day. Since suitable percolation conditions for rice growth differ according to the type of soil and water quality, it is necessary to establish rates for local conditions.

Optimum percolation rate was discussed by many researchers in the workshops on "Physical Aspects of Soil Management in Rice-Based Cropping Systems" at the International Rice Research Institute (Tabuchi et al., 1986) and on "Soil and Water Engineering for Paddy Field Management" at the Asian Institute of Technology (Woodhead, 1985; Adachi, 1992; Humphreys et al., 1992; Inoue, 1992; Koga, 1992; Murty and Koga, 1992). It was concluded that percolation may not be essential for wetland rice under tropical conditions having organic matter lower than 5% and not receiving high additions of organic amendments.

7.2.4 Soil Properties of Paddy Fields

The soil in a paddy field has a cultivated layer of surface soil, an impervious layer, and the subsoil. Roots of the rice plant grow in the topsoil, which is usually 10 to 20 cm deep. The surface soil goes through several wetting and drying cycles each year. Before planting, the saturated surface soil is mechanically puddled or remolded. The hydraulic conductivity of puddled clayey surface soil is small, on the order of 10^{-5} to 10^{-6} cm/sec (Table 7.6), which corresponds to about 1 to 0.1 cm/day. When the surface water is drained during the midsummer drainage period or during the harvest period, cracks appear in the soil making it more pervious. The conductivity increases to 10^{-3} cm/sec (100 cm/day) (Table 7.6).

The impervious soil below the surface soil is not affected by cultivation, and its properties are almost constant. An impervious hardpan is gradually formed by yearly cultivation and is often created by artificial compaction to prevent the percolation loss of water (Tabuchi et al., 1987).

Data on paddy field soils are given in Table 7.7. The data for fields 1 to 3, with a high groundwater level, indicate that the volume of air is nearly zero and that soil pores are saturated. The texture of these soils is clayey, and saturated hydraulic conductivity is very small, on the order of 10^{-5} to 10^{-7} cm/sec. Paddy fields 4 and 5 have a low groundwater level. The former is on an alluvial cone where the subsoil consists of gravel, and has an air volume of about 10%. The latter is a paddy field on a diluvial hill, whose parent material is volcanic ash. The air volume exceeds 10%; the soil is unsaturated even during the irrigation period.

Table 7.6. Changes of physical properties of puddled clayey soil in a rice paddy field after drainage. (From Tabuchi et al., 1966c)

Property	Before drainage	After drainage
Bulk density (g/cm^3)	0.55	0.54
Porosity (%)	79	79
Saturation (%)	100	92
Hydraulic conductivity (cm/sec)	10^{-5}-10^{-6}	10^{-2}-10^{-3}

Table 7.7. Soil physical properties of rice paddy fields in Japan during the irrigation period.

Field	Soil type, place	Depth (cm)	Texture	Bulk density (g/cm^3)	Volume %			Hydraulic conductivity K_s (cm/sec)
					Solid	Water	Air	
1	Alluvial plain, Ōhama	0-20	SiC	0.95	36	64	0	10^{-4}
	(Tabuchi et al., 1966a)	20-40	SiC	1.26	49	50	1	10^{-6}-10^{-7}
2	Alluvial plain, Nagaoka	0-5	LiC	0.55	21	79	0	10^{-5}
	(Tabuchi et al., 1966c)	20-25	LiC	0.87	35	65	0	10^{-5}-10^{-6}
3	Sea bottom reclamation, Ariake	0-10	HC	0.72	28	72	0	10^{-7}
	(Nagaishi, 1959)	10-12	HC	0.79	31	69	0	10^{-7}
		40+	HC	0.55	21	79	0	10^{-7}
4	Alluvial cone, Rokugo	0-14	SL	--	--	--	--	--
	(Yamazaki et al., 1961)	14-25	SL	--	26	67	7	10^{-3}
		25-40	LiC	--	30	58	12	10^{-2}-10^{-3}
		40-60	SL	--	38	53	9	10^{-2}-10^{-3}
		60-80	SL	--	33	60	7	10^{-2}
5	Volcanic ash, Utsunomiya,	0-12	--	0.72	28	--	--	10^{-4}
	(Yamazaki et al., 1960)	12-23	--	0.58	25	65	10	10^{-3}-10^{-4}
		23-55	LiC	0.49	20	65	15	10^{-3}
		55-110	LiC	0.55	21	59	20	10^{-3}

7.2.5 Control of Percolation in Paddy Fields

Percolation rates vary widely, even in the same paddy field. Figure 7.12 shows the percolation rates at 64 points selected at intervals of 5 m, measured by the rapid percometer (Yamazaki et al., 1960). The minimum value was 0.2 cm/day and the maximum value was 35.5 cm/day. Large values were recorded at points near the borders where mechanical compaction by the bulldozer was inadequate. Ishikawa et al. (1963) reported percolation rates from 0.4 to 38 cm/day in an area of about 50 paddy fields in a district situated on a volcanic ash plateau. The standard deviations of percolation rates in newly-constructed paddy fields varied from 1.4 to 51.4 cm/day; in old paddy fields constructed 60 years ago, standard deviations are 0.6 to 12.8 cm/day. The range of percolation rates within a paddy field gradually decreases with time.

Percolation rates within a paddy field are lower and more uniform when the paddy is formed mechanically by bulldozer than when done by hand. Many years were required to decrease percolation to an adequate value in old paddy fields formed by hand labor.

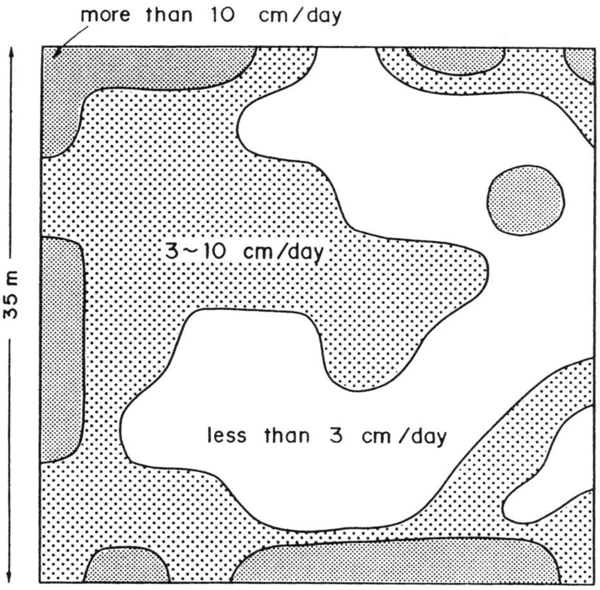

Fig. 7.12. The variability of percolation rates of a paddy field on volcanic ash soil. (From Yamazaki et al., 1962)

Bulldozer compaction or rolling is not very effective on volcanic ash soil unless the natural structure is first destroyed. The microcapillary pore structure of volcanic ash soil is so strong that it cannot be broken by the direct passage of a bulldozer. It can be broken by the crushing action of cultivation or by excavating the soil. Field tests carried out by Ishikawa et al. (1964) showed that percolation can be reduced below 2.0 cm/day and hydraulic conductivity reduced below 10^{-5} cm/sec by crushing before compaction. To achieve such a low percolation, the thickness of the crushed-banked soil should be 20 to 30 cm and it should be rolled at least seven to nine times. This method of paddy field reclamation, called the "Gandai operation" in Japan, has been used in districts with volcanic ash soil (Yamazaki, 1969; 1971).

Humphreys et al. (1990; 1992) in Australia, tested percolation control by rotary hoeing in wet or flooded soil (puddling) and by a vibrating roller on wet soil (compaction). Average deep percolation over five months was reduced from 16 mm/d to 1 mm/d (rotary hoe) or 3 mm/d (vibrating roller).

Additions of impermeable soil are useful for seepage control in leaky paddy fields. Amounts of 400 to 600 m^3/ha are used for seepage control; this is a layer 4 to 6 cm deep. In addition to seepage control, soil dressing may improve soil texture, increase the depth of cultivated soil, and increase the supply of iron.

The values of hydraulic conductivity of puddled soil change with the clay content (Fukioka and Nagahori, 1964). At less than 10% clay, K exceeded 10^{-4} cm/sec, but decreased to about 2×10^{-5} cm/sec at more than 25% clay. To gain effective seepage control, the clay content of the soil should be maintained at more than 25%. It has been reported that percolation decreases as much as one-half to one-third with the addition of 10 ton/ha of bentonite, and the seepage loss in the border regions also decreases through use of bentonite (Tsukidate and Tokunaga, 1962).

Vinyl sheets are used in sand dune districts for the construction of paddy fields. After removal of the surface soil, a vinyl sheet is laid and underdrain pipes are set on it. Percolation can be controlled by the elevation of the outlet of the drainage pipe. The groundwater level and water content can thus be controlled in a paddy field.

7.2.6 Puddling

Puddling of paddy fields is important to make soil soft for transplanting and impervious for percolation control. The soil surface should remain flat to keep a uniform water depth and to drain surface water rapidly at harvest time. Puddling breaks down large soil clods and soil aggregates, and decreases transmission pores. The permeability of the puddled soil generally depends on soil texture, especially clay content (Fujioka and Nagahori, 1964), but also depends on puddling method and frequency (Osari, 1988). The disadvantages of puddling (Sharma and De Datta, 1985; Koga, 1992) include a high labor requirement, a high water requirement for puddling, difficult regeneration of soil structure, and impeded root development.

The soils of the plowed layer are stratified by puddling. The coarse particles settle first, and the fine soil particles concentrate in the surface layer. Changes in bulk density, soil water pressure profile, particle size distribution, and hydraulic conductivity (Fig. 7.13) were investigated by Adachi (1988). For the fine-textured soil, the least permeable layer was just below the puddled layer, which was blocked with fine particles dispersed by puddling.

Settlement and permeability of puddled soil are influenced by changes in water pressure (Lei and Tada, 1987; 1988). After puddling, the soil in the lower part is packed more densely than that of the upper part due to its own weight and seepage forces. The percolation rate decreased with development of the consolidation layer. The mechanism of this process was analyzed by Koga (1989) and Adachi (1990, 1992).

Water pressure distribution in the paddy field and nitrogen concentration of percolating water were measured under flooded conditions (Tabuchi et al., 1990). Pressure of percolating water becomes negative in the subsoil due to impervious puddled soil and a hardpan as shown in Fig. 7.14. However in the paddy field without puddling, water pressure was not negative during the first period of flooding. Percolation rate increased to 7 cm/d, and the ammonium concentration in the percolating water also increased due to excess percolation. The loss of chemical fertilizer by percolation can be controlled by puddling.

7.3 DRAINAGE OF CLAYEY RICE PADDY FIELDS

In Japan, there are many clayey and swampy paddy fields where it is impossible to use combine harvesters and tractors due to low bearing capacity of the soils. Low bearing capacity results in decreased trafficability. To obtain sufficient drying of the paddy fields during the nonirrigation periods of cultivating and harvesting, it is necessary not only to install drainage canals, but also to improve surface field conditions.

7.3.1 General Pattern of Drainage in a Clayey Paddy Field

Drainage of a clayey paddy field is delayed by the large water holding capacity and low hydraulic conductivity. The hydraulic conductivity of clayey soil is less than 10^{-5} cm/sec, although once the soil is dried, water flows easily through

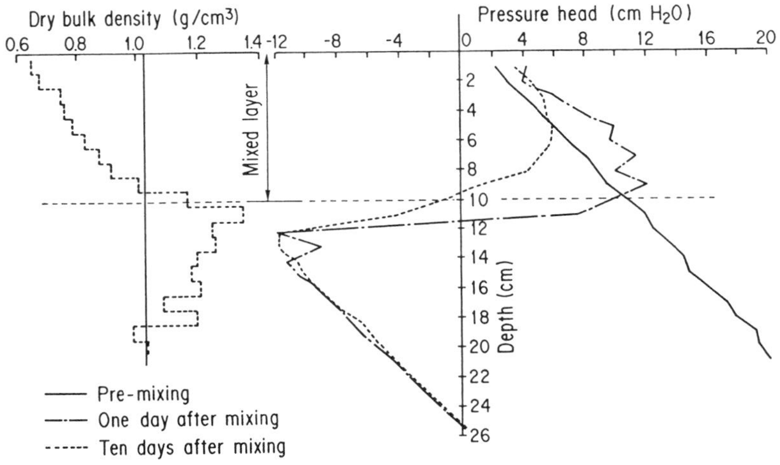

Fig. 7.13. Effect of puddling on dry bulk density and water pressure distribution in heavy clayey soil. (From Adachi, 1988; 1992)

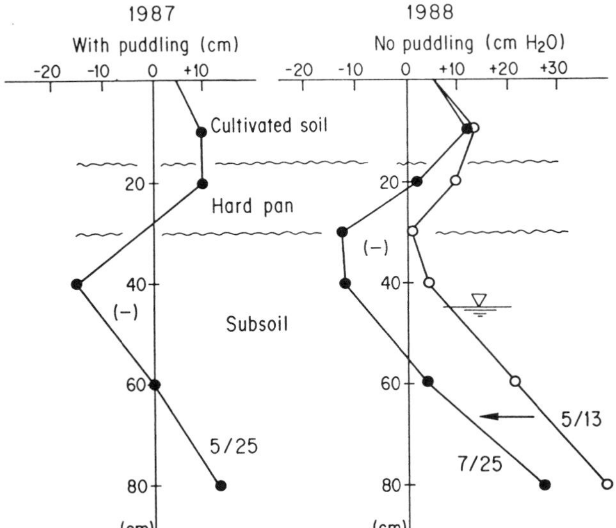

Fig. 7.14. Puddling effect on water pressure distribution measured in a rice field. (From Tabuchi et al., 1990)

cracks. Accordingly, water movement in clayey soil depends to a very great extent on whether or not it has cracks.

In a district with little rainfall, water movement in a clayey paddy field is as shown in Fig. 7.15a. After puddling, the surface soil becomes impervious, and percolation is low. When cracks appear at the surface as a result of midsummer drainage, the surface soil becomes pervious. After midsummer drainage, paddy fields are flooded again, and the cracks usually close. During the harvest period, surface water is drained to the drainage canal and drying of the surface soil will start again. When the soil is dried to the extent that cracks appear, it is easy to drain rainwater from the surface of a paddy field to the drainage canal or the underdrain. The drier the soil, the deeper the cracks become and the more easily water drains.

In a district with heavy rainfall (Fig. 7.15b) the surface does not dry sufficiently to form cracks. Water cannot percolate through the soil to the underdrain. Surface flow of the water is the only route for drainage in these paddy fields (Tabuchi, 1968).

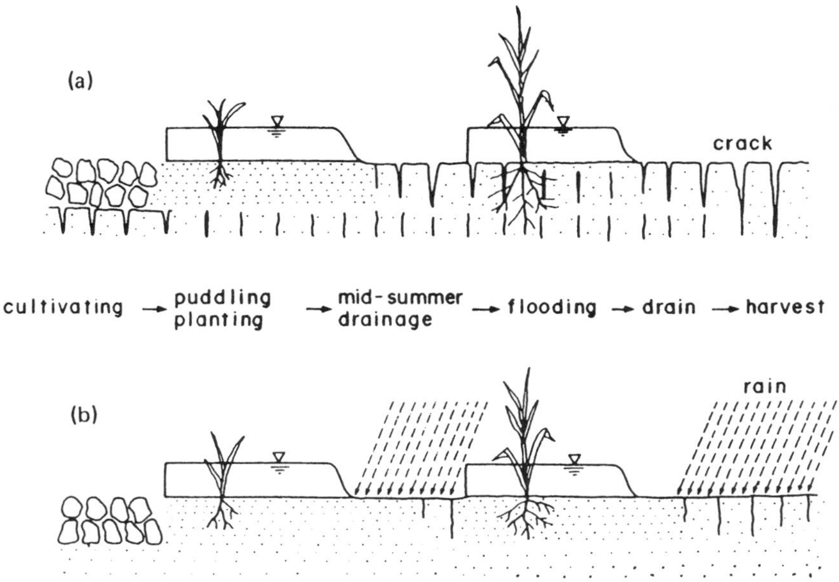

Fig. 7.15. Seasonal changes of surface conditions on clayey paddy fields: (a) low rainfall; (b) high rainfall.

7.3.2 Drainage Conditions in a Clayey Paddy Field

Surface outflow of water to the drainage canal is the first stage of drainage. Figure 7.16 shows daily changes of actual surface conditions in a paddy field. Two days after the start of midsummer drainage, much of the paddy field wasstill covered by water. The total amount of surface outflow was 5 cm, which corresponded to 70% of the flooding water. After 9 days, when the period of midsummer drainage had finished, 30% of the water remained in the lower parts of the paddy field due to unevenness of the surface. Low spots were usually 3 to 5 cm deep; in a large paddy field they sometimes attained a depth of 10 cm. Leveling of the surface of a clayey paddy field is therefore necessary for drainage.

The water depth decrease is about 4 to 7 mm/day, which means that water loss is due mainly to evapotranspiration. Only 10 mm of rainfall occurred during this period. It is impossible in this condition to use a large combine harvester in the paddy field.

7.3.3 Effect of Subsurface Drains

Many rice paddy fields in Japan have subsurface drains. However, they are often ineffective because percolation into the drains is limited. Accordingly, it becomes necessary in the first stage of drainage to change the impervious property of the clayey soil, in other words, to effect cracks. This requires precise leveling of paddy fields, a long drainage period, and low rainfall (Fig. 7.17). Field observations indicate 30 to 40 mm water loss is needed before cracks have formed sufficiently to drain water. Fujioka and Sato (1968) found that after soil water loss of more than 30 mm, the cracks were 2 cm wide and 20 cm deep.

Once cracks appear in the surface soil, water from rainfall flows rapidly into the underdrains through the cracks and the backfilled soil layer above the drain (Fig. 7.18). Table 7.8 indicates that the hydraulic conductivities of the backfilled soils are much larger than for undisturbed subsoil. In some cases sand or rice husks are used instead of soil for backfilling, to maintain a high permeability.

Water will flow through cracks in the subsoil (Tabuchi et al., 1968). Inoue (1988, 1992) studied the role of subsoil cracks in a clayey field. The subsoil structure was massive and hydraulic conductivity was very low. However, cracks developed in the subsoil at intervals of 10 to 20 cm. Accordingly, most of the rain water was drained quickly through the drains. The flow through cracks obeyed Darcy's law with hydraulic conductivity of about 10^{-2} cm/s.

Mole drains, subsoiling, and shallow ditches are used to increase the effect of underdrains (Nagahama et al., 1968; Ezaki, 1973; Tabuchi, 1985). A mole drain is constructed at a depth of 30 to 40 cm in the direction crossing the

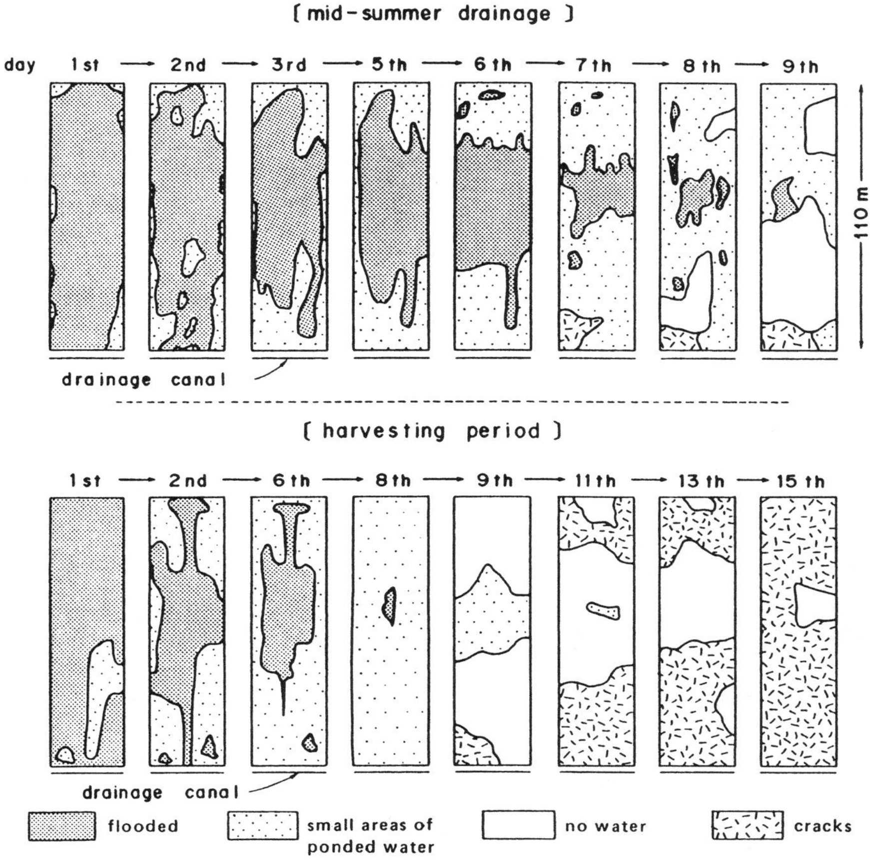

Fig. 7.16. Sketches of the water conditions on the surface of a clayey paddy field during two drainage periods. (From Tabuchi et al., 1966a,b)

underdrain; however, the effect of mole drains in the wet clay soil does not last long because they are often destroyed by the operation of tractors (Nagahori and Ogino, 1972).

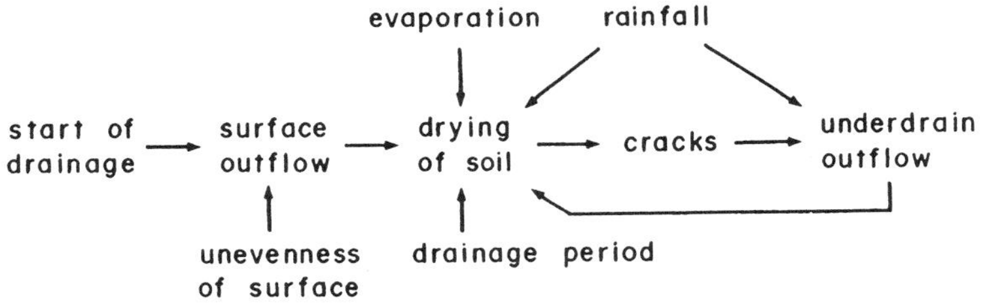

Fig. 7.17. Some factors relating to the effectiveness of underdrains.

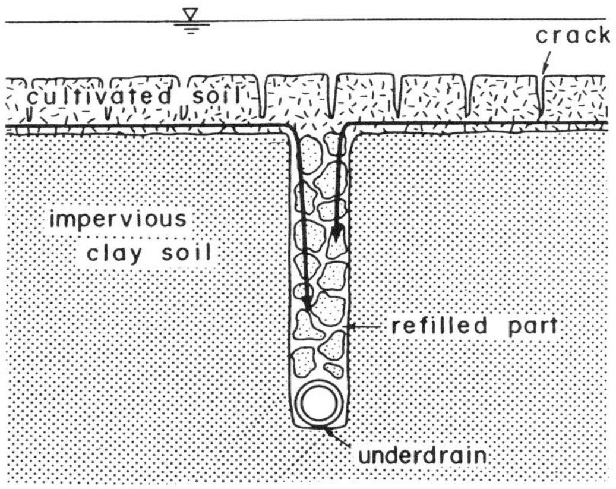

Fig. 7.18. Role of the backfilled soil zone for drains in a clayey paddy field.

Table 7.8. Hydraulic conductivity of the backfilled soil above the underdrain and in undisturbed subsoil (cm/sec). (From Fujioka and Maruyama, 1971)

Soil	Backfilled soil	Undisturbed subsoil
Clayey	3×10^{-2}	3×10^{-5}
Loamy	2×10^{-3}	4×10^{-5}
Sandy	2×10^{-3}	4×10^{-4}

7.4 INTERACTION OF CHEMICALS AND WATER IN PADDY FIELDS

7.4.1 Improvement of Paddy Fields Contaminated by Cadmium

A survey of 4094 paddy fields in Japan in 1973 showed that the Cd content of the soil increased from 0.4 ppm on the average to 0.9 ppm in contaminated districts. The maximum value was 13.8 ppm. The average Cd content of the rice plant was 0.1 ppm, but a maximum value of more than 3 ppm appeared in the contaminated districts. It is considered necessary for public health to have Cd content of the rice plant less than 1 ppm (Iimura et al., 1977).

The methods for improvement of contaminated soils are: removal of contaminated soil, adding uncontaminated soil as a topdressing, use of a soil conditioner such as calcium or phosphoric acid, and improvement of water management. Kondō et al. (1975) reported that topdressing with soil under constant waterlogging was most effective. By this method, the Cd content of unpolished rice could be decreased to a value less than 0.1 ppm. Itō and Iimura (1975) found that uptake of Cd by rice under constant waterlogging was so low that the Cd content was less than 1 ppm even with a high content in the soil. In control plots, the Cd content of the rice exceeded 1 ppm.

Consequently, seepage control has become important to maintain waterlogging and constant water depth in Cd-contaminated paddy fields. Tokunaga et al. (1975) reported on a paddy field with a percolation rate of 50 to 100 mm/day, a Cd content of 1.3 to 2.6 ppm in soil, and 0.3 to 3.4 ppm in the rice. The surface soil is thin and the subsoil is sand or gravel. They improved this paddy field by (1) crushing and compacting the surface soil to construct an impervious layer, (2) adding soil above this layer and rolling to make a hard pan, and (3) a secondary soil dressing of 15 to 20 cm soil depth to give a surface cultivated zone as shown in Fig. 7.19. After this improvement, the water loss was 11 mm/day and the Cd content of rice was 0.056 ppm. The low conductivity of the impervious layer is shown in the water pressure distribution curve (Fig. 7.19). Head loss in the impervious zone (20 to 50 cm depth) was larger than that in other zones. Water pressure was negative in the subsoil, which indicates unsaturated flow in an open system, typical of percolation in layered soils as discussed in Chapter 5.

7.4.2 Nitrogen and Phosphorus Concentrations in a Paddy Field

The routes of outflow and inflow of fertilizers in a paddy field are shown in Fig. 7.20. A budget of nitrogen and phosphorus would require measurements of volumes of water flowing through all routes and concentrations of nutrients.

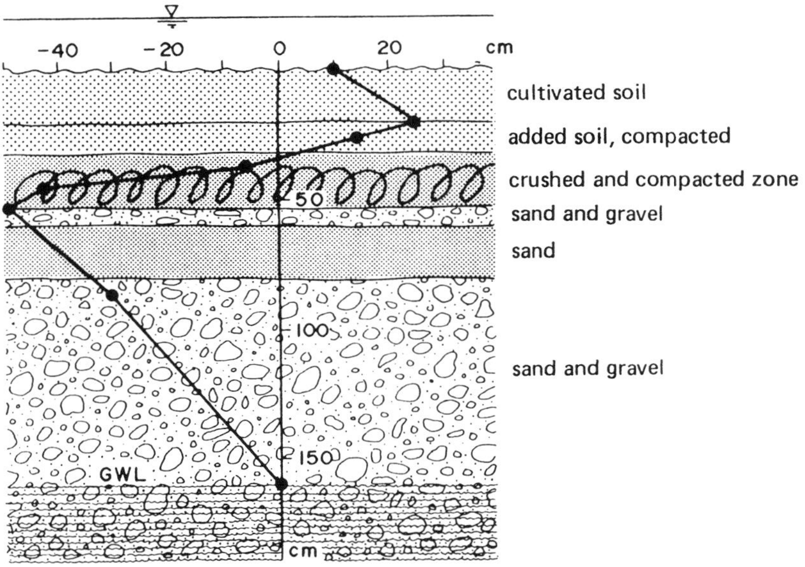

Fig. 7.19. Soil profile and the distribution of water pressure in a paddy field after improvement. (From Tokunaga et al., 1975)

Figure 2.21 is an example of the annual variation in the concentration of total nitrogen (TN) in a rice paddy field with a high groundwater table. The flooding surface water contained 64 ppm immediately after the basal fertilizer application. After about 2 weeks, the concentration of TN in the flooding surface water was reduced to the level of the unfertilized plot. A supplemental fertilizer dressing in late June increased TN to 17 ppm. The dominant form of nitrogen in the flooding surface water was ammonium; the concentration of nitrate was less than 0.6 ppm throughout the irrigation period.

The TN of percolating water was highest at 20 cm below the surface, with a maximum value of 7.3 ppm. The percolating waters trapped at 40 cm and 60 cm below the surface showed only slight increases in TN concentration. The dominant form was ammonium. However, nitrate was the dominant form in other paddy fields with low groundwater and unsaturated water percolation.

The concentration of total phosphorus (TP) in the flooding surface water increased to 26.5 ppm at the time of fertilizer application. The phosphorus concentration decreased more rapidly with time than the nitrogen concentration. The application of fertilizer did not affect the concentration of TP in the percolating water.

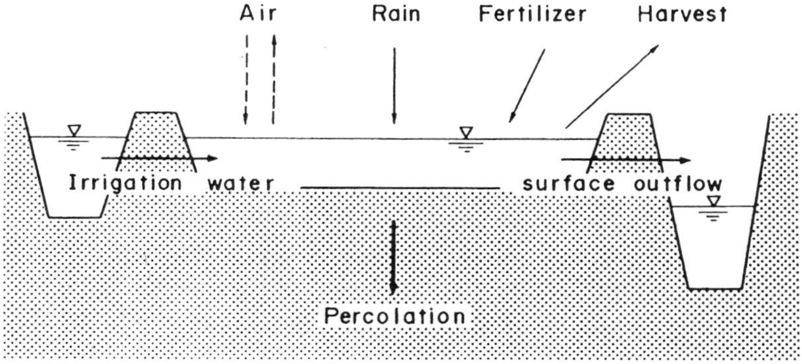

Fig. 7.20. Routes of input and output of nitrogen in a paddy field.

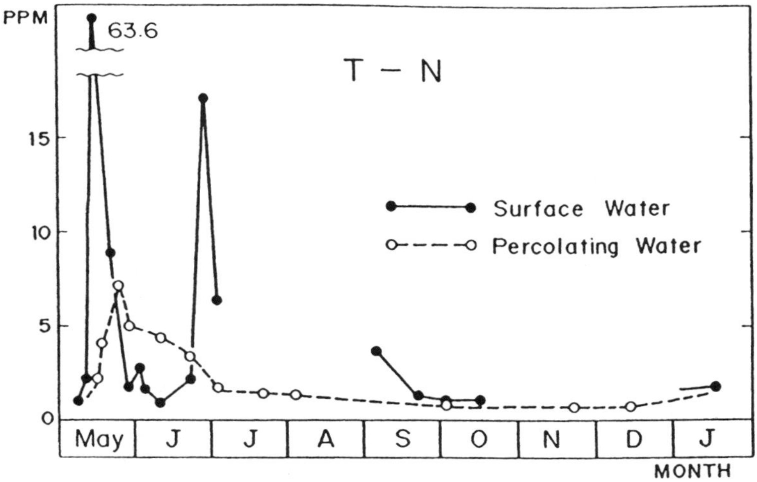

Fig. 7.21. Annual variation of total nitrogen concentration in water in a poorly drained paddy field. (From Tabuchi, 1976)

7.4.3 Nitrogen and Phosphorous Budgets in Paddy Fields

The nutrient load flowing into or out of a paddy field can be calculated from measured concentrations of nutrients and the water budget. An example of seasonal changes for total nitrogen is shown in Fig. 7.22. Fertilizer is the largest input, which appears in May and June with basal and top dressings. The irrigation water shows a peak in May, while the load from rainwater exists in all seasons.

The largest output is due to harvest in September. Output by surface flow is large during the irrigation period, especially in May and July. Drainage of flooding water after fertilization causes a large amount of fertilizer outflow, as shown in the output for May. Percolation output is small due to the low percolation rate of the clayey soil.

Table 7.9 shows annual budgets of total nitrogen in poorly-drained, imperfectly-drained, and well-drained rice paddy fields in the Kasumigaura Lake basin (Takamura et al., 1976; 1977a,b). The nitrogen input totaled 89 to 163 kg/ha, of which more than 77% was supplied as fertilizer. The output of nitrogen totaled 99 to 133 kg/ha, of which more than 78% was harvested rice. The phosphorus input totaled 33 to 96 kg/ha, of which more than 97% was fertilizer. Output totaled 11 to 16 kg/ha, of which more than 91% was harvested rice.

The nitrogen lost by surface outflow and percolation was 18 to 30 kg/ha in fertilized fields. The extra nitrogen lost from fertilized over unfertilized fields was 8 to 21 kg/ha. The output of phosphorous was only one-third of the input; most of the fertilizer phosphate added was retained in the soil. The loss of nitrogen from poorly-drained rice paddy fields is due to surface outflow, while that from a well-drained paddy field is due to percolation. In a nonfertilized plot, part of the nitrogen supplied by irrigation and rainwater does not flow out.

The budget for nutrients in actual paddy fields is variable, especially for nitrogen, in response to differences in rainfall, water management, fertilizer application method, concentration in irrigation water and soil properties. From measured values reported in Japan, we can determine that the load due to irrigation water is in the wide range of 3 to 40 kg/ha of nitrogen, which is affected by amount of irrigation water and nitrogen concentration. When irrigation water is polluted by sewage, the load of nitrogen becomes large. The sum of inflow load for both irrigation and rainwater is in the range 10 to 50 kg/ha.

The loss by surface outflow or percolation is also variable. The relation between inflow load and outflow is examined in Fig. 7.23, which contains data on load of nitrogen measured in many districts. There is a tendency for values of outflow to increase with increase of inflow; the data plot as a nearly straight line with a slope of 45°. Accordingly, the balanced losses fluctuated over such a small range that their values are in the range -20 to +20 kg/ha. Efforts are

Table 7.9. Measured annual nitrogen budgets in paddy fields in the Lake Kasumigaura Basin (kg/ha). (From Takamura et al., 1976, 1977a,b)

	Poorly-drained[a] (1974)		Imperfectly-drained (1975)		Well-drained[b] (1976)	
	Fertilizer[c]	None[d]	Fertilizer	None	Fertilizer	None
Inputs						
Fertilizer	68.9	--	141.0	--	120.4	--
Irrigation	6.8	6.3	15.3	15.3	7.1	7.1
Rainfall	12.8	12.8	6.4	6.4	8.1	8.1
Surface inflow	0.2	0.5	0.0	0.0	3.3	3.3
Total	88.7	19.6	162.7	21.7	138.9	18.5
Outputs						
Surface outflow	16.6	5.0	13.1	6.5	4.4	1.5
Percolation	4.7	2.5	4.9	3.1	25.2	7.4
Harvested crops	77.7	63.9	112.7	65.4	103.2	55.0
Total	99.0	71.4	130.7	75.0	132.8	63.9

[a]High groundwater table
[b]Low groundwater table
[c]Fertilizer, fertilized plot
[d]None, unfertilized plot

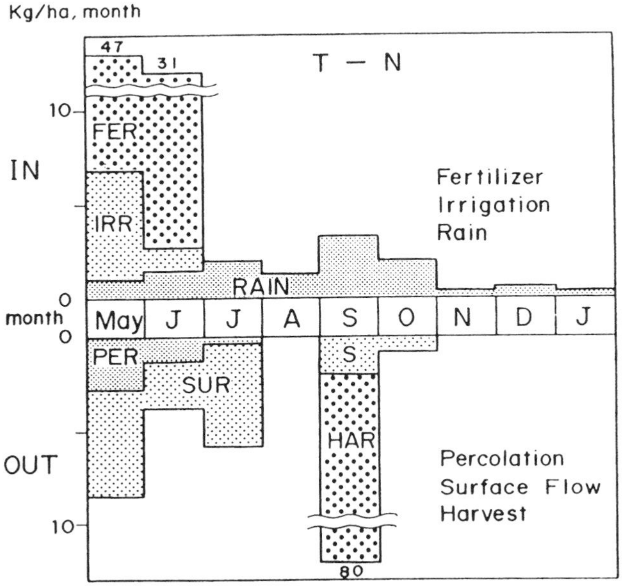

Fig. 7.22. Annual total nitrogen budget in a poorly drained paddy field. (From Tabuchi, 1976)

now being made to change paddy fields to absorbers, where outflow is less than inflow. This requires improvement of field water management, different methods of applying fertilizer and percolation control.

Side-dressing resulted in high efficiency of utilization of fertilizer. Compared with the standard method, the amount of basic fertilizer decreased by 20% (Nagano Agr. Exp. Station, 1990). Recycling of drained water to irrigation also decreased nitrogen losses. The value of balanced loss changed from +2.6 kg/ha to -2.1 kg/ha on recycling (Hasegawa, 1992).

7.4.4 Nitrogen Removal in Paddy Fields

Control of nitrogen outflow from agricultural areas is a major problem in Japan. Nitrate percolates to groundwater from large amounts of fertilizer applied to upland fields. It finally flows to lowland areas, where nitrate can be removed by plant uptake and denitrification under anaerobic conditions in paddy fields. This natural geographic sequence formed by upland fields and lowland rice fields has potential for the control of nitrogen losses.

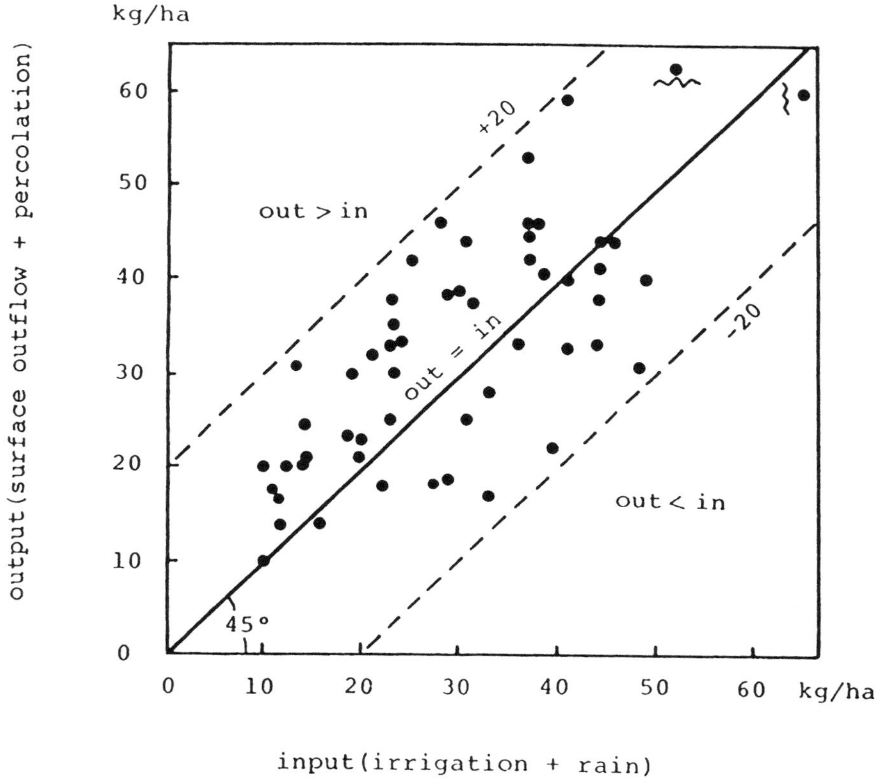

Fig. 7.23. Relation between nitrogen inflow and outflow in paddy fields. (From Tabuchi and Takamura, 1985)

The changes of nitrogen concentration in flooded surface water of paddy fields were investigated by Tabuchi et al., (1983, 1987) as shown in Fig. 7.24. Nitrogen concentration, mainly in the form of nitrate, decreased from 6.3 mg/l to 2 mg/l. This corresponded to a nitrogen removal of 0.1 g/m^2 per day. The amount of nitrogen removed depended on nitrate concentration in the water and on temperature.

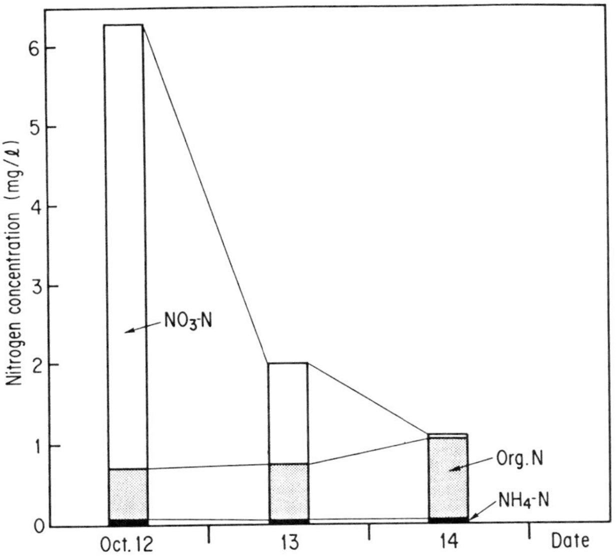

Fig. 7.24. Decrease of nitrogen concentration in surface flooded water in a paddy field. (From Tabuchi et al., 1983)

REFERENCES

Adachi, K., Experimental studies of the effects of puddling on percolation control, *Trans. Jpn. Soc. Irrig. Drain. Reclam. Eng.* 135:1-8 (1988). (Japanese with English summary)

Adachi, K., Effect of puddling on sedimentation of puddled soil and flow rate, *Trans. Jpn. Soc. Irrig. Drain. Reclam. Eng.* 148:67-73 (1990). (Japanese with English summary)

Adachi, K., Effect of puddling on rice-soil physics. *In* Soil and Water Engineering for Paddy Field Management, Asian Institute of Technology, pp. 220-231 (1992).

Arai, M., Unpublished (private communication) (1976). Committee on Underdrains, Report on planning, construction and management of underdrains, *J. Jpn. Soc. Irrig. Drain. Reclam. Eng.* 41:9-30 (1973). (Japanese)

Ezaki, K., On the furtherance of drying in marshy heavy clay soil ground, *Trans. Jpn. Soc. Irrig. Drain. Reclam. Eng.* 44:34-42 (1973). (Japanese with English summary)

Fujioka, Y., and T. Maruyama, Studies on underdrainage of clayey paddy field: 1. Drainage effect of cracks in plow-layers and backfilled trench. *Trans. Jpn. Soc. Irrig. Drain. Reclam. Eng.* 35:48-53 (1971). (Japanese with English summary)

Fujioka, Y., and K. Nagahori, A rational method for deciding the quantity of bentonite corresponding to the condition of excessive percolative paddy field, *J. Agric. Eng. Soc. Jpn.* 31:393-397 (1964). (Japanese with English summary)

Fujioka, Y., and K. Sato, On the cracks with drying of clayey paddy field soil: III, *Trans. Jpn. Soc. Irrig. Drain. Reclam. Eng.* 26:8-14 (1968). (Japanese with English summary)

Hasegawa, F., Report of Ibaraki Agricultural Experimental Station (1963). (Japanese)

Hasegawa, K., Studies on the behavior and balance of nitrogen in paddy field near the lake and its influence on environment especially on the water quality, Shiga Pref. Agric. Exp. Station, Report 17 (1992). (Japanese with English summary)

Hayasaka, T., Behavior of water in upland volcanic ash soil, *Soil Phys. Cond. Plant Growth Jpn.* 37:17-21 (1978). (Japanese)

Humphreys, E., W. A. Muirhead, and C. M. Steer, Reducing deep percolation on rice soils, *Trans. 14th Int. Cong. of Soil Science* 1:354-355 (1990).

Humphreys, E., W. A. Muirhead, and B. J. Fawcett, The effect of puddling and compaction on deep percolation and rice yield in temperate Australia, *In* Soil and Water Engineering for Paddy Field Management, Asian Institute of Technology, pp. 212-219 (1992).

Iimura, K., H. Itō, M. Chino, T. Morishita, and H. Hirata, Behavior of contaminant heavy metals in soil-plant system, *Proc. Int. Sem. SEFMIA*, Soc. of Sci. of Soil and Manure, Japan, pp. 357-368 (1977).

Inoue, H., Hydrological properties of pipe drainage and physical properties of water flow in the soil of agricultural fields with shrinkage cracks, *Trans. Jpn. Soc. Irrig. Drain. Reclam. Eng.* 137:25-33 (1988). (Japanese with English summary)

Inoue, H., Percolation and pipe drainage. *In* Soil and Water Engineering for Paddy Field Management, Asian Institute of Technology, pp. 99-108 (1992).

Ishihara, K., Percolation and rice growing, *Soil Phys. Cond. Plant Growth Jpn.* 16:22-26 (1967). (Japanese)

Ishikawa, T., K. Tokunaga, and K. Tsukidate, Studies on the reclamation scheme of rice paddies in the Piedmont district of Mt. Iwate: II. *J. Agric. Eng. Soc. Jpn.* 31:80-88 (1963). (Japanese with English summary)

Ishikawa, T., K. Tokunaga, and K. Tsukidate, Studies on the reclamation scheme of rice paddies in the Piedmont district of Mt. Iwate: III. On the permeability decrease of volcanic ash subsoil due to cultivating and rolling, *J. Agric. Eng. Soc. Jpn.* 31:263-269 (1964). (Japanese with English summary)

Isozaki, H., On the moderate percolation of paddy fields, *J. Agric. Eng. Soc. Jpn.* 24:311-312 (1957). (Japanese)

Itō, H., and K. Iimura, Absorption of cadmium by rice plants in response to change of oxidation-reduction conditions of soils, *J. Sci. Soil Manure Jpn.* 46:82-88 (1975). (Japanese)

Iwata, S., The characteristics of water movement in soils during drainage with special reference to field moisture capacity, *Bull. Natl., Inst. Agric. Sci.* B16:149-176 (1966). (Japanese with English summary)

Kanō, T., S. Nakagawa, H. Onishi, T. Maruyama, and T. Furuki, The changes of the duty of water on a paddy field by underdrain, *J. Agric. Eng. Soc. Jpn.* 28:425-431 (1961). (Japanese)

Kawasaki, T., Physical properties of soil water requirements in paddy fields after direct drilling on ponding or upland condition, *Trans. Jpn. Soc. Irrig. Drain. Reclam. Eng.* 59:15-22 (1975). (Japanese with English summary)

Kira, Y., K. Soma, and H. Takenaka, On the relationships between moisture fluctuation and evapotranspiration in the Kantō-loam volcanic ash soil, *Trans. Jpn. Soc. Irrig. Drain. Reclam. Eng.* 7:81-86 (1963). (Japanese with English summary)

Kitajima, S., and H. Hashimoto, The relationship between wilting point and water content at pF 4.2 in a volcanic ash soil, *Rep. Kyushu Agric. Exp. Stn.* (1967). (Japanese)

Koga, K., Some phenomena in prolonged water percolation through soils, *Soil Phys. Cond. Plant Growth Jpn.* 59:17-27 (1989). (Japanese with English summary)

Koga, K., *Introduction to Paddy Field Engineering*, Asian Institute of Technology, pp. 30-31, Bangkok (1992).

Kondō, H., K. Suzuki, C. Kokubo, and N. Kanno, An actual condition of the paddy field contaminated by cadmium in the town of Bandai and the engineering method for reclamation, *J. Jpn. Soc. Irrig. Drain. Reclam. Eng.* 43:655-659 (1975). (Japanese)

Lei, P., and A. Tada, Effect of downward percolation on settlement of puddled soil, *Trans. Jpn. Soc. Irrig. Drain. Reclam. Eng.* 132:35-42 (1987). (Japanese with English summary)

Lei, P., and A. Tada, The effect of drainage level and initial void ratio on settlement and permeability of puddled soils, *Trans. Jpn. Soc. Irrig. Drain. Reclam. Eng.* 133:69-77 (1988). (Japanese with English summary)

Masujima, H., Percolation of paddy fields, *Soil Phys. Cond. Plant Growth Jpn.* 15:12-14 (1966). (Japanese)

Ministry of Agriculture and Forests, Government of Japan, *Volcanic Ash Soils in Japan* (1964).

Misono, S., and S. Terasawa, Studies on the soil water system of volcanic ash soils in Japan, *Bull. Natl. Inst. Agric. Sci.* B7:77-103 (1957). (Japanese)

Misono, S., S. Terasawa, A. Kishita, and S. Sudo, Studies on the soil water system of volcanic ash soils in Japan, *Bull. Natl. Inst. Agric. Sci.* B2:95-124 (1953). (Japanese)

Murty, V. V. N., and K. Koga (eds.), *Soil and Water Engineering for Paddy Field Management*, Proc. of Int. Workshop, Asian Institute of Technology (1992).

Nagahama, K., S. Tejima, M. Tomita, and H. Taniguchi, On the substance of the mechanism in the combined mole-pipe drainage system, *Trans. Jpn. Soc. Irrig. Drain. Reclam. Eng.* 26:29-34 (1968). (Japanese with English summary)

Nagahori, K., and Y. Ogino, On the behavior of moledrains constructed in the subsoil of the poldered paddy fields caused by the operation of a tractor. *Trans. Jpn. Soc. Irrig. Drain. Reclam. Eng.* 41:7-12 (1972). (Japanese with English summary)

Nagaishi, Y., On the depth of water requirement in paddy fields created by impoldering, *J. Agric. Eng. Soc. Jpn.* 27:60-65 (1959). (Japanese with English summary)

Nagano Agricultural Experimental Station, Side-dressing for lowland rice, Advances in Soils and Fertilizers in Japan, 14th Int. Cong. Soil Science, pp. 104-105 (1990).

Nagata, N., On the nonuniformity of air and water permeability of sandy soils, *Trans. Jpn. Soc. Irrig. Drain. Reclam. Eng.* 36:60-66 (1971). (Japanese with English summary)

Nakagawa, S., *The Method of Planning for Irrigation of Paddy Fields*, Hatachi-nogyo Shinkokai, Tokyo (1967). (Japanese)

Nakajimada, M., Unpublished (private communication) (1976).

Nakano, M., Studies on water movement in layered volcanic ash soil: 2, *Trans. Jpn. Soc. Irrig. Drain. Reclam. Eng.* 31:10-16 (1970). (Japanese with English summary)

Osari, H., Analysis of stirred and compacted operation on puddling by a tractor, *Trans. Jpn. Soc. Irrig. Drain. Reclam. Eng.* 134:1-7 (1988). (Japanese with English summary)

Sato, K., S. Aoki, and H. Akashi, On the duty of water in the paddy fields with pipe system, *J. Jpn. Soc. Irrig. Drain. Reclam. Eng.* 43:783-788 (1975). (Japanese)

Sharma, P. K., and S. K. De Datta, Effects of puddling on soil physical properties and processes, *Soil Physics and Rice*, Int. Rice Res. Institute, pp. 217-234 (1985).

Tabuchi, K., Studies on soil pores through microscopical observation of thin sections, *Trans. Jpn. Soc. Irrig. Drain. Reclam. Eng.* 7:21-31 (1963). (Japanese with English summary)

Tabuchi, T., Studies on drainage in clayey paddy fields: 8. A method of calculation of drainage and its fluctuations due to the precipitation, *Trans. Jpn. Soc. Irrig. Drain. Reclam. Eng.* 25:50-56 (1968). (Japanese with English summary)

Tabuchi, T., Outflow of fertilizers from the rice paddy field, *Soil Phys. Cond. Plant Growth Jpn.* 33:16-20 (1976). (Japanese)

Tabuchi, T., S. Suzuki, and Y. Takamura, Nitrogen removal in paddy fields during the nonrice growing period, *Trans. Agric. Eng. Soc. Jpn.* 104:9-15 (1983). (Japanese with English summary)

Tabuchi, T., Underdrainage of lowland rice fields, *Soil Physics and Rice*, Int. Rice Res. Institute, Manila, pp. 147-159 (1985).

Tabuchi, T., and Y. Takamura, *Nitrogen and Phosphorous Outflow from Catchment Area*, Tokyo University Press, Tokyo (1985). (Japanese)

Tabuchi, T., S. Hasegawa, and S. Iwata, Report from the Workshop PASMIRCS (IRRI), *Irrigation Eng. and Rural Planning*, 10:70-76 (1986).

Tabuchi, T., M. Nakano, and S. Suzuki, Studies on drainage in clayey paddy fields: 2, *Trans. Agric. Eng. Soc. Jpn.* 18:12-17 (1966a). (Japanese with English summary)

Tabuchi, T., M. Nakano, and S. Suzuki, Studies on drainage in clayey paddy fields: 3, *Trans. Agric. Eng. Soc. Jpn.* 18:18-24 (1966b). (Japanese with English summary)

Tabuchi, T., M. Nakano, A. Sumida, and I. Maurta, Studies on drainage in clayey paddy fields: 5, *Trans. Agric. Eng. Soc. Jpn.* 18:31-38 (1966c). (Japanese with English summary)

Tabuchi, T., M. Nakano, A. Kondō, H. Matsumura, and I. Maruta, Studies on drainage in clayey paddy fields: 7. On the effect of underdrainage, *Trans. Jpn. Soc. Irrig. Drain. Reclam. Eng.* 25:42-49 (1968). (Japanese with English summary)

Tabuchi, T., N. Suemasa, and M. Takanashi, Nitrate removal in the flooded paddy field, *J. Jpn. Soc. Irrig. Drain. Reclam. Eng.* 55:53-58 (1987). (Japanese)

Tabuchi, T., S. Iwata, S. Hasegawa, T. Woodhead, and E. Maurer (eds.), *Physical Measurements in Flooded Rice Soils*, Int. Rice Res. Institute, Manila (1987).

Tabuchi, T., I. Yamafuji, and H. Kuroda, Effect of puddling on percolation rate and nitrogen concentration in percolating water, *Trans. 14th Int. Cong. Soil Sci.*, Vol. I, pp. 287-288 (1990).

Tada, A., H. Takenaka, K. Soma, T. Kurobe, and Y. Hayama, On the characteristics of the particles of the Kantō loam volcanic ash soil, *Trans. Agric. Eng. Soc. Jpn.* 7:14-21 (1963). (Japanese with English summary)

Takamura, Y., T. Tabuchi, S. Suzuki, Y. Harigae, T. Ueno, and H. Kubota, The fates and balance sheets of fertilizer nitrogen and phosphorous applied to a rice paddy field in the Kasumigaura basin, *J. Sci. Soil Manure Jpn.* 47:398-405 (1976). (Japanese)

Takamura, Y., T. Tabuchi, and H. Kubota, Behaviour and balance of applied nitrogen and phosphorus under rice field conditions, *Proc. Int. Sem. SEFMIA*, Soc. of Sci. of Soil and Manure, Japan pp. 342-349 (1977a).

Takamura, Y., T. Tabuchi, Y. Harigae, H. Otsuki, and H. Kubota, The behavior and balance for nitrogen and phosphorus in the badly-drained paddy field in the Shintone river basin, *J. Sci. Soil Manure Jpn.* 48:431-436 (1977b). (Japanese)

Terasawa, S., Studies on the movement of moisture in upland soils, *J. Sci. Soil Manure Jpn.* 33:456-460 (1962). (Japanese)

Terasawa, S., Studies on the movement of moisture in upland soils, *Bull. Natl. Inst. Agric. Sci.* B13:1-115 (1963). (Japanese)

Terasawa, S., and K. Ueda, Vertical percolation and rice plant growth under the ponding paddy field of compacted soil column, *Trans. Jpn. Soc. Irrig. Drain. Reclam. Eng.* 34:10-16 (1970). (Japanese with English summary)

Tokunaga, K., H. Baba, T. Ishikawa, M. Ishihata, and M. Ishikawa, Regeneration method of the paddy field contaminated by cadmium, *J. Jpn. Soc. Irrig. Drain. Reclam. Eng.* 43:660-666 (1975). (Japanese)

Tsukidate, K., and K. Tokunaga, Investigation on the vertical migration of bentonite particles from paddy field surfaces, *J. Agric. Eng. Soc. Jpn.* 29:355-361 (1962). (Japanese with English summary)

Woodhead, T., E. A. Tout, and G. S. Argosino (eds.), *Soil Physics and Rice*, Intl. Rice Res. Inst., Manila (1985).

Yamazaki, F. ed., *Soil Physics*, Yokendo, Tokyo (1969). (Japanese)

Yamazaki, F., *Land Reclamation Engineering, Vol. I*, Tokyo University Press, Tokyo (1971). (Japanese)

Yamazaki, F., *Paddy Field Engineering*, Asian Inst. Tech. (1988). (Translated from Japanese)

Yamazaki, F., T. Yawata, and T. Tabuchi, Percolation of the paddy fields on the upland, *J. Agric. Eng. Soc. Jpn.* 28:137-141 (1960) (Japanese with English summary)

Yamazaki, F., T. Yawata, T. Tabuchi, T. Ishikawa, and A. Nagasaki, Studies on the seepage in paddy fields with gravel layer in their subsoils, *J. Agric. Eng. Soc. Jpn.* 29:9-18 (1961) (Japanese with English summary)

Yamazaki, F., T. Yawata, T. Tabuchi, and T. Ishikawa, Percolation through ridge and inside area of paddy fields *J. Agric. Eng. Soc. Jpn.* 29:309-314 (1962) (Japanese with English summary)

Yamazaki, F., T. Yawata, and S. Sudo, Physical properties of the Kantō loam, *Trans. Agric. Eng. Soc. Jpn.* 7:1-13 (1963) (Japanese with English summary)

Yokoi, Y., Studies on soil water of irrigated mineral soils, *Bull. Tokai-Kinki Agric. Exp. Stn.* 12:1-26 (1965) (Japanese)

Index

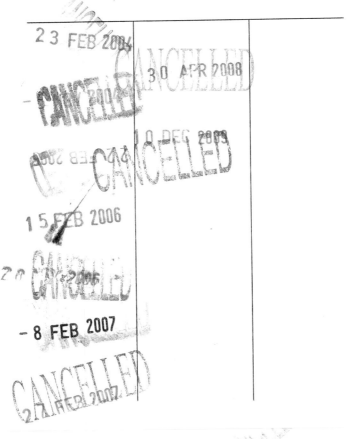